Science, Technology, *and* Commercialization *of* Powder Synthesis *and* Shape Forming Processes

Related titles published by the American Ceramic Society:

Phase Equilibria Diagrams, Phase Diagrams for Ceramists, Volume XI: Oxides
Edited by Robert S. Roth
© 1995, ISBN 0-944904-90-4

Crystal Growth of Novel Electronic Materials (Ceramic Transactions Volume 60)
Edited by R.K. Pandey and Ruyan Guo
© 1995, ISBN 1-57498-003-3

Microwaves: Theory and Application in Materials Processing III (Ceramic Transactions Volume 59)
Edited by David E. Clark, Diane C. Folz, Steven J. Oda, and Richard Silberglitt
© 1995, ISBN 1-57498-002-5

Advanced Synthesis and Processing of Composites and Advanced Ceramics (Ceramic Transactions Volume 56)
Edited by Kathryn V. Logan
© 1995, ISBN 1-57498-000-9

Sol-Gel Science and Technology (Ceramic Transactions Volume 55)
Edited by Edward J.A. Pope, Sumio Sakka, and Lisa Klein
© 1995, ISBN 0-944904-97-1

Science, Technology, and Applications of Colloidal Suspensions (Ceramic Transactions Volume 54)
Edited by James H. Adair, Jon A. Casey, Clive A. Randall, and Sridhar Venigalla
© 1995, ISBN 0-944904-96-3

Low-Expansion Materials (Ceramic Transactions Volume 52)
Edited by David P. Stinton and Santosh Y. Limaye
© 1995, ISBN 0-944904-92-0

Ceramic Processing Science and Technology (Ceramic Transactions Volume 51)
Edited by Hans Hausner, Gary L. Messing, and Shin-ichi Hirano
© 1995, ISBN 0-944904-89-0

Design for Manufacturability of Ceramic Components (Ceramic Transactions Volume 50)
Edited by Asish Ghosh, Basavaraj Hiremath, and John Halloran
© 1995, ISBN 0-944904-88-2

Silicon-Based Structural Ceramics (Ceramic Transactions Volume 42)
Edited by Brian W. Sheldon and Stephen C. Danforth
© 1994, ISBN 0-944904-76-9

Grain Boundaries and Interfacial Phenomena in Electronic Ceramics (Ceramic Transactions, Volume 41)
Edited by Lionel M. Levinson and Shin-ichi Hirano
© 1994, ISBN 0-944904-73-4

Science, Technology, *and* Commercialization *of* Powder Synthesis *and* Shape Forming Processes

▼▲▼

Edited by

J.J. Kingsley
Battelle Pacific Northwest Laboratory

Christopher H. Schilling
Iowa State University

James H. Adair
University of Florida

▲▼▲

Ceramic Transactions
Volume 62

Published by
The American Ceramic Society
735 Ceramic Place
Westerville, Ohio 43081

Proceedings of the Symposium on Science, Technology, and Commercialization of Powder Synthesis and Shape Forming Processes, held at the 97th Annual Meeting of the American Ceramic Society, Cincinnati, Ohio, April 30–May 3, 1995.

For information on ordering titles published by the American Ceramic Society, or to request a publications catalog, please call 614-890-4700.

Printed in the United States of America.

1 2 3 4–00 99 98 97 96

ISSN 1042-1122
ISBN 1-57498-005-X

Contents

Preface ix

Overviews

Advanced Ceramic Powder and Nano-Sized Ceramic Powder: An Industry and Market Overview 3
Thomas Abraham

Commercialization of Powder Synthesis Processes 15
Greg Geiger

The Critical Role of Raw Material Suppliers in Market Development ... 21
Keith A. Blakely

Issues in Powder Forming Technologies 29
Laurel M. Sheppard

Powder Synthesis

Synthesis of Ceramic Powders by Novel Microwave-Hydrothermal Processing 37
Sridhar Komarneni, V.C. Menon, and Qing Hua Li

Synthesis of Monodisperse Spherical $ZrTiO_4$ Powders by Thermal Hydrolysis of Metal Salts 47
Hong Kyu Park, Do Kyung Kim, and Chong Hee Kim

A Simple Solution-Polymerization Route for Oxide Powder Synthesis 57
M.A. Gülgün and W.M. Kriven

Preceramic Polycyclomethylsilazanes: Chemistry of Processing 67
Yigal D. Blum, Sylvia M. Johnson, and Gregory A. McDermott

Thermodynamic Analysis of the Hydrothermal Synthesis of the $Mn_{0.5}Zn_{0.5}Fe_2O_4$ Ferrite 75
Tsuneharu Ogasawara and Ana Maria Rangel de Figueiredo Teixeira

The Effect of Diffusion Flame Configuration in Aerosol Synthesis of Nanosize Titania and Silica Powders 83
Wenhua Zhu and Sotiris E. Pratsinis

Nitriding of Large-Sized Aluminum. 91
Shigeru Ito, Kazuharu Yoshida, Shin-ichi Saito, and Kazuo Akashi

Manufacturing of Iron Carbides in Fluidized-Bed Reactors 99
Seetharama C. Deevi, Sarojini Deevi, and Mohammad R. Hajaligol

Synthesis of Ultra-Fine Metallic Powders in a DC Plasma Reactor by Chemical Vapor Deposition 107
G.Y. Zhao and V. Hlavacek

Powder Processing—Dispersion/Rheology

Modification of Silicon Nitride Slip Properties by Poly(Acrylic Acid) . . 117
Vincent A. Hackley and Subhas G. Malghan

Adsorption of Organic Additives on Powder Surfaces 125
Amanda K. Berry, Leonard E. Bogan, and Sharon E. Agostine

Improved Aqueous Dispersion of Silicon Nitride with Aminosilanes 133
Michael A. Buchta and Wei-Heng Shih

Organic Additive Systems for Spray Drying and Dry Pressing Silicon Nitride. 141
W.J. Walker Jr., J.S. Reed, and S.K. Verma

Surface Modification of Silicon Nitride Powders by Metal-Organic Surfactants. 149
H.M. Wang and R.A. McCauley

Acrylic Emulsion Binder for Tape Casting. 157
Xun Tang, Scott A. Ibbitson, Antonio T. Donato, and Hayes Byrd Jr.

Viscoelasticity of Nano-Alumina Dispersions. 165
B. Rand and R. Fries

Suspension Parameters of SiC_w/Al_2O_3 Composites 173
Melissa J. Crimp, Mingli Zhang, and Dean A. Oppermann

Rheological Characterization of Temperature Effects in Particle-Filled Polymer Fluids 183
J.A. Horn, B.R. Patterson, and J. Zhou

Using Polymer Blends to Fine-Tune Binder Performance 193
David W. Whitman, Xun Tang, Donald I. Cumbers, and Scott A. Ibbitson

Powder Processing—Forming and Consolidation

A Comparison of Pressure Compaction and Diametral Compression Tests for Determining Granule Strengths . . . 203
S. Jill Glass and Clay Newton

Springback in Dry Pressing: Time, Additive, and Processing Effects . . . 217
Kevin J. Moeggenborg

Control of Dry Pressing Performance via Binder Optimization . . . 225
David W. Whitman, Xianliang K. Wu, and William C. Finch

Green-State Deformation of Boehmite-Coated Silicon Nitride Compacts . . . 233
W.-H. Shih, D.J. Farrell, and W.Y. Shih

Plastic Shaping of Aqueous Alumina Suspensions with Saccharides and Dicarboxylic Acids . . . 241
H. Goel, C.H. Schilling, S.B. Biner, J.A. Moore, and B.K. Lograsso

Gelcasting of Silicon Compositions for SRBSN . . . 255
S.D. Nunn, J.O. Kiggans Jr., R.E. Simpson II, and J.P. Maria

Filter-Pressing of Alumina Dispersions . . . 263
R. Fries and B. Rand

Mechanism and Kinetics of Electrophoretic Deposition of Al_2O_3 . . . 271
Partho Sarkar and Patrick S. Nicholson

Acrylic Binders for Alumina and Other Non-Clay Ceramic Extrusions . . . 281
X.L. Kevin Wu and Tony T. Donato

Binders for Flexible Ceramic Sheet Extrusion . . . 289
X.L. Kevin Wu and Tony T. Donato

Effects of Particle Size Distribution and Extrusion Rate on the Dimensional Stability of Extruded Alumina Rods . . . 297
K.J. Woodard and D.R. Dinger

Injection Molding of Ceramics Using a Polyacetal-Based Binder System . . . 305
J.S. Ebenhöch, J.H.H ter Maat, W. Hesse, and P. Truebenbach

Chemically Bonded Ceramic Processing of Mono-Calcium Aluminate . . . 313
Bradley R. Johnson, Mehmet A. Gülgün, and Waltraud M. Kriven

Thermal Stability of Y-α'-Sialon at High Temperature . . . 321
Akihiko Suda and Yoshio Ukyo

Index . . . 329

Preface

An understanding of the sciences of powder synthesis and processing is a critical component in the fabrication of ceramic articles with specified, engineering-oriented properties. The objective of this volume is to present a collection of papers on advanced powder synthesis and processing techniques as well as several analyses of the commercial viability of some of these processes. The papers were contributed from participants at the Symposium on the Science, Technology, and Commercialization of Powder Synthesis and Shape Forming Processes held at the 97th Annual Meeting of the American Ceramic Society in Cincinnati, Ohio, April 30–May 3, 1995.

The objective of the symposium was to bring together researchers and industrial colleagues with common interest in the application of scientific principles and the commercialization of advanced ceramic powder synthesis and processing techniques. As a part of the effort to establish criteria for the commercial viability of advanced powder synthesis and processing techniques, two workshops were held during the symposium: Commercialization of Powder Synthesis: Barriers and Opportunities; and Commercialization of Ceramic Shape-Forming Processes: Barriers and Opportunities. Panel members composed of research and industrial leaders provided overviews in each workshop followed by lively discussions among the audience and panel members. The discussions held during the workshop are summarized in the three papers presented in the Overview section of this volume. The symposium organizers would like to express our appreciation to the participants in each of the workshops by recognizing each of the panel members: for the synthesis workshop, Tom Abraham, Keith Blakely, Gary Messing, William Payne, Bill Rafaniello, and Stephen Costantino with James Adair as the moderator; and for panel members on the shape-forming workshop, Ilhan Aksay, John Jones, Mark Janney, Fred Lange, George Onoda, John Pollinger, and Vimal Pujari with Arvid Pasto as the moderator. The goal of the symposium organizers is that these workshops will form a suitable model for future symposia to ensure that the developers and potential users of advanced technology are in communication.

This Ceramic Transactions volume is organized by topic with collections of papers relevant to each of the following aspects of powder synthesis and processing: overviews related to commercialization and workshop reports, powder synthesis, powder processing—dispersion and rheology, and powder processing—forming and consolidation.

Any undertaking of this nature requires the combined effort of many individuals and organizations. The editors are grateful for the sponsorship of the symposium by the Basic Science, Engineering Ceramics, and Electronics Divisions of the American Ceramic Society, the National Institute of Ceramic

Engineers, University of Florida Foundation, Pacific Northwest Laboratory, Ames Laboratory, and the Department of Materials Science and Engineering at Iowa State University for financial support in organizing the symposium. We are also grateful for the many individuals who contributed their time and expertise as reviewers of the manuscripts and would like to recognize the authors of each manuscript for the excellent quality of each submitted document. Finally, we would like to thank Ms. Lori Kozey, Rich Zumpone, and other individuals in the publications department of the American Ceramic Society for their patience and support during the review and editing process.

It is the sincere hope of the editors that the readers of this volume will find it informative and useful in both their research and educational endeavors.

J.J. Kingsley
Christopher H. Schilling
James H. Adair

Overviews

ADVANCED CERAMIC POWDER AND NANO-SIZED CERAMIC POWDER: AN INDUSTRY AND MARKET OVERVIEW

Dr. Thomas Abraham
Business Communications Co., Inc.
25 Van Zant St., Norwalk, CT 06855
Phone: (203) 853-4266, Fax: (203) 853-0348

ABSTRACT

Advanced ceramic materials constitute an emerging technology with a very broad base of current and potential applications and an ever growing list of material compositions. Ceramic powder is a necessary ingredient for most structural ceramics, ceramic composites, electronic ceramics and ceramic coatings. For most advanced ceramic components, starting powder is usually of critical importance. In the last few years, several new powder synthesis techniques have been developed as alternatives to established methods. A recent development is the availability of nano-sized powders. This paper analyzes the industry and focuses on the current and emerging applications, major issues, current and future markets.

INTRODUCTION

Advanced ceramics encompass a variety of materials, markets, and applications. The materials include: oxides, carbides, nitrides, borides, silicates and glass ceramics. The most commonly used advanced ceramic materials are: alumina, zirconia, silicon carbide, silicon nitride, sialon, ferrites and titanates.

Ceramic powder is a necessary ingredient for most of the structural ceramics, ceramic composites, electronic ceramics and ceramic coatings. The most commonly used ceramic powders are oxides, carbides, nitrides and borides. New technologies are emerging for the production of ultrapure and fine ceramic powders. These new developments have helped to improve the quality of ceramic components. The impact of these new developments will be felt as the advanced ceramics industry scales up manufacturing processes and the powders produced by the new processes become competitive with those produced by established techniques.

ADVANCED CERAMIC APPLICATIONS AND MARKET SEGMENTS

The major applications and market segments can be broadly categorized as follows:

STRUCTURAL CERAMICS
- Heat Engine Components
- Cutting Tools
- Wear-Resistant Parts and Industrial Applications
- Energy Related
- Aerospace and Defense
- Bioceramics

ELECTRONIC CERAMICS
- Insulators, Substrates and IC Packages
- Piezoelectric Ceramics
- Ferrite Magnets
- Capacitors
- Superconductors

CERAMIC COATINGS
- Auto, diesel & Land-Based Turbine Engines
- Aircraft and Aerospace Engines
- Heat Exchangers
- Wear Parts & Industrial Applications
- Cutting Tool Inserts

U.S. MARKETS FOR ADVANCED CERAMICS

Business Communications Co. has been studying the various advanced ceramics market segments for the last twelve years. In a study[1], BCC analyzed the current and future U.S. markets for advanced ceramics. Table I summarizes BCC's estimate of the size of the 1994 and future market potential for the various advanced ceramics market segments in the U.S. The total value of the U.S.

advanced ceramic components market for 1994 is estimated to be $4.9 billion. This will increase to $8.5 billion by the year 2000 with a growth rate of 9.8% from 1994 to 2000. In terms of market share, electronics constitute 79% of the market now. Figure 1 shows share of the different market segments for 1994.

Table I. U.S. Markets for Advanced Ceramic Components from 1994 -2000[1]

	1994 ($ mil.)	2000 ($ mil.)	AAGR 1994-2000 (%)
Structural Ceramics	475	1,020	13.5
Electronic Ceramics	3,864	6,573	9.3
Ceramic Coatings	525	940	10.2
Total	4,864	8,533	9.8

AAGR - Average Annual Growth Rate
Note: All dollar values are in current U.S. dollars
Source: Business Communications Co., Inc.

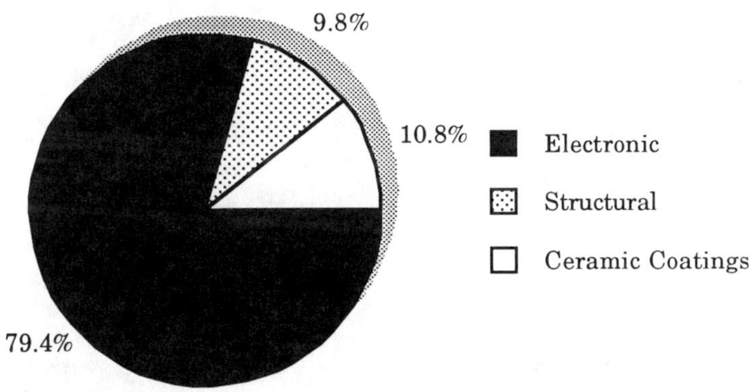

Total Market Value for 1994: $4,864 million

Figure 1. Summary Chart Showing Share of U.S. Advanced Ceramic Market Segments for 1994[1]

Source: Business Communications Co., Inc.

ADVANCED CERAMIC POWDER SYNTHESIS

The basic raw materials for ceramics occur abundantly in nature. However, they must be refined, processed or synthesized before they can be used to fabricate structures. The ceramic powder production technologies are listed in Table II.

Table II. Advanced Ceramics Powder Production Technologies

CONVENTIONAL	Alumina from Bauxite Zirconia from Zircon
POWDER SYNTHESIS TECHNOLOGIES	
CARBOTHERMAL REDUCTION	AlN, AlON, SiC, TiB_2, ZrB_2
NITRIDATION	AlN, Si_3N_4
VAPOR PHASE REACTION	Si_3N_4, AlN and SiC
PLASMA PROCESSES	SiC, TiC, B_4C, AlN, Si_3N_4, BN Al_2O_3, ZrO_2, Cr_2O_3
POLYMER PYROLYSIS	SiC, Si_3N_4
SOL GEL	Al_2O_3, $BaTiO_3$, ZrO_2
PRECIPITATION	$BaTiO_3$, $SrTiO_3$, MgO-ZrO_2 Y-Ba-Cu Oxalate(to $Y_1Ba_2Cu_3O_{7-\partial}$) $Si(NH)_2$ (to Si_3N_4)
HYDROTHERMAL PROCESS	Al_2O_3, $BaTiO_3$, ZrO_2
EMULSION PROCESS	Oxide, titanate & superconductor
LASER SYNTHESIS	SiC, Si_3N_4, B_4C, Al_2O_3, TiB_2
SHS	TiC, Si_3N_4, TiB_2

U.S. INDUSTRY STRUCTURE

The U.S. advanced ceramic powder industry is characterized by about 40 producers and suppliers. There are about two dozen suppliers of oxide ceramic powders. There are four major producers of alumina powders and an additional three major suppliers. Zirconia is produced by two companies while an additional four companies supply foreign produced zirconia. The U.S. carbide powder business is characterized by over a dozen players. Most of them supply silicon carbide powders and a few supply boron carbide powders.

There are not many producers of silicon nitride and sialon powders in the U.S. and most of the demand is met by imports. The major producers for silicon nitride are in Europe and Japan. There are two producers each in boron nitride, aluminum nitride and titanium diboride powders.

There are a dozen companies producing developmental or prototype quantities of nano-sized ceramic powders. Two companies produce nano-sized powders in large quantities. About one half dozen companies are in a position to produce large quantities, if there are sufficient orders.

U.S. MARKETS

Business Communications Co. recently published a study[2], "Advanced Ceramic Powders & Nano-Sized Ceramic Powders." Table III shows a summary of the 1994 U.S. market for advanced ceramic powders and nano-sized ceramic powders and the market forecast for the year 2000. The U.S. market for advanced ceramic powders (including nano-sized ceramic powders) was valued at $684 million in 1994. By 2000, the market is expected to reach almost $1.1 billion, reflecting 8.1 percent average annual growth. Figure 2 shows an illustration of the market share of the three broad categories of the applications.

Table III. U.S. Market for Advanced Ceramic Powders by Application[2]
(Millions of Current Dollars)

	1994	2000	AAGR(%) (1994-2000)
Electronic	613	977	8.2
Structural	45	83	10.7
Thermal Spray	26	31	3.6
TOTAL	684	1,092	8.1

Note: AAGR - Annual Average Growth Rate

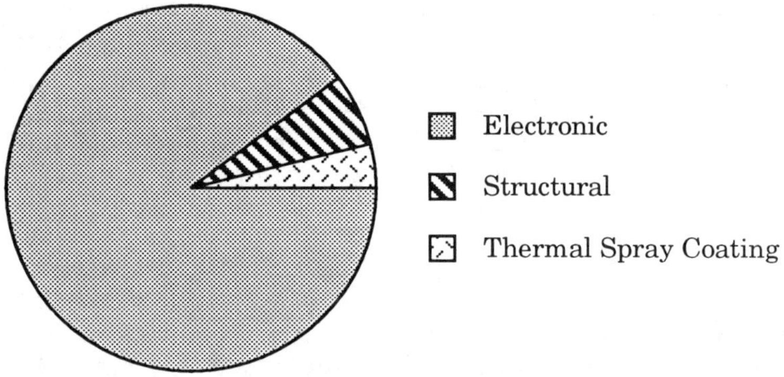

Total Market for 1994 - $684 million

Figure 2. Market Share of Advanced Ceramic Powder Applications, 1994[2]

Electronic applications such as insulators, substrates, IC packages, capacitors, and magnets will dominate the advanced ceramic powder marketplace, accounting for over 90 percent of total powder usage. In 1994, electronic applications used advanced ceramic powders valued at $613 million. By 2000, the market value of ceramic powders used in electronic applications will reach an estimated $977 million as a result of 8.1 percent average annual growth.

Advanced ceramic powders used in structural applications such as wear-resistant parts, cutting tool inserts and bioceramics are expected to experience the most rapid growth in usage. Accounting for advanced ceramic powder sales of $45 million in 1994, structural applications are expected to utilize approximately $83 million worth of advanced ceramic powders by 2000, reflecting 10.7 percent average annual growth. Use of advanced ceramic powders in thermal spray coatings (such as those used in aircraft engines) will grow only modestly in value at 3.6 percent annually to reach $31 million by 2000. This smaller growth rate is due to the reason that there has been a decrease in demand for new military aircraft engines where thermal spray coatings are used. Also, ceramic coatings used for wear resistant applications use more of CVD coatings.

In the material types, oxide ceramic powders have as much as 97% of the market and this is expected to continue till 2000. This is due to the fact that the present oxide markets will continue and, secondly, more non-oxide ceramic components will be produced by non-powder routes.

Among the oxides, electronic applications such as substrate materials for integrated circuits and insulators have emerged as the largest market for alumina ceramic powders. Ceramic ferrite powders used for making permanent and soft magnetic materials constitute the second largest market. Ceramic titanates are the basis for electronic capacitors as well as piezoelectric components and constitute the third largest market.

Commercial applications of superconductor mixed oxide ceramics are yet to take off. Many users formulate their own powders. Several companies, however, supply these powders in developmental and prototype quantities. There are a few companies currently using ceramic superconductor powders in prototype and limited commercial quantities. The market is still small; however, there is a tremendous potential for superconductor ceramic powders.

Among carbides, silicon carbide constitute 96% of the market, the rest being boron carbide. These powders are used mostly for structural ceramic applications such as mechanical seals, sliding bearings, cutting tools, linings and other wear resistant products together with rocket nozzles, engine components, heat exchangers and other high temperature applications.

Nitride powders commercially used are silicon nitride and sialon, boron nitride and aluminum nitride. The largest use is for silicon nitride and sialon for applications such as cutting tools and wear parts. Aluminum nitride is yet to be developed into a major market segment while boron nitride has an established niche market for evaporator crucibles and boats.

Titanium diboride moistly constitutes the boride ceramics market and is used in niche applications. Current major uses are for making evaporation boat and industrial armor.

There has been much hyperbole in the market about nano-sized ceramic powders. The industry is now beginning to recognize the various technical and business issues. The current market is less than half a million dollars which include experimental, developmental and prototype quantities. Current applications include the materials used as catalysts in solid rocket fuel which is an established use. However, this market has been declining because of the cuts in defense. Other uses are materials for making developmental and prototype ceramic components and for making ferrofluids. BCC does not expect a large market as compared to other predictions since several technical and business issues are yet to be resolved before massive commercialization can take place.

COMMERCIALIZATION ISSUES FOR NEW SYNTHESIS TECHNOLOGIES

In the processing of advanced ceramics, there is a growing need to develop synthesis techniques capable of producing submicron, chemically pure powders with a tailored size distribution. However, new synthetic processing techniques are comparatively more expensive than the currently established powder manufacturing methods.

Over the last ten years, U.S. researchers have been attempting to develop technologies for the production of ultrapure and nano-sized ceramic powders. As the advanced ceramic industry upgrades manufacturing processes, these new powders should become competitive with those produced using conventional production techniques. Despite inflated estimates of huge future market demand for nano-sized ceramic powders, it is more likely that consumption will remain relatively small compared to that of conventional ceramic powders. Even with an estimate of large growth rate of about 60% per year, BCC study shows that the nano-sized ceramic powders are expected to reach only less than one percent of the total market share of the advanced ceramic powders in the year 2000. However, there has been a recent surge of interest in nano-sized ceramic powders.

Presently, over a dozen companies are supplying developmental quantities, while a few will soon be able to supply prototype and commercial quantities. However, many technical and business issues (including proper packaging and handling, processing powders into components, and high cost) must be resolved before nano-sized powders can be used in commercial applications. Also, environmental and health related issues in the production of nano-sized powders have not been assessed.

GLOBAL COMPETITION

In terms of global competition, tremendous inroads have been made by Japanese and European companies in the production of advanced ceramic powders. While the U.S. still maintains a lead in some ceramic powder markets, foreign companies are dominating the U.S. companies in zirconia and nitrides. Also, in the last six years, the Japanese and European companies have penetrated the U.S. advanced ceramic industry via acquisitions, mergers, joint ventures and licensing agreements (see Table IV). With these new developments, the U.S. ceramic industry is highly dependent on foreign owned companies as well as imported products.

Table IV. Recent Major Acquisitions of the U.S. Ceramic Companies by Foreign Companies

Foreign Company	Acquired U.S. Company	Ceramic Business	Year
British Petroleum (U.K.)	Standard Oil Engineered Materials (Carborundum)	Structural	1988
Cookson (U.K.) (Tam Ceramics)	Du Pont Electronics	Electronic (Dielectric Powder)	1991
Didier Werke (Germany)	Corning's Zircoa Products	Structural & Powder	1988
Elkem Metals (Norway)	Ceramatec	Structural	1989
Feldmuhle (Germany)	Alsimag	Structural & Electronic	1991
Hertel AG (Germany)	Adamas Carbide	Structural	1988
Kyocera (Japan)	AVX	Electronic	1989
Morgan Matroc (U.K.)	Duramic Products	Structural	1988
	Vernitrorn	Electronic	1989
	GTE Wesgo	Structural	1991
Osram/Siemens (Germany)	GTE Products	Structural	1992
SECO/Sandvik (Sweden)	Carboloy Div. of GE	Structural	1987
Saint-Gobain (France)	Norton	Structural	1990
TDK (Japan)	Allen-Bradley	Electronic	1992

CONCLUSIONS

- Advanced ceramics is expected to continue its growth in the 1990s and into the next century.

- Electronic ceramics will continue to hold the largest share of the ceramic powder market, although the largest growth rate is expected to be for the structural ceramics market.

• To promote structural ceramics applications, there is a need for better quality powder at affordable cost.

• There are still many technological and business issues to be resolved before nano-sized ceramic powder could become commercially viable.

References

1. "A Complete Guide to the U.S. Advanced Ceramics Industry," Published by BCC, 1995.

2. "Advanced Ceramic Powders & Nano-Sized Ceramic Powders - Technologies, Applications, New Developments and Market Opportunities," Published by BCC, 1994.

COMMERCIALIZATION OF POWDER SYNTHESIS PROCESSES

Greg Geiger
The American Ceramic Society
735 Ceramic Place
Westerville, OH 43081-8720

Ceramic powders are a critical first step in the production of most advanced ceramic products. The synthesis of these powders has been an area of much interest to engineers and scientist in industry, academia, and government. Basic research is aimed at lowering powder production and shape-forming costs, improving powder properties, increasing product reliability and improving characterization techniques.

To explore how technology derived from basic research in powder synthesis affects commercialization, a panel discussion at the technical session "Commercialization of Powder Synthesis and Shape-Forming Processes," was held during the 97th ACerS Annual Meeting.

This session was part of the "Science, Technology, and Commercialization of Powder Synthesis and Shape-Forming Processes Symposium" organized by J. J. Kingsley, Pacific Northwest Laboratory, C. H. Schilling, Iowa State University, and J. H. Adair, University of Florida.

Panel members of the powder synthesis discussion included T. Abraham, Business Communication Co. Inc. (BCC); K. A. Blakely, Advanced Refractory Technologies; G. L. Messing, Pennsylvania State University; W. H. Payne, Visatech Corp.; W. Rafaniello, Dow Chemical Co.; and S. Costantino, Cabot Corp. The discussion was moderated by Adair. Below is an overview of this panel discussion.

INTRODUCTION

The market for advanced ceramics is large and growing. BCC estimates the total value of the U.S. advanced ceramics components market for 1994 to be \$4.9 billion. This is expected to increase to \$8.5 billion by 2000. BCC estimates the value of the U.S. advanced ceramics powders market for 1994 to be \$648 million—increasing to almost \$1.1 billion by 2000 [1].

Advanced ceramics are generally divided into three categories—structural, electronic, and coatings. Structural ceramics, the fastest growing segment, includes applications such as industrial wear parts, bioceramics, cutting tools and engine components. Electronic ceramics, which has the largest share of the advanced ceramics market, includes dielectrics (capacitors, insulators and substrates), integrated circuit packages, piezoelectrics, superconductors and magnets. Ceramic coating applications include engine components, cutting tools and industrial wear components.

The advanced ceramic fabrication process generally includes powder production, powder conditioning, forming, and densification. The starting powders are critical to the quality of the final product because flaws that occur at this stage cannot be corrected in the following processing steps. Powder characteristics that are of importance include particle size, particle shape, particle size distribution, chemical purity, surface chemistry, loss on ignition, morphology, and degree of agglomeration [2].

The starting powders for advanced ceramic applications normally involve some type of chemical synthesis process. Some of these processes include organometallic synthesis, sol–gel processing, polymer pyrolysis, plasma processing, metal nitridation, carbothermal processes (i.e., electric arc, thermite, SHS), precipitation, hydrothermal processing, aqueous processing, and laser synthesis [1,3].

ROLE OF BASIC SCIENCE

Panel members were asked how the emphasis on basic science in the area of powder synthesis had contributed to ceramic manufacturing. This prompted some lively and interesting discussion from both the panel members and the audience on the relationship between academia and industry.

While agreeing that there has to be some connection between industry and universities in the area of direction for R&D, one participant commented that universities are not always as well aware of industry's needs as they should be. On the other hand, universities have the ability to do "blue sky" research for little cost, he said. This type of research will result in periodic breakthroughs to assist industry.

An audience member asked if there was an over-emphasis on "novel and innovative" approaches to powder processing and whether these approaches were always helpful to industry. He asked what kind of basic science would have the greatest impact on industry.

One panelist commented that the academic community can best assist in commercialization of advanced ceramics powders by advancing the fundamental understanding of how heat and mass transfer affect the reaction kinetics. This was later echoed by another panelist, who said kinetic studies are needed to increase production of commercial powders. The increased efficiency would lower powder cost, he said.

It also was pointed out that synthesis research thus far has focused mainly on the chemistry of the precursor with little attention to the characteristics of the final powder. Ceramic powder manufacturers need detailed information on powder characteristics in order to achieve reproducibility and cost competitiveness, the panelist said.

Another need cited is to conduct research that couples reaction engineering with precursor chemistry in support of powders synthesized with targeted properties. This also is known as "smart processing." Other needs and directions in powder synthesis pointed out by one of the panelist included:

Needs:

- Synthesis processes resulting in <u>dense</u> particles;
- Comminution, separation and classification of <0.5 μm powders;
- Synthesis of dispersible powders (with dispersants);
- Cost effectiveness.

Direction:

- Seeded or controlled nucleation during synthesis;
- Processes for synthesis of composite particles;
- Processes for synthesis of anisotropically shaped particles;
- Powders <100 nm in diameter.

One panelist commented that, in many cases, there is a true discontinuity between R&D that is "blue sky" and R&D that is directly transferable to industry. It is more "sexy" to do R&D on an complex technique that will produce, in the lab, if processed carefully, unique characteristics, than it is to take the same powder and make it applicable for processing high volumes of a finished product in the plant, he said.

A representative from a powder manufacturer commented that you need to be careful of your expectations of what academia can provide. You should not look to university researchers to become industrial process engineers that are going to solve plant problems, he said. Nor should we think that we are going to pick up a process technology from a university, license it and turn it into a commercial product.

Companies need to look primarily for injections of very specific technical capabilities or insights from the universities, he said. As an example, he described the development of an aqueous, processable powder that was commercialized as a result of utilizing basic research by a university chemist.

IMPACT ON THE ECONOMY

When asked how new technology coming from basic research had impacted the economy, most panelists agreed that there has not yet been a major impact on the economy from new technology or that it was too early to tell what amount of growth might occur.

One panelist pointed out that there has been progress and renewed interest in nano-sized powders. Despite this, there are still many technological and business issues to overcome before nano-sized powders become commercially viable, the panelist said. Some of these barriers include packaging and handling, component processing technology, high cost, and environmental and health issues. According to BCC, nano-sized powders are expected to reach <1% of the total market share of advanced ceramic powders by the year 2000.

In the area of electronic ceramics, one panelist pointed out that chemically precipitated barium titanate powders are the fastest growing market for multilayer ceramic capacitors (MLCCs) application. The current world market for MLCCs is ~$3 billion/year. The annual use of barium titanate in MLCCs is ~10 million lb, while the market for chemically precipitated barium titanate powders is currently the smallest (~25%). Because of these new powders, MLCCs with more than 100 layers are somewhat common today.

Chemically synthesized ceramic dielectric powders are now available with controlled particle size distribution, controlled chemistry and improved homogeneity. These powders enable capacitor manufacturers to make thinner (≤6 μm) dielectric layers, using both doctor blade and new coating processes.

Reducing the dielectric thickness of MLCCs increases the capacitance efficiency of the device. This results in both improving the cost effectiveness of MLCCs and increasing the market share of ceramic capacitors at the expense of electrolytic capacitors.

COMMERCIALIZATION BARRIERS

The commercialization barriers for implementing the new technologies was also discussed during the session. Improvements in shape forming, characterization techniques, nondestructive testing, machining and development of a materials property database were all cited as having a direct impact on the commercialization of advanced ceramic powders.

Researchers from industry, government and academia continue to make advances in all of these areas. ASTM Committee C28 Advanced Ceramics, for example, has been developing a standard system for classification of advanced ceramics that will pave the way for a usable materials property database.

Advances in shape-forming technology were evident from the many papers presented on this topic during the symposium.

Another key factor for increasing commercialization of powders, according to one panelist representing a powder manufacturer, is going the extra mile for your customers. Developing enabling technology that allows your customers to use your powders was one example of a value-added service.

From a technology standpoint, there are things that academia can do to help industry, but most of the problems with the advanced ceramics market are problems that must be solved by industry, one panelist commented. Value, he said, is the main issue—"Where are you getting the value from the product? Until industry can focus enough on applications where the benefits of new materials can be exploited, we will always be in a quagmire where industry will not grow at the predicted rates."

Another panelist agreed that value is a strong consideration. "This implies that there is an open dialogue between the end user, the processor and the raw-material supplier." The raw-material supplier needs to hear what the real requirements are in the end application, he said. He also commented that our real customers—the end users—are not at this meeting. The advanced ceramics producers need to be educating the end user of the advantageous properties and potential applications that ceramics can provide.

Continuing the dialogue on value, another panelist commented that value to those participating in the session equals technology push. According to the panelist, one of the problems with the advanced ceramics market, particularly structural ceramics, is that the right questions are not being asked; e.g., "What is the market pull, or what will sell a product?" Such questions as, "What is the value of having a ceramic component in an automobile engine in the eyes of the buyer?" need to be considered, he said.

COST CONSIDERATIONS

The cost of advanced ceramic powders varies widely. For example, electronic-grade alumina (96%) powders range from $1–2/lb while high-purity-grade alumina (>99.9%) and silicon nitride powders can cost as high as $70/lb and $150/lb, respectively. Taking everything into consideration, the high cost of powders is still the major factor holding back the large scale use of advanced ceramics said one panelist.

One panelist commented that there is a growing need to develop synthesis techniques capable of producing submicron, chemically pure powders with a narrow size distribution. These new synthetic routes to advanced ceramic powders are more expensive than currently established powder-manufacturing methods, however.

Speaking from a non-oxide powder supplier's point of view, one panelist stated that the availability of high-purity, aqueous, processable powders at low cost will enable advanced ceramics manufacturers to make more mistakes. It is these mistakes that will lead to increased knowledge in the area of advanced ceramics processing, which, in turn, will lead to products that are reliable and insertable into real-world applications.

It also was pointed out that overall processing and systems costs must be taken into consideration when producing advanced ceramic products. There is a linkage between powder cost and machining cost he said, referring to an earlier comment that machining contributes up to 50% of the overall cost of a finished component. If producers drive down powder specifications purely for economic requirements, the result will be higher machining cost, he said.

Tremendous progress has been made in advanced ceramics technology in the last decade said one panelist. The advanced powders coming out of these new technologies, however, are not always good for the commercialization of advanced ceramics. This is due to the high costs of these powders, and most importantly, because they are not needed for the applications in which they will be used, he said.

The panelist commented that there is a need to find areas to sacrifice in order to lower costs of the powder, while at the same time not compromising the finished product. Another participant agreed. There are a million ways to make ceramic powders, but you have to marry the process to the application and cost, he said.

One audience participant commented that, in the case of advanced structural ceramics, the failure of industry to market certain products is more related to cost than to the relationship between basic science and industry. He believes that, because powder and shape forming technologies take such a long time to develop, only large companies are able to afford the investment. Smaller companies are often not able to sustain the development needed to see a profit due to the amount of time and money required.

References:

1. T. Abraham, "Advanced Ceramic Powder and Nano-Sized Ceramic Powder: An Industry and Market Overview;" Paper No. SXIX-46-95 presented at ACerS 97th Annual Meeting, May 2, 1995.
2. S. Hirano, G. L. Messing and H. Hausner, Ceramic Transactions, Vol. 51, Ceramic Processing Science and Technology, p. 3, The American Ceramic Society, Westerville, Ohio, 1995.
3. M. Schwartz, Handbook of Structural Ceramics; 4.1–4.10. McGraw-Hill, New York, 1992.

THE CRITICAL ROLE OF RAW MATERIAL SUPPLIERS IN MARKET DEVELOPMENT

Keith A. Blakely, Advanced Refractory Technologies, Inc., Buffalo, NY

EVOLUTION OF THE MATERIAL SUPPLIER'S ROLE

Historically, the view which most people have had of the role of the raw material supplier has been rather straightforward - the supplier should make a product which is high quality and low cost and be sure that it never varies. From a customer's point of view, having a supplier meet these "simple" requirements enabled them to produce components which were high quality, affordable, and reliable or uniform from batch to batch.

As the world moves toward a truly global economy and designers are challenged to increase a component's performance while at the same time reducing its cost, the role of the supplier of raw materials, including advanced ceramic materials, is changing. Both economics and product enhancements are critical "musts" for the advanced ceramic materials supplier to be successful today. These elements combined with the supplier's ability to be responsive to the needs of the designer and his/her application will result in market expansion for advanced ceramic materials.

THE IMPORTANCE OF ECONOMICS

There is no doubt that economics are a significant driver in materials selection and, as importantly, in the growth of applications and market segments. ART's own estimate of the relationship between powder cost and market size for aluminum nitride is depicted graphically in Figures 1 and 2. At $50 per pound, the only applications which can clearly afford to utilize AlN are those in the electronics segment - and even there the use is limited. To date, the economics of most AlN powders has borne out this observation and the domestic market for high performance, high cost AlN is roughly $1 million (powder value, not component value). The structural market for this material, however, is much larger - we

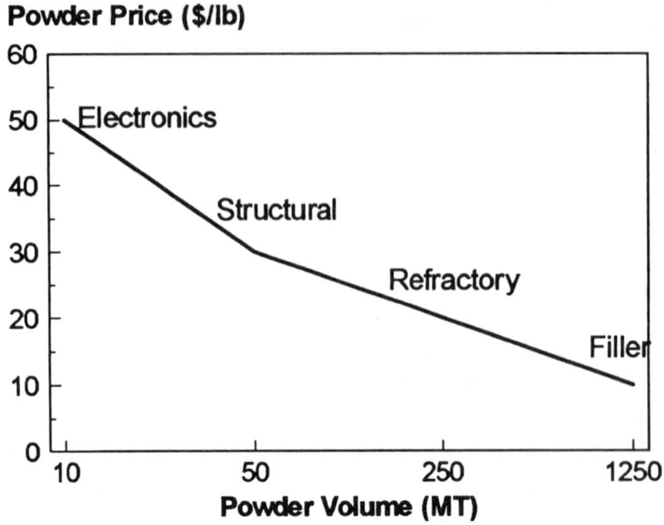

Figure 1. Price-Volume Sensitivity for AlN Markets

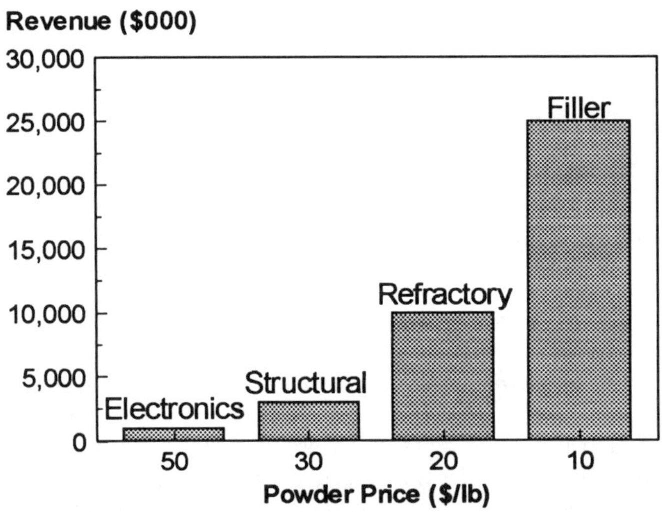

Figure 2. Opportunity for AlN in Market Segments

estimate it to be $3 million or more - but requires pricing which is competitive with other structural non-oxide powders, such as silicon carbide and silicon nitride, which run roughly $30 per pound or less. An even greater potential for AlN exists in the refractory market segment, where the corrosion resistance and compatibility with molten metal makes AlN a very interesting material. However, the economic limitations of this market segment would require AlN powders which are $20 per pound or less. Getting the economics to that level could open up market applications which would consume upwards of 250 MT annually, for a total market value of $10 million or more. And finally, an existing market for fillers for plastics and other materials exists which consumes an estimated 1,250 MT of alumina and silica powders annually. The high thermal conductivity of AlN makes it very attractive for this application, but the economics of the final filled plastics demand that the filler material not exceed $10 per pound in price.

ADDITIONAL REQUIREMENTS

It would appear, from this quick analysis, that the simple job of the raw material supplier, then, is to concentrate on cost reduction and let the customers and market forces handle everything from that point forward. Unfortunately, in reality, it is much more complicated than that. Particularly in an emerging area such as high performance ceramics, the role of the raw materials supplier involves delivering a broad range of "services" and "products" - better called product enhancements - only one of which may actually be specified on the purchase order! Without these other deliverables, the reduced pricing of the raw material may not (in fact, almost never will) be enough to answer market needs or create the market usage which is projected.

While the following list is not intended to be all-inclusive, it covers many of the more critical deliverables which every successful raw material supplier must be prepared to offer the customer. Note that these go well beyond the product itself:

- Technical Support
- Engineering Assistance
- Quality & Quality Assurance
- Processing Aids
- Applications Ideas
- Customer Contacts
- Technology Protection
- Economic Advantages

THE IMPORTANCE OF PRODUCT ENHANCEMENTS

The supply of these peripheral elements of the product are critical to the development of the commercial market for these new materials. A raw material supplier cannot be content to make his product, announce its availability to the world, and expect the proverbial path to be beaten to his door. To understand why, it is helpful to view the typical, highly simplified, user channel associated with an advanced ceramic material.

Raw Material Supplier(s) → Fabricator → Machine Shop → Assembler → System Integrator → End User (the "Market")

What drives the user to specify an advanced ceramic? Generally, they don't! What they (the market) demand is higher performance which can only be achieved using high performance materials, such as ceramics. The design engineers at the system integrator must therefore be aware of what ceramics can do to meet the performance demands of the marketplace.

In a user channel such as this, with a technically challenging material, such as ceramics, the probability is high that there will not be adequate understanding somewhere along the chain and, as a result, the use of ceramics will NOT be specified and an opportunity may be lost. This is where the role of the raw material supplier becomes critical in the development of the advanced ceramic marketplace. Although many have described this role as "pushing a wet noodle up a hill", it is very important that the issues and opportunities associated with using or incorporating advanced ceramics into products be well-communicated through the entire user channel.

AN EXAMPLE

Perhaps the best example of this can be found in ART's development of the lowest cost aluminum nitride powder currently commercially available - called "A500" by the company. This product was developed, in fact, to help create the market awareness and potential utilization in high volume applications which had been identified by the company in various market surveys conducted over the past few years. Although the material met the specifications required by many high end users, including a number of electronics companies (see Table 1), it was manufactured in such as way as to permit pricing at less than $10 per pound in large volumes. One of the first responses to this new material came from a customer who was primarily interested in the refractory properties of AlN and

Table I. Specification for A500 Aluminum Nitride

Property	Units	Value
Surface Area	m^2/gram	1 - 3
Oxygen	%	< 1.3
Carbon	%	< 0.05
Trace Metals	%	< 0.1
Price at 10 MT	$/lb.	< 10
Production Capacity	MT	> 100

whose application could ultimately lead to hundreds of tons of product demand annually. This opportunity demonstrates the extensive role of the raw material supplier in market development.

The specific project ART became involved in was for advanced batteries. The high operating temperatures, corrosive electrolytes, and high power densities used in these batteries meant that conventional materials of construction would not be adequate and design engineers were searching for new materials that would resolve their problems. The development of that solution is an interesting example of how the raw material supplier is called upon to help create the market opportunity for advanced ceramics.

The original inquiry to ART came in the form of a simple question: Can you supply cheap aluminum nitride? The answer, naturally, was yes. Fortunately, however, the conversation did not end there. The customer DID need aluminum nitride, but what they really needed was a solution to the following problem:

The Customer's Problem: Develop a porous separator for a high power density battery which can survive high temperature, corrosive environments for extended periods without change.

The answer to that was more involved than simply providing aluminum nitride powder. The customer's experience was in using other ceramic powders in the form of a pressed plate, unsintered, consisting of the ceramic separator material (a non-conducting, corrosion resistant material which keeps the electrodes from touching and shorting out the battery) pressed together with the battery electrolyte (which thus is in contact with and conducts electricity from electrode to electrode). The problem with this was simple: the pressed ceramic/electrolyte plate would deform over time, the result of thermal expansion and contraction as the battery cycled through charging and discharging. Eventually, this deformation would lead to shorting in the cell.

So, what the customer really wanted wasn't just aluminum nitride, but AlN in a form that would resist this deformation and provide longer battery life and higher operating performance. They also needed the solution at a price that would not make the battery uneconomical. The entire battery needed to sell for under $300. With roughly 16 separator plates per battery, the typical price for sintered AlN plates ($1 per square inch) would put the overall cost of the battery separators at $500! After some discussion of system economics, it was decided that the separator plates needed to be made for roughly one tenth the conventional price. Clearly, a different solution was needed than conventional ceramic processing. Further complicating matters is that aluminum nitride powders are normally reactive with water, tending to hydrolyze and form ammonia gas. So simple, water-based processing wasn't going to be an obvious answer either.

What ultimately emerged as the most cost effective solution to this customer's challenging problem was a combination of technologies, processing techniques and equipment, and a novel raw material that offered the performance and economics desired. But even with all of that, the funding required to develop and demonstrate all of that was arranged by ART - both for itself and for its customer.

In the final analysis, ART, as the raw material supplier, was instrumental in providing the following to the customer:

- Raw Material - A500 Grade
  - -superior economics and appropriate purity and grain size
- Water Resistant Surface Treatment
  - -permitting aqueous based processing of AlN powders
- High speed compaction technology for green tape
  - -involving proprietary binders and processing conditions
- Dimensionally stable composition and firing conditions
- Continuous high temperature inert atmosphere furnaces
- Development and demonstration contract funding
  - -through the Department of Defense, Department of Energy & New York State Energy Research and Development Agency
- Promotional and marketing support
- Customer introductions

RELEVANCE TO MARKET EXPANSION

How much of this type of support is typically required for the sales of non-oxide powders? Throughout its almost fifteen years of business, ART has found that it is NEVER enough to simply sell the raw material. Without strong technical support, engineering assistance, patent advice or interpretation, customer or sponsor identification, processing aids, analytical testing services, furnacing services, and, occasionally, even the furnace design itself, these high performance materials are less likely to succeed. Since it is only through our success in commercial applications and commercial markets that ART can develop a market for its raw materials, the company will continue to provide these services and more as a means to expanding the market for advanced ceramic materials.

Why bother? The first reason is that customer-supplier interactive relationships often lead to strategic commercial partnerships with both parties having a strong commitment to succeeding and a significant motivation for doing so, as well. Such partnerships have, in our experience, been the most productive means for developing commercial products, identifying new applications, and creating long term value for the companies involved. The second reason is that such projects almost always lead to new understandings of the product, process, or marketplace and, as a result, spin-off opportunities are abundant. Even if the outcome of the initial initiative is not favorable, working closely with companies throughout the entire user channel will often lead to secondary opportunities - some of which may be more lucrative than the original project. The third reason is that, as the raw material supplier, if you don't do it, it either won't get done, or your competition will do it. In any case, it's how you sell powder.

ISSUES IN POWDER FORMING TECHNOLOGIES

Laurel M. Sheppard
Lash Publications International
7438 Sawmill Road, Suite 456
Columbus, OH 43235

The continued commercialization of new powder shape forming methods is dependent on solving process safety and economy issues, overcoming limitations with particular manufacturing technologies, and scaling up methods developed in the laboratory into viable manufacturing processes. These issues were addressed in a special panel discussion of industry experts as part of this symposium.

Ceramic powder is a necessary ingredient for most advanced ceramic applications, with the total market estimated at $684 million in 1994, according to Business Communications Company. This market is expected o grow at an annual rate of 8.1% and reach almost $1.1 billion by 2000. The market for components and coatings made from such powders is expected to grow to $8.5 billion at an annual rate of 9.8% by 2000. Such growth rates will depend on a number of factors, including solving some of the problems mentioned above.

Improvements in Binder Technology

Optimizing the binder and other additives is an important step in any ceramic forming process. For handling of cast green tapes during cutting, printing, laminating, and final dicing of electronic substrates, the mechanical performance is partially dependent on the binders. An ideal binder will produce a flexible green tape with reasonably high tensile strength and elasticity. Rohm and Haas Company (Spring House, PA) has introduced an aqueous acrylic emulsion binder, Duramax B-1050, with these properties that also does not produce formaldehyde when burned.

The improvement in elasticity also reduces green defects, such as wrinkling, ridging, and stretching, as well as fired defects caused by shrinkage and warpage. A relatively high solids content can be achieved (48%) and the tensile strength can be doubled if a crosslinking polymer is used. In addition to these advantages, the binder has a low viscosity (100-1000 cP), can be blended with other binders, and burns out cleanly. Tape thicknesses from 2 to 6 mil are possible, though adjusting the glass transition temperature is required for lamination.

In addition to binders, other process parameters must be controlled to minimize defects. For tape casting of lead zirconate titanate, deairing time and fish oil content has been found to be the most important factors to control large pores (>20 microns). The optimum content is 0.3 wt%; the number of pores increases with a content >1.5 wt%. An increase in deairing time reduces the number of pores.

For thicker tapes, extrusion is preferred and is currently used for small scale production of substrates, resistors, and sensors. An optimum binder for such a process must mix readily, be highly plastic and homogeneous, retain both wet and dry extrudiate dimensions, and burn out cleanly in either an oxidizing or reducing atmosphere. A two-part binder system has been developed by Rohm and Haas (Table I) that consists of several aqueous emulsions (30-50% polymer content) whose ratio can be adjusted to tailor the ductility and plasticity. The binder system has been successfully used to extrude tapes 1/16 inch thick by 1/2 inch wide. Fast drying without cracking is another advantage.

Table I. Typical Binder Formulations for Extrusion

Component (wt %)	High modulus	Low modulus
Alumina	100	100
PEZ-400	0.5	0.5
AMP 95	1.0	1.0
High green strength binder*	2.7	----
Medium green strength binder**	1.3	1.3
High plasticity binder***	---	2.7
Water	20.5	19.1

*Duramax B-1039, **B-1051, **B-1052, all manufactured by Rohm & Haas

Such polymer blends can fine tune binder properties since the ideal glass transition temperature (T_g, the softening temperature of the polymer) for an application may not be readily available. Such an approach can also be applied to dry pressing and tape casting. To determine if binders are compatible, they must produce a clear film after mixing together in a solvent. Such blends can result in a higher green strength, no evaporation of toxic volatiles during drying, and less variability Because there are two T_gs, the microscopic properties are preserved and the macroscopic properties change smoothly. A general rule is to add a high

T_g material to increase green strength and to add a low T_g material to increase green density and flexibility.

Improved binder systems have also been developed for injection molding, which is attractive for making small complex parts. BASF Corp. (Wyandotte, MI) has developed a catalytic debinding process for polyacetal binders, which results in short molding cycle times, high green strength, and fast debinding without deformation. Ready-to-mold compounds are commercially available for various oxides and non-oxides, as well as the equipment needed for this process.

The Search for the Perfect Forming Technique

The ideal forming technique for ceramic components used in such applications as turbine engines must be able to produce complex shapes at high rates (1-5 min/part) and high yields (>90%) with dimensional control of <1%. A colloidal processing method is preferred due to its ability to control flaws and should be able to produce parts with an as-processed strength of 80%. The ideal forming method should also be easily adapted to automation and be able to produce near net shapes so that machining costs are minimal. One of the most important requirements is that the method be "robust," which means optimization by experimental design.

It is also vital to have a reliable reproducible raw material. Unfortunately, some commercial powders still have lot-to-lot variation and must undergo a treatment to normalize the powder. Otherwise a uniform green density and high green strength cannot be obtained. Of course, such treatments will add to the overall cost even though the powder may be affordable otherwise.

Though conventional forming techniques, such as slip casting and injection molding, can meet some of the above requirements, the search for an ideal method continues. One method that may be the solution is gel casting, developed at Oak Ridge National Laboratory in the mid-80s and first licensed to Coors in 1989. This method uses a slip containing from 1-4 wt% monomer capable of in situ polymerization forming a polymer solvent gel, a crosslinking agent or catalyst (ammonium sulfate), and a mold, which is heated to moderate temperatures. Molds can be made out of aluminum, wax, and plastics (except for polyurethane and silicone rubber). After the part is removed from the mold, it is dried and fired, with or without green machining.

One major advantage of this process is that the mold filling step occurs separately from the setting mechanism, avoiding problems encountered with injection molding where these steps occur simultaneously. Gellation generally occurs in 10 minutes to half an hour from room temperature up to 60°C, with times as low as 5 minutes with microwave heating. Packing density gradients are minimal (0.7%) and green strengths are higher than conventional processes (600 psi). A 50-80% reduction in process temperature is possible.

Drying and sintering cycles are similar to that of slip cast parts, whereas binder burnout cycles are similar to dry pressing. However, binder burnout and pyrolysis have been shown to be a function of both ceramic and type of polymer; alumina burnouts by 550°C with pyrolysis occurring in several stages, and ferrites burnout by 380°C within a narrow processing window. Polymerization is accelerated for SiC and retarded when firing alumina. Silicon nitride requires the highest binder burnout temperature.

Cost sharing between government and industry has been important in commercializing this process. A three-year ATP program, sponsored by DOE and DOC, has resulted in transferring the technology to other companies. The technology has been licensed to Lo Tec and a number of CRADAs have been signed. Potential applications include rapid prototyping and UV curing of thin layers. Gel casting of silicon nitride and $MoSi_2$ composites is also being investigated.

A two-year program 50% funded by ARPA will now attempt to scale up the process to high volumes. Allied Signal is sharing the cost of this program and has formed a "virtual company" consisting of 7 turbine engine companies, two universities, a silicon nitride powder producer, and a machining company. A low-toxicity gel system has been developed and demonstrated (based on polyvinyl alcohol), and a six-inch turbine wheel has been made. The transfer of personnel between government and industry has also played a key role in making such progress.

Though gel casting is a colloidal process, it still has problems with hetereogeneities (agglomeration). Though these heterogeneities can be removed by using such techniques as filtration, the ideal approach is to use a powder that behaves like clay. This is possible by changing the repulsive potential of the particles to short range. Weakly attractive networks are formed, which prevent mass segregation. Other approaches to improve dispersion properties of slips include using coupling agents (aminopropylsilane) or acrylic-based polyelectrolytes (polyacrylic acid for silicon nitride). Colloidal processing is also being applied to SiC-whisker reinforced alumina composites and other nonoxides via electrophoretic deposition and osmotic consolidation methods.

Emerging Processes and Market Issues

A number of forming processes under development have demonstrated that unique materials can be made. For instance, co-extrusion of two or more materials can create structures with features on the order of 10 microns. Another method called orientation selective processing has demonstrated it is possible to orient a matrix of monolithic ceramic using anisometric particles incorporated into a material of the same phase. This seeded texturing of one ceramic by a composite addition of an oriented minor phase provides a unique way for microstructural control for composites such as AlN reinforced with SiC platelets.

Superplasticity behavior of ceramics is now being applied to forming processes. Processes that have been demonstrated include extrusion, closed-die deformation (isothermal forging), punch forming, and gas pressure forming of sheets and tubes. However, for these and other processes to be commercialized, they must be able to produce affordable components for new markets that focus on real world applications. If such a market driven approach is taken by both universities and industries, user commitment will be ensured and these technologies will end up in manufacturing; otherwise it will be a long road to commercialization. Advanced ceramics will then eventually become the traditional ceramics of the future.

Author's note: This article is based on papers found elsewhere in this proceedings and comments made by the panel discussion. A similar article has been published in *Ceramic Industry* magazine.

Powder Synthesis

SYNTHESIS OF CERAMIC POWDERS BY NOVEL MICROWAVE-HYDROTHERMAL PROCESSING

Sridhar Komarneni, V. C. Menon and Qing Hua Li
Intercollege Materials Research Laboratory
The Pennsylvania State University
University Park, PA 16802, USA

Abstract

Microwave-hydrothermal processing was compared with conventional-hydrothermal processing in the crystallization of hematite from 1M ferric nitrate and hydroxyapatite from calcium oxide and phosphoric acid. Microwave-hydrothermal processing led to the crystallization of hematite while convention-hydrothermal process led to the crystallization of a mixture of goethite and hematite under the same temperature conditions. Thus the crystallization of hematite which is the stable iron oxide phase at high temperatures has been catalyzed by microwaves. Microwave-hydrothermal processing also led to highly crystalline hydroxyapatite powders compared to the conventional hydrothermal processing which also points to the catalytic role of microwaves. The microwave induced effects may be attributed to the generation of localized high temperatures at the reaction sites which enhance reaction rates.

INTRODUCTION

The use of microwaves in solid-state synthesis and sintering (1-5) and dissolution of inorganic and organic materials (6-9) is now an established practice. The use of microwaves, however, for synthesis in solutions is new. There have been several reports of the utilization of microwaves in catalyzing organic reactions (10-15) and in the rapid synthesis of organometallic (16-18) and intercalation compounds (19). The rapid synthesis of inorganic zeolite phases in solution under microwave field has also been shown recently (20-25). We are, however, the first group to show the rapid synthesis of anhydrous ceramic oxides (26-30), hydroxylated phases (28), metal powders (29, 30) and metal intercalated clays (32). Following our early work, there are several recent reports of inorganic syntheses in solutions under the microwave field (33-35). All these studies show the catalytic role of microwaves in solution syntheses. We, however, showed that new metastable phases (27,29) can be synthesized using microwaves under hydrothermal conditions. The term "Microwave-hydrothermal" processing was coined by us (26) for reactions taking place above 100°C and under the microwave field. In this paper we demonstrate the application of microwave dielectric loss heating effects for accelerating the synthesis of one unary oxide and a binary hydroxylated phase under hydrothermal conditions.

EXPERIMENTAL

A microwave digestion system, MDS-2000 (CEM Corporation) was used to conduct the microwave-hydrothermal experiments. This equipment is normally used for acid dissolution of inorgaic and organic samples for chemical analysis. The system operates at a 2.45 GHz frequency and power can be varied from 0-100% of maximum power (630 ± 50 watts). Hematite was prepared from 1M $Fe(NO_3)_3 \cdot 9H_2O$ solution and hydroxyapatite was prepared from stoichiometric mixture of $CaO + H_3PO_4$. The above solution and mixture of solid and solution were treated with 45% of full power in double-walled digestion vessels which consist of a Teflon PFA inner liner and cover surrounded by a higher strength shell of Ultem polyetherimide. The system is controlled by pressure and operates at a maximum of 200 psi which is equivalent to approximately 194°C when pure water is heated. This temperature is based on steam tables (36). We varied only two parameters, i.e., pressure (and thus, temperature) and duration of treatment in these synthesis experiments. A few conventional hydrothermal experiments were conducted in Teflon-lined Parr bombs (Parr Instrument Company, Moline, IL) which were heated in an oven for comparison with the microwave-hydrothermal experiments. The solid and solution phases were separated by centrifugation after the treatments under microwave-hydrothermal and conventional hydrothermal methods. The solid phases were washed with deionized water and ethanol to remove any soluble salts and dried in an oven at 40°C in air. The solid samples were characterized by powder x-ray diffraction (XRD) using a Scintag diffractometer with CuKα radiation. Morphology and particle size were determined by scanning electron microscopy (SEM) using an ISI-DS 130 instrument.

RESULTS AND DISCUSSION

Hematite Powders

Microwave-hydrothermal (M-H) treatment of 1M $Fe(NO_3)_3 \cdot 9H_2O$ resulted in the formation of hematite at all the temperatures used i.e., 116° (25 psi), 138° (50 psi), 164° (100 psi) and 194°C (200 psi) and all the durations used (Table 1). Conventional-hydrothermal (C-H) method was used for comparison at 138°C for different durations i.e., 2.5, 4, 8, 24 and 72 hours. The above conventional treatment led to the formation of pure hematite only after 72 hours of treatment. The formation of pure hematite rapidly under microwave hydrothermal conditions but not under conventional hydrothermal conditions clearly demonstrates the value of microwave catalysis. This result clearly shows that the M-H method leads to enhanced kinetics by one- to two-orders of magnitude compared to C-H processing for the crystallization of hematite which is the equilibrium phase at high temperatures. Scanning electron micrographs show large agglomerated crystallites of hematite under M-H conditions (Fig. 1A) and very small aggregates of goethite and hematite under C-H conditions. The size of the agglomerates is another indication of rapid kinetics under M-H conditions due to microwave catalysis. The catalytic role of microwaves may be attributed to the generation of localized high temperatures at the reaction sites which enhance reaction rates in an analogous manner to that of ultrasonic waves (37) where both high temperatures and pressures have been reported. Further experiments are necessary to prove this hypothesis, however.

Table I. XRD Analyses of Hematite Produced by Microwave- and Conventional-Hydrothermal Techniques Using 1M $Fe(NO_3)_3 \cdot 9H_2O$.

Temp. (°C)	Time (Hr)	Reaction Products by XRD
Microwave-Hydrothermal		
194	2	HEMATITE
194	1	HEMATITE
194	0.5	HEMATITE
194	0.25	HEMATITE
194	0.08 (5 MIN)	HEMATITE
164	2	HEMATITE
138	2	HEMATITE
116	2	POORLY CRYSTALLINE HEMATITE
116	1	VERY POORLY CRYSTALLINE HEMATITE
116	0.5	VERY POORLY CRYSTALLINE HEMATITE
Conventional-Hydrothermal		
138	72	HEMATITE
138	24	HEMATITE + GOETHITE
138	8	GOETHITE + HEMATITE
138	4	GOETHITE + HEMATITE
138	2.5	GOETHITE + HEMATITE

Figure 1. Scanning electron micrographs of iron oxide and oxyhydroxide powders: (A) Large agglomerates of hematite prepared by M-H process at 138°C/2 hours and (B) Small aggregates of goethite and hematite mixture prepared by C-H process at 138°C/8 hours.

Hydroxyapatite Powders

Hydroxyapatite powders were prepared from calcium oxide and phosphoric acid under all the temperature (or pressure) and duration conditions of M-H treatments used (Table 2). Hydroxapatite was also crystallized at 116° and 195°C in 2.5 and 8

Table II. XRD Analyses of Hydroxyapatite Produced by Microwave- and Conventional-Hydrothermal Techniques Using $CaO+H_3PO_4$.

Temp. (°C)	Time (Hr)	Reaction Products by XRD
Microwave-Hydrothermal		
194	2	HYDROXYAPATITE
194	1	HYDROXYAPATITE
194	0.5	HYDROXYAPATITE
194	0.25	HYDROXYAPATITE
164	2	HYDROXYAPATITE
138	2	HYDROXYAPATITE
116	2	HYDROXYAPATITE
Conventional-Hydrothermal		
195	2.5	HYDROXYAPATITE
195	3	HYDROXYAPATITE
116	8	HYDROXYAPATITE
116	3	POORLY CRYSTALLINE HYDROXYAPATITE

hours under C-H conditions. Powder XRD patterns of hydroxyapatite crystallized at 194°C clearly show the difference between the M-H and C-H crystallized powders (Fig. 2). The hydroxyapatite crystallized under M-H conditions shows sharp and well separated peaks while the one crystallized under C-H conditions shows broad peaks. Thus highly crystalline powders of hydroxyapatite were obtained by microwave catalysis. Figure 3 shows the SEM pictures of hydroxyapatite crystallized under M-H and C-H conditions. The agglomerated hydroxyapatite crystallized under M-H conditions consists of larger crystals (Fig. 3A) compared to those from aggregates of hydroxyapatite under C-H conditions (Fig. 3B). The SEM results substantiate the XRD results and the catalytic role of microwaves in crystallization. The enhanced rate of crystallization may be attributed to the localized heating effects as discussed above in the case of hematite crystallization.

CONCLUSION

Microwaves catalyze the crystallization of hematite and hydroxapatite under hydrothermal conditions. One possible hypothesis for the catalytic role of

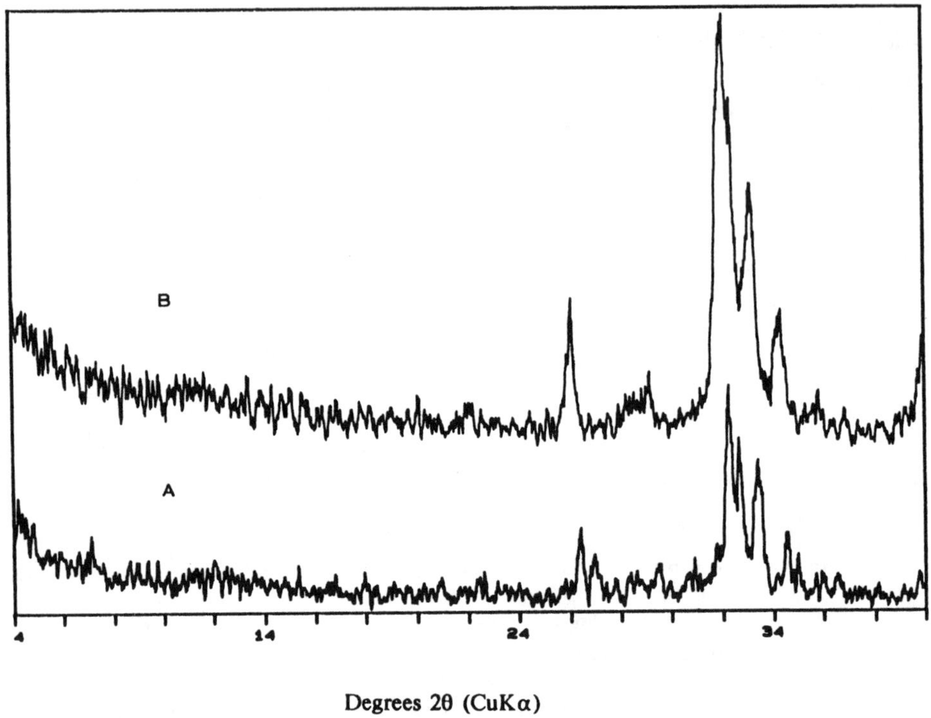

Figure 2. Powder XRD patterns of hydroxyapatite: (A) Highly crystalline powder prepared by M-H process at 194°C/2 hours and (B) Powder prepared by C-H process at 195°C/3 hours showing broad peaks.

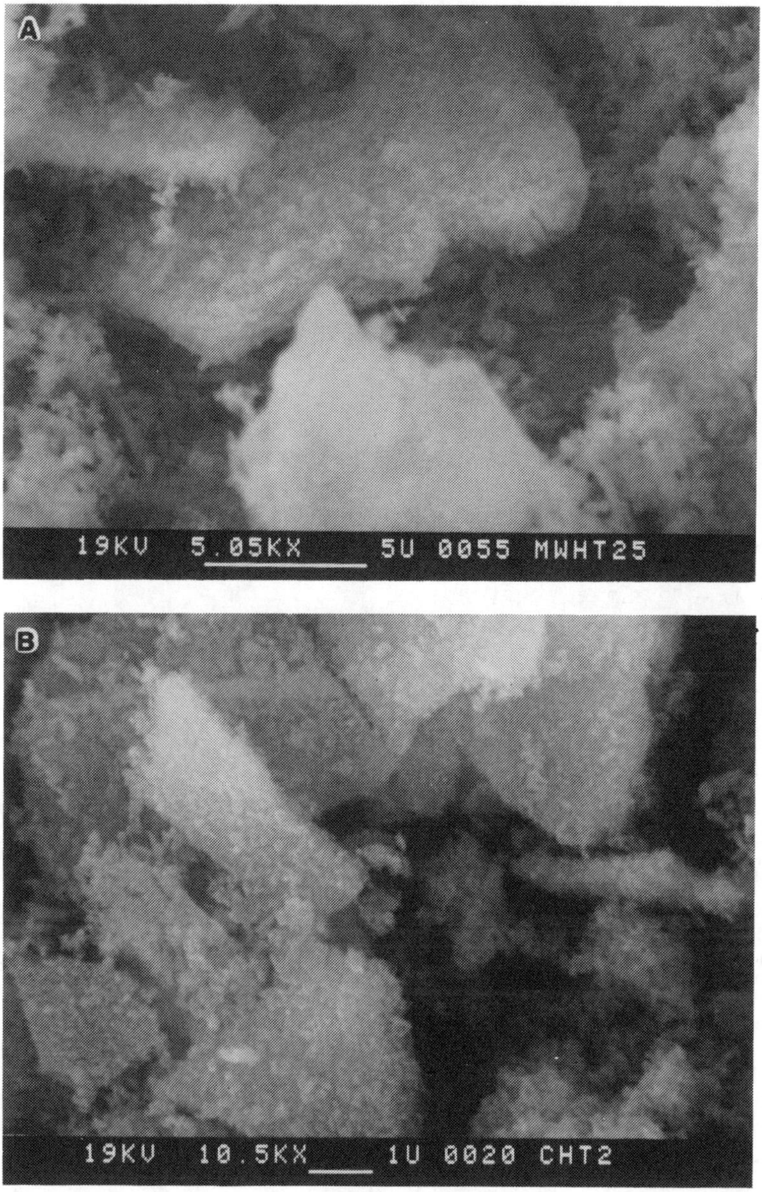

Figure 3. Scanning electron micrographs of hydroxyapatite powders: (A) Large agglomerates of hydroxyapatite consisting of individual crystals prepared by M-H process at 194°C/2 hours and (B) Large agglomerates of hydroxyapatite consisting of smaller individual crystals prepared by C-H process at 195°C/3 hours.

microwaves is the creation of localized high temperatures at the reaction sites which enhance reaction rates.

ACKNOWLEDGEMENTS

This research was supported by the Intercollege Materials Research Laboratory Consortium on Chemically Bonded Ceramics. The senior author thanks the organizing committee for inviting him to present this paper at the 1995 American Ceramic Society meeting.

REFERENCES

1. R. Roy, S. Komarneni and L. J. Yang, "Controlled Microwave Heating and Melting of Gels", J. Am. Ceram. Soc. 68: 392-395 (1985).

2. S. Komarneni and R. Roy, "Anomalous Microwave Melting of Zeolites", Mat. Lett. 4:107-110 (1986).

3. S. Komarneni, E. Breval and R. Roy, "Microwave Preparation of Mullite Powders" in Microwave Processing of Materials (Eds. M. H. Brooks, I. J. Chabinsky and W. H. Sutton), Mat. Res. Soc. Proc., Pittsburgh, PA , Vol. 124, pp. 235-238 (1988).

4. W. H. Sutton, "Microwave Processing of Ceramic Materials", Am. Ceram. Soc. Bull., 68 [2] 376-386 (1989).

5. M. K. Krage, "Microwave Sintering of Ferrites", Am. Ceram. Soc. Bull., 60 [11] 1323-1234 (1981).

6. A. Abu-Sharma, J. S. Morris and S. R. Koirtyohann, "Wet Ashing of Some Biological Samples in Microwave Oven", Anal. Chem. 47:1475-1477 (1975).

7. R. A. Nadkarni, "Multitechnique Multielement Analysis of Coal and Fly Ash", Anal. Chem. 52:929-935 (1980).

8. L. B. Fischer, "Microwave Digestion of Geologic Material: Application to Isotopic Dilution Analysis", Anal. Chem. 58:261-263 (1986).

9. H. M. Kingston and L. B. Jassie, "Microwave Energy for Acid Decomposition at Elevated Temperatures and Pressures Using Biological and Botanical Samples," Anal. Chem. 58:2534-2541 (1986).

10. R. Gedye, F. Smith, K. Westaway, H. Ali, L. Baldisera, L. Laberge and J. Rousell, "The Use of Microwave Ovens for Rapid Organic Synthesis," Tetrahedron Lett. 27 [3] 279-282 (1986).

11. R. J. Gigure, T. L. Bray, S. M. Duncan and G. Majetich, "Application of Commercial Microwave Ovens to Organic Synthesis" Tetrahedron Lett. 27 [41] 4945-4948 (1986).

12. D. R. Baghurst, S. R. Cooper, D. L. Greene, D. M. P. Mingos, and S. M. Reynolds, "Application of Microwave Dielectric Loss Heating Effects for the Rapid and Convenient Synthesis of Coordination Compounds", Polyhedron 9 [6] 893-895 (1990).

13. D. M. P. Mingos and D. R. Baghurst, "Application of Microwave Dielectric Heating Effects to Synthetic Problems in Chemistry," Chem. Soc. Rev. 20:1-47 (1991).

14. D. R. Baghurst and D. M. P. Mingos, "A New Reaction Vessel for Accelerated Syntheses Using Microwave Dielectric Super-heating Effects", J. Chem. Soc. Dalton Trans. 1151-1155 (1992).

15. D. R. Baghurst and D. M. P. Mingos, "Superheating Effects Associated with Microwave Dielectric Heating", J. Chem. Soc. Chem. Commun. 674-677 (1992).

16. M. Ali, S. P. Bond, S. A. Mbogo, W. R. McWhinnie and P. M. Watts, "Use of a Domestic Microwave Oven in Organometallic Chemistry", J. Organometallic Chem. 371:11-13 (1989).

17. D. R. Baghurst, D. M. P. Mingos and M. J. Watson, "Application of Microwave Dielectric Loss Heating Effects for the Rapid and Convenient Synthesis of Organometallic Compounds", J. Organometallic Chem. 368:C43-C45 (1989).

18. D. L. Greene and D. M. P. Mingos, "Application of Microwave Dielectric Loss Heating Effects for the Rapid and Convenient Synthesis of Ruthenium (II) Polypyridine Complexes", Transition Met. Chem. 16:71-72 (1991).

19. K. Chatakondu, M. L. H. Green, D. M. P. Mingos and S. M. Reynolds, "Application of Microwave Dielectric Loss Heating Effects for the Rapid and Convenient Synthesis of Intercalation Compounds", J. Chem. Soc. Chem. Commun. 1515-1517 (1989).

20. V. C. Virtuli, P. Chu and F. G. Dwyer, "Crystallization Method Using Microwave Radiation", U. S. Patent Application, 4,778,666 (1988).

21. P. Chu, F. G. Dwyer and V. J. Clarke, Eur. Patent 358,827 (1990).

22. J. C. Jansen, A. Arafat, A. K. Barakat and H. van Bekkum, "Microwave Techniques in Zeolite Synthesis" in Synthesis of Microporous Materials (Eds. M. L. Occelli and H. E. Robson), Van Nostrand Reinhold, New York, Vol. 1, pp. 507-521 (1992).

23. A. Arafat, J. C. Jensen, A. R. Ebaid and H. van Bekkum, "Microwave Preparation of Zeolite Y and ZSM-5", Zeolites 13: 162-165 (1993).

24. I. Girnus, K. Hoffman, F. Marlow, J. Caro and G. Doring, Microporous Materials: 2:537 (1994).

25. I. Girnus, K. Jancke, R. Vetter, J. Richter-Mendau and J. Caro, "Large $AlPO_4$-5 Crystals by Microwave heating", Zeolites 15:33-39 (1995).

26. S. Komarneni, R. Roy and Q. H. Li, "Microwave-Hydrothermal Synthesis of Ceramic Powders", Mat. Res. Bull. 27 [12] 1393-1405 (1992).

27. S. Komarneni, Q. H. Li, K. M. Stefansson and R. Roy, "Microwave-Hydrothermal Processing for Synthesis of Electroceramic Powders", J. Mat. Res. 8 [12] 3176-3183 (1993).

28. S. Komarneni, Q. H. Li and R. Roy, "Microwave-Hydrothermal Processing for Layered and Network Phosphates," J. Mat. Chem. 4: 1903-1906 (1994).

29. S. Komarneni, R. Pidugu, Q. H. Li and R. Roy, "Microwave-Hydrothermal Processing of Metal Powders", J. Mat. Res. 10 [7] (in press, 1995).

30. S. Komarneni, "Novel Microwave-Hydrothermal Processing for Synthesis of Ceramic and Metal Powders" in Novel Techniques in Synthesis and Processing of Advanced Materials (Eds. J. Singh and S. M. Copley), published by TMS, Warrendale, PA, pp. 103-117 (1995).

31. S. Komarneni, "Environmentally Benign Microwave-Hydrothermal Processing for Synthesis of Ceramic Powders", in Proceedings of the International Symposium on Environmental Issues of Ceramics (in press, 1995).

32. S. Komarneni, "Microporous Metal Intercalated Clay Nanocomposites", Proceedings of the International Forum for Materials Engineers at Sanjo-Tsubame on Porous Materials, Ceramic Transactions, Vol. 31 (Eds. K. Ishizaki et al.) publ. by Am. Ceram. Soc., Columbus, OH, pp. 155-168 (1993).

33. Y. T. Moon, D. K. Kim and C. H. Kim, "Preparation of Monodisperse ZrO_2 by the Microwave Heating of Zirconyl Chloride Solutions", J. Am. Ceram. Soc., 78 [4] 1103-1106 (1995).

34. D. Daichuan, H. Pinjie and D. Shushan, "Preparation of Uniform α-Fe_2O_3 Particles by Microwave-Induced Hydrolysis of Ferric Salts", Mat. Res. Bull. 30 [5] 531-535 (1995).

35. D. Daichuan, H. Pinjie and D. Shushan, "Preparation of Uniform β-FeO (OH) Colloidal Particles by Hydrolysis of Ferric Salts Under Microwave Irradiation", Mat. Res. Bull. 30 [5] 537-541 (1995).

36. R. C. Weast (Ed.) in CRC Handbook of Chemistry and Physics, Publ. by CRC Press Inc., Boca Raton, FL, pp. E16-E21 (1988).

37. K. S. Suslick, "Sonochemistry", Science 247: 1439-1445 (1990).

SYNTHESIS OF MONODISPERSE SPHERICAL $ZrTiO_4$ POWDERS BY THERMAL HYDROLYSIS OF METAL SALTS

Hong Kyu Park, Do Kyung Kim, and Chong Hee Kim
Department of Materials Science and Engineering
Korea Advanced Institute of Science and Technology, Taejon, 305-701
South Korea

ABSTRACT

The $ZrTiO_4$ powders were synthesized by thermal hydrolysis of $ZrOCl_2 \cdot 8H_2O$ and $TiCl_4$ in the mixed solvent of 1-PrOH and water. The morphology of the powders was observed to be spherical and the mean particle diameter was in the range of 0.75 μm. The thermal behavior analysis by XRD and TG/DTA showed that the precursor powders transformed into crystalline $ZrTiO_4$ after calcination at about 600°C. The calcined powders retained their spherical shapes. Sinterability of the powders without any additives was examined. The density of the $ZrTiO_4$ sintered at 1500°C for 3 h was more than 97% of the theoretical.

INTRODUCTION

In communicative and electronic engineering industry, the advanced ceramics of high performance and reliability are required to meet the high electrical properties. $ZrTiO_4$ ceramics are one of the most applicable materials as a resonator because of their high microwave dielectric properties, such as high dielectric constant, high Q factors, and low frequency temperature coefficient.[1-4] $ZrTiO_4$ is synthesized by the solid state reaction of TiO_2 and ZrO_2 at 1300°C and requires a high sintering temperature of above 1600°C to obtain the densified ceramics of 98% in its relative density.[5]
Recently, a number of researches have been accomplished to prepare fine spherical powders by the chemical route in order to lower the sintering

temperature. Belyaev *et al.*[6] reported that $ZrTiO_4$ powders were synthesized by coprecipitation of zirconium hydroxide and titanium hydroxide. After calcination at about 1000°C, the $ZrTiO_4$ powders with the compositional homogeneity were obtained. The powder morphology, however, could not be controlled because of the uncontrollable precipitation behavior of zirconium hydroxide and titanium hydroxide and the formation of hard agglomerates at high calcination temperatures. Shnoeller *et al.*[7] synthesized $[Zr_{0.8},Sn_{0.2}]TiO_4$ powders from the mixed solution of $Zr(O^n\text{-}Pr)_4$, $Ti(O^n\text{-}Pr)_4$ and $Sn(OAc)_2$. The powder morphology was very sensitive to the reaction conditions, such as temperature, water-to-alkoxide ratio, and the rate of hydrolysis and condensation of metal alkoxides. Moon *et al.*[8] introduced a new synthesis method of spherical ZrO_2 powders. They obtained monodisperse spherical ZrO_2 powders by heating a zirconyl chloride solution containing the 1-PrOH or 2-PrOH. They showed that precipitation was due to the decrease of solubility caused by the decrease of the dielectric constant of the solution.

In this study, the novel route is examined for the synthesis of the monodisperse spherical $ZrTiO_4$ powders using metal salts, such as $ZrOCl_2{\cdot}8H_2O$, $TiCl_4$, as starting reagents. And the precipitation behavior and the chemical composition of the precursor powders are investigated. The thermal behavior and the sinterability of the powders are also examined.

EXPERIMENTAL PROCEDURE

Starting reagents, zirconyl chloride octahydrate (98% $ZrOCl_2{\cdot}8H_2O$, Aldrich Chemical, Milwaukee, WI.) and titanium tetrachloride (98% $TiCl_4$, Wako Chemical Industries, Ltd., Japan), were dissolved in the mixed solvent of 1-PrOH (98% Junsei Chemical Co., Ltd., Japan) and distilled, deionized water. Hydroxypropyl cellulose (HPC) (molecular weight ~100 000, Aldrich Chemical, Milwaukee, WI.) of 0.016 g/cm^3 was used as a steric dispersant. The concentration of the solution was maintained at 0.12 M for this study. When the prepared solutions were kept at 70°C for 6 min in isothermal water bath, white products were precipitated. The precipitates were separated from the mother liquor by centrifuging at 10 000 rpm. The precipitates, washed twice with distilled water and then finally with 1-PrOH, were dried in vacuum oven at 50°C for 24 h.

Phase transformation and thermal behavior of the precursor powders were determined by X-ray diffractometry (XRD) and thermogravimetry and differential thermal analysis (TG/DTA). The composition of the powders was investigated by energy-dispersive spectroscopy (EDS) and wavelength-dispersive spectroscopy (WDS). After calcination at 800°C, the powders were sintered at 1300°, 1400°, 1500°, and 1550°C for 3h, respectively. The sintered density was measured by using the water displacement technique.

RESULTS AND DISCUSSION

Figure 1 shows the precipitation behavior of zirconium hydroxide and titanium hydroxide from $ZrOCl_2 \cdot 8H_{2z}O$ and $TiCl_4$, respectively. Precipitation reactions were performed in the mixed solvent of 1-PrOH and water with the various volume ratios of 1-PrOH to water (P/H ratio).

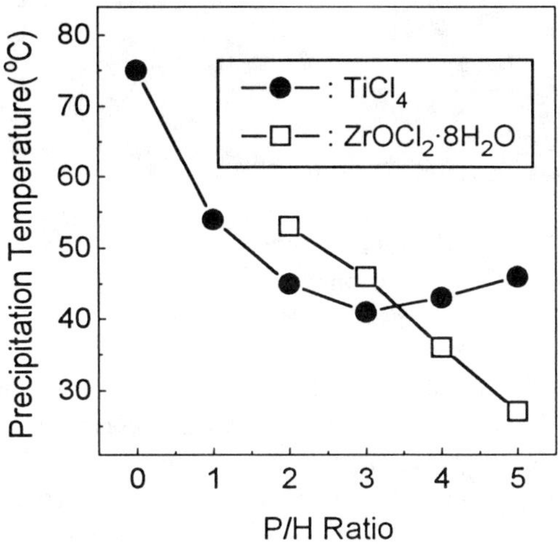

Figure 1. Precipitation temperature of the precursor powders in the mixed solvent of 1-PrOH and water. The starting reagents were $ZrOCl_2 \cdot 8H_2O$ and $TiCl_4$.

In both cases, the precipitation temperature was decreased, as the P/H ratio was increased. In the case of $TiCl_4$, the precipitation temperature slightly increased at the P/H ratio of 4 and 5. According to Moon *et al.*,[8] the decrease of the precipitation temperature in the mixed solution of alcohol and water was due to the decrease of solubility caused by the decrease of the dielectric constant of the solution. In this study, it is thought that the increase of P/H ratio promoted the precipiation reaction. At the P/H ratio of 3.4, zirconium hydroxide and titanium hydroxide show the same value of the precipitation temperature. From these results, we chose the P/H ratio as 3.5 to prevent the compositional inhomogeneity caused by the difference of the precipitation temperature.

Figure 2 (A) shows the SEM micrographs of the $ZrTiO_4$ precursor powders derived by thermal hydrolysis of the mixed salts of $ZrOCl_2 \cdot 8H_2O$ and $TiCl_4$ without addition of any dispersants.

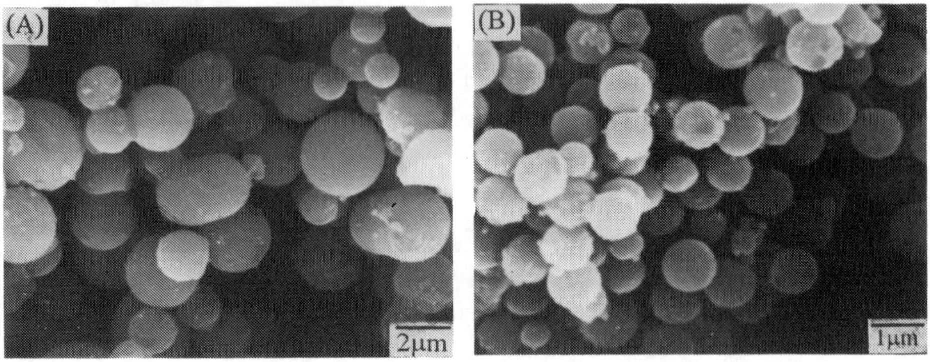

Figure 2. SEM micrographs of the precursor powders derived from the solutions without (A) and with (B) HPC. The concentration of HPC was 0.016 g/cm^3.

The powders were spherical but agglomerated. When HPC was added as a steric dispersant, the agglomeration of the powders decreased and monodisperse, spherical $ZrTiO_4$ powders were obtained as shown in Fig. 2 (B). These results indicate that HPC added as a dispersant prevents the particle agglomeration by the steric stabilization.

The compositional homogeneity in multicomponent ceramic powders is very important to fabricate ceramics with high performance and reliability. Therefore, the chemical homogeneity of the precursor powders plays a key role to enhance the final properties. Figure 3 shows the result of the compositional analysis of the precursor powders by EDS. The particles are composed of zirconium, titanium, and oxygen. No elements, especially, chlorine which was the anion source of starting reagents, were not detected by EDS analysis.

Figure 4 shows the micrograph and the result of the quantitative analysis by WDS. The atomic percent of zirconium and titanium of the particle is 18 and 17.7, respectively. This result shows that the particle has the stoichiometric amounts of zirconium and titanium.

Figure 5 shows the X-ray spectra of the precursor powder, which shows the precursor powders to be amorphous. After calcination at 600°C for 1 h, peaks of

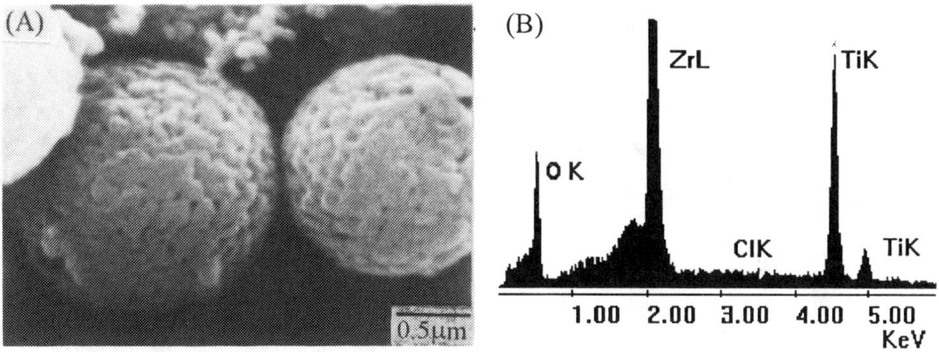

Figure 3. SEM micrograph (A) and EDS spectrum (B) of the precursor powder.

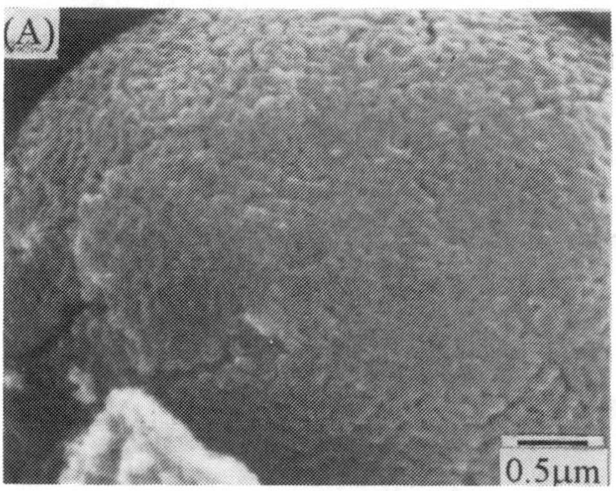

(B)

Element	Composition	
	at. %	Error range
Zr	18.0043	±0.1106
Ti	17.6692	±0.2992
O	64.3265	±0.4930

Figure 4. SEM micrograph (A) and the chemical composition analyzed by WDS (B) of the precursor powder.

$ZrTiO_4$ phase appeared. It is well known that the formation of $ZrTiO_4$ by solid state reaction between TiO_2 and ZrO_2 occurs above 1250°C after several hours. Thermal behavior of the precursor powders was also investigated by TG/DTA as shown in Fig. 6. The broad peak formed in the range of 25 to 200°C is believed to be due to the dehydration of adsorbed water and the evaporation of organic solvent. A sharp peak is observed at around 600°C. Comparing with the result of XRD analysis, this peak is thought to be due to the crystallization of amorphous $ZrTiO_4$ powders. The phase transformation of $ZrTiO_4$ at low temperature of 600°C means that titanium and zirconium atoms are intimately mixed in the

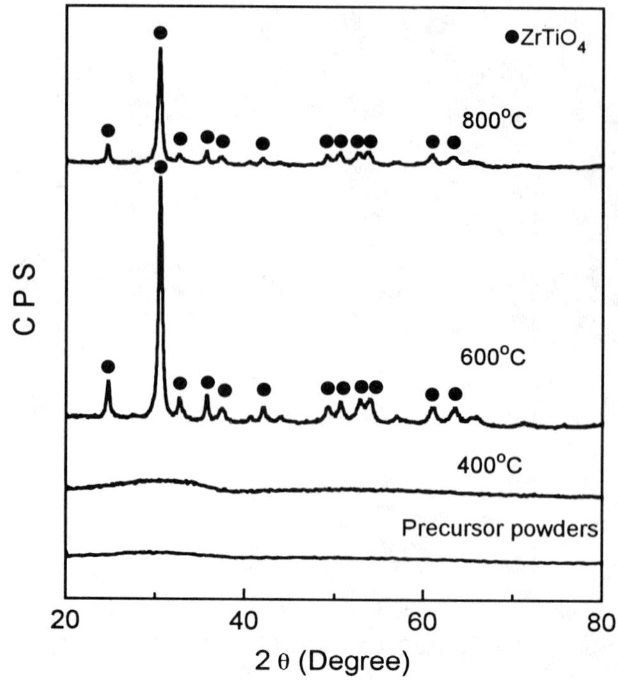

Figure 5. XRD patterns of the $ZrTiO_4$ powders calcined at various temperatures for 1 h in air.

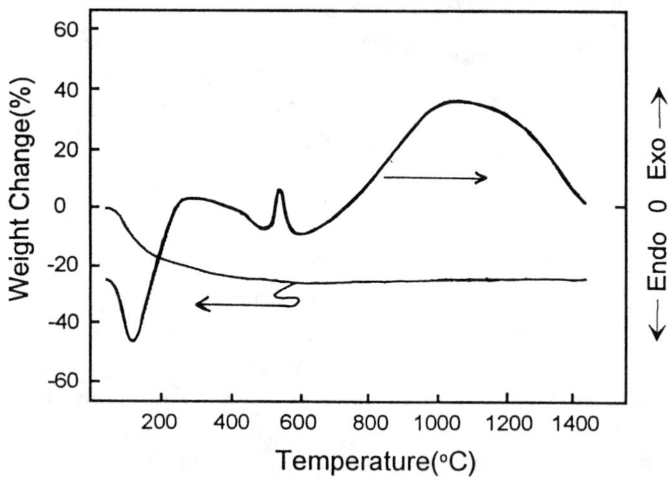

Figure 6. TG/DTA curves of the precursor powders heated at a rate of 10 °C/min in air.

precursor powder. Thus, the diffusion distances for each atom are sufficiently short to form the compound of $ZrTiO_4$ easily.

The morphologies of the precursor and calcined powders are shown in Fig. 7. After calcination at 800°C for 1 h, the powders preserved their spherical morphology with a reduction in their size, which was believed to be due to the dehydration of adsorbed water and the decomposition of organic solvent.

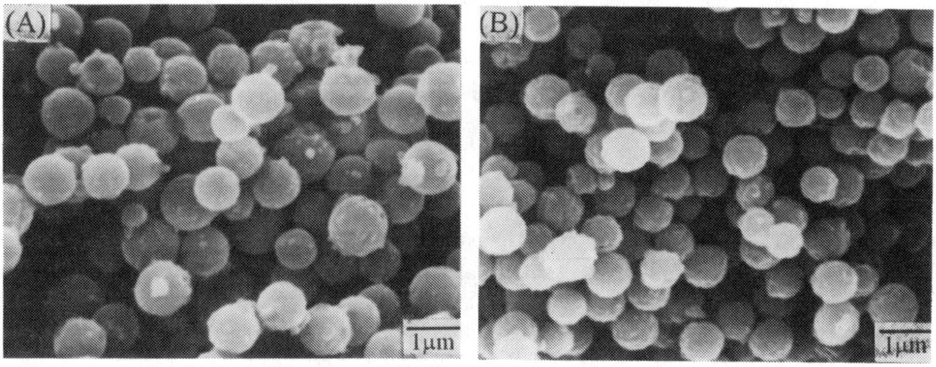

Figure 7. SEM micrographs of the precursor powders (A) and calcined powders at 800°C for 1 h in air (B).

Figure 8 shows the relative density of the specimens sintered at various temperatures. The relative density of about 97% was obtained for the specimen sintered at 1500°C for 3 h. On the other hand, the specimen obtained by solid state reaction of ZrO_2 and TiO_2 powders had a density of about 88%, despite of the higher sintering temperature of 1550°C.

CONCLUSIONS

The precursor powders of $ZrTiO_4$ were synthesized by thermal hydrolysis of $ZrOCl_2 \cdot 8H_2O$ and $TiCl_4$ in the mixed solvent of 1-PrOH and water. Monodisperse spherical $ZrTiO_4$ precursor powders with a mean particle diameter of 0.75μm were obtained using HPC as a steric dispersant. The chemical composition of these powders well corresponded to the stoichiometry of $ZrTiO_4$. TG/DTA analysis showed that the precursor powders were transformed into crystalline $ZrTiO_4$ powders at about 600°C. The calcined powders had the size of 0.6μm and could be sintered to a dense ceramic with 97% of theoretical density at 1500°C for 3 h in air.

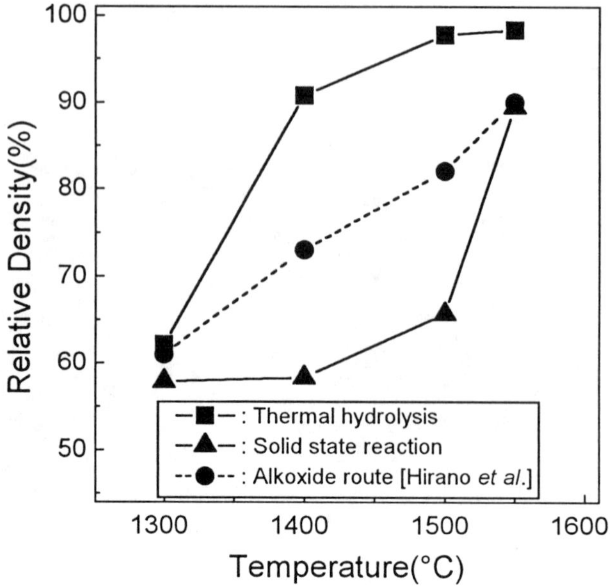

Figure 8. Relative density of $ZrTiO_4$ specimens sintered at various temperatures for 3 h in air.

REFERENCES

1. G. Wolfram and H. E. Gobel, "Existence Range, Structural and Dielectric properties of $Zr_xTi_ySn_zO_4$ Ceramics (x+y+z=2)," *Mat.Res. Bull.*, Vol. 16, 1455–63 (1981).
2. K. Wakino, K. Minali, and H. Tamura, "Microwave Characteristics of $(Zr,Sn)TiO_4$ and BaO-PbO-Nd_2O_3-TiO_2 Dielectric Resonators," *J. Am. Ceram. Soc.*, 67 [4] 278–81 (1984).
3. A. E. McHale and R. S. Roth, "Investigation of the Phase Transition in $ZrTiO_4$ and $ZrTiO_4$-SnO_2 Solid Solutions," *J. Am. Ceram. Soc.*, 66 [2] C-18–C-20 (1983).
4. R. Christoffersen, P. K. Davies, and S. Wei, "Effect of Sn Substitution on Cation Ordering in $(Zr_{1-x}Sn_x)TiO_4$ Microwave Dielectric Ceramics," *J. Am. Ceram. Soc.*, 77 [6] 1441–50 (1994).

5. S. Hirano, T. Hayashi, and A. Hattori, "Chemical Processing and Microwave Characteristics of [Zr,Sn]TiO_4 Microwave Dielectrics," *J. Am. Ceram. Soc.*, 74 [6] 1320–24 (1991).
6. N. Belyaev, N. P. Popovskaya and T. I. Ivleva, *Izv. Vyssh, Uchebin. Zaved., Khim Tekhnol*, 22 [3] 382 (1979).
7. M. Schnoeller and W. Wersing, pp.45 in *Maerials Research Society Symp. Proc.*, Vol. 155, Edited by I. A. Aksay, G. L. McVay and D. R. Ulrich, Materials Research Society, Pittsburgh, PA., 1989.
8. Y. T. Moon, H. K. Park, D. K. Kim, and C. H. Kim, "Preparation of Monodisperse Spherical Zirconia Powders by Heating of Alcohol–Aqueous Salt Solutions," *J. Am. Ceram. Soc.*, in press (1995).

A SIMPLE SOLUTION-POLYMERIZATION ROUTE FOR OXIDE POWDER SYNTHESIS

M. A. Gülgün and W. M. Kriven, University of Illinois at Urbana-Champaign, 105 S. Goodwin Ave., Urbana, Illinois 61801.

ABSTRACT

A simple solution polymerization route based on polyvinyl alcohol (PVA) as the organic carrier for the cations is used to process pure, mixed oxide powders. Simultaneous DTA/TGA analyses revealed that the organics are burnt out below 600°C. The new system produces X-ray amorphous, fine, white powders at temperatures as low as 650°C. The powders thus prepared have a 12 m^2/g as-calcined specific surface area. The succesful use of a non-chelating polymer for the solution polymerization method indicated that cation chelation is not the only route for creating stable ceramic precursors. The process has a net yield of 1.8 for the weight ratio of the mixed oxide powder to the organics. The organic carrier has a significant influence on the morphology of the final ceramic powder.

INTRODUCTION

As the requirements of purity, phase distribution and other physical characteristics of ceramic powders become more stringent, chemical synthesis methods are becoming more popular over conventional high temperature synthesis routes. Chemically prepared powders are pure and homogeneous at a molecular level [1]. Their chemical and physical characteristics can be closely controlled. They are synthesized at much lower temperatures than conventional ceramic powders. In addition, the costly grinding-refiring steps are eliminated.

One such chemical synthesis route for producing pure, mixed oxide powders is the solution-polymerization technique, better known as the "Pechini Method" [2, 3]. The technique employs the Pechini resin as the polymeric carrier of the pre-ceramic powders. The resin consists of citric acid, as the chelating agent, and ethylene glycol, to promote polymerization during the esterification process. The cation sources are usually nitrate salts of the metals, mixed in stoichiometric proportions in the aqueous resin solution. Modifications have been suggested to move from costly nitrate salts to more readily available sources, specially for silicon containing systems [4]. An important aspect in the selection of these cation precursor salts is the aqueous solubility. Alternatives have been suggested to use other resins in order to reduce the cost [5, 6].

In the literature [6-10], the following reactions are proposed to occur during the Pechini process (Figure 1). The heart of the process is chelation of the dissolved metal ions by the carboxylic acid end of the citric acid which has a configuration ideal for chelating (step I). Polyesterification follows at higher temperatures (step II).

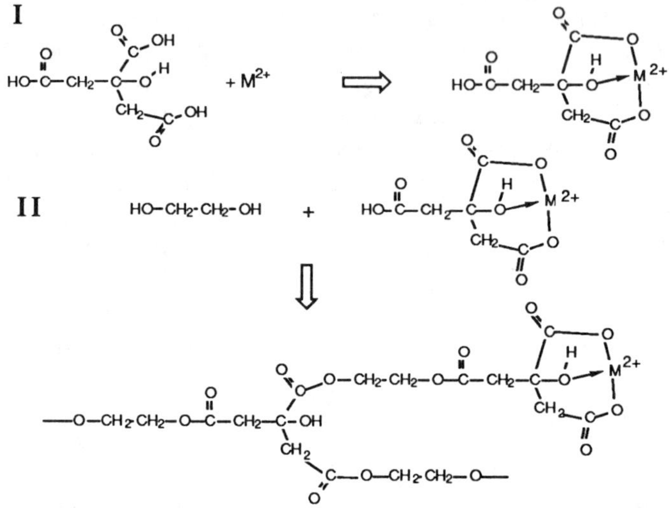

Figure 1. Proposed chemical reactions for the Pechini Process.

The chelating action of the organics is suggested to be responsible for the formation of a stable, atomistically homogeneous, pre-ceramic organometallic. Previous studies with calcium aluminate powders have shown however, that the resin can accommodate higher amounts of cations than it can chelate in the solution [11]. Thus, it is possible that the stabilization of the metal ions in the organics is due to other effects in addition to chelation.

In this study, a rather simple-structured and inexpensive polymer, polyvinyl alcohol (PVA), is used as the organic carrier. The effects of this new organic carrier on the chemical, physical and thermal characteristics of the precursor-to-ceramic conversion are reported.

EXPERIMENTAL PROCEDURE

$Ca(NO_3)_2{\cdot}4H_2O$ and $Al(NO_3)_3{\cdot}9H_2O$ were used as the cation sources. The polymeric precursors were prepared by mixing stoichiometric proportions of the nitrate salts into a 5 wt% aqueous solution of PVA. The PVA solutions were made from three types of PVA with three different degrees of polymerization (DP), i.e. DP 1700 (Type A), DP 580 (Type B) and DP 330 (Type C). Type A PVA was dissolved by stirring on a hot plate at 68°C. The other two types were dissolved by stirring at room temperature. The proportions of the PVA to cation sources in the solution were adjusted in such a way that there were 4 times more positively (+) charged valences from the cations than the negatively (-) charged functional ends of the organics. Solutions were then heated on a hot plate until the water of solution evaporated, and a crisp, light brown, aerated gel formed. This gel was ground and calcined in air at various temperatures for 1 h. The heating rate for the calcination process was 10°C/min. Some powders were annealed for dwell times of up to 90 h to promote crystal growth at lower temperatures.

The phase formation in the powders was analyzed as a function of calcination temperature using a powder X-ray spectrometer and Cu K_α radiation (40 kV and 40 mA). ^{27}Al MAS-NMR analyses were performed on powders calcined at various temperatures, using a home-built spectrometer based on a Nicolet 1280 data-acquisition system at 11.7 tesla. The samples were spun at about 10 KHz. The FTIR measurements were done on some samples, using a Nicolet spectrometer (resolution of 2 cm^{-1}).

Pyrolysis of the ground gel samples was monitored by simultaneous differential thermal analysis and thermogravimetric analysis (DTA/TGA) at a heating rate of 10°C/min.

The effect of calcination temperature on the specific surface area of powders was studied by nitrogen gas adsorption. The surface area data were obtained by five point BET analysis. The particle size distribution of powders was studied using a LASER absorption spectrometer (Sedigraph). The powders were suspended in a calibrated non-aqueous dispersing liquid (Sedisperse A-12). The suspension was sonically agitated with a quartz tip ultrasonic processor for 5 minutes before analysis.

The morphology of powders was studied by scanning electron microscopy. A drop of powder suspended and sonically agitated in isopropyl alcohol was deposited on to an aluminum stub, followed by carbon coating.

RESULTS

Simultaneous DTA/TGA analysis of the precursors revealed that they undergo a four step decomposition process (Fig. 2). The first weight loss is observed in the temperature range 80°C to about 200°C, which is

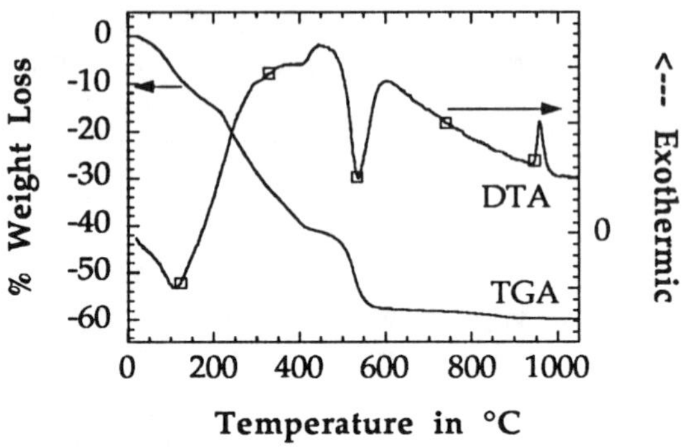

Figure 2. Simultaneous DTA/TGA analysis of the pre-ceramic precursor.

associated with an endotherm in the simultaneous DTA curve. The second major weight loss in TGA and a corresponding exotherm in the DTA curve occur between 220°C and 400°C, which are followed by the third large reduction in weight between 450°C and 550°C. This third weight loss is associated with a strong endotherm in the DTA trace. Later, powders gradually lost about 1% of the their weight between 600°C and 900°C. The DTA trace also exhibits an exotherm due to crystallization at about 940°C which occurs in a region with constant weight.

Powders calcined for one hour at the specified temperatures are x-ray amorphous up to 800°C (Fig. 3). After one hour of annealing at 900°C the x-ray spectrum of the powders reveal crystalline peaks of $CaAl_2O_4$. Powders calcined at 750°C for 60 hrs and at 850°C for 1 h showed evidence of incipient crystallization. The very broad x-ray peak

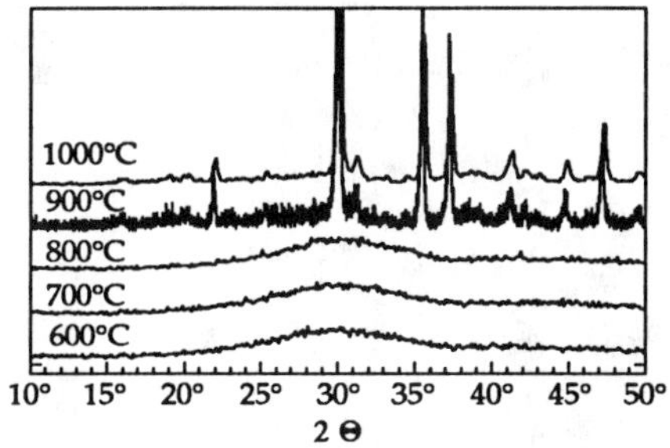

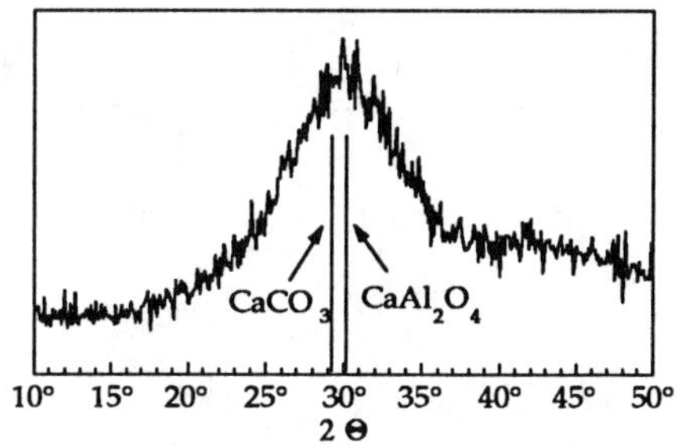

Figure 3a. X-ray spectra of powders calcined 1 h at specified temperatures.

Figure 3b. X-ray spectrum of the powders calcined 60 h at 750°C.

centered around a 2Θ value of 30° covers the range of the most intense peaks of calcium aluminate and also of calcium carbonate. In order to gain more insight into the precursor-to-ceramic conversion, as well as local ordering in these x-ray amorphous powders, FTIR and NMR analyses were performed. The FTIR analysis of the powder calcined at 600° and 800°C revealed absorption bands usually assigned to carbon-oxygen stretching in inorganic carbonates. A series of ^{27}Al-NMR analyses on the powders calcined at temperatures 500°C - 900°C revealed broad peaks at about 10, 40 and 75 ppm (Fig. 4). The peak at 40 ppm was not detected in powders calcined above 550°C. The peak at 10 ppm persisted up to the calcination temperature of 700°C. The peak at 75 ppm progressively became sharper with increasing calcination temperature.

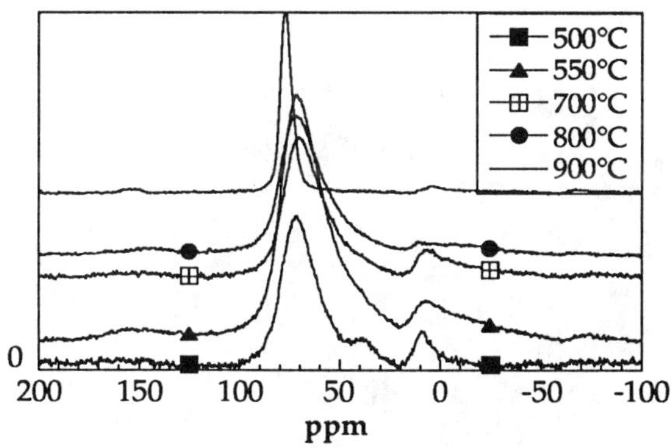

Figure 4. ^{27}Al MAS-NMR spectra from powders calcined at different temperatures.

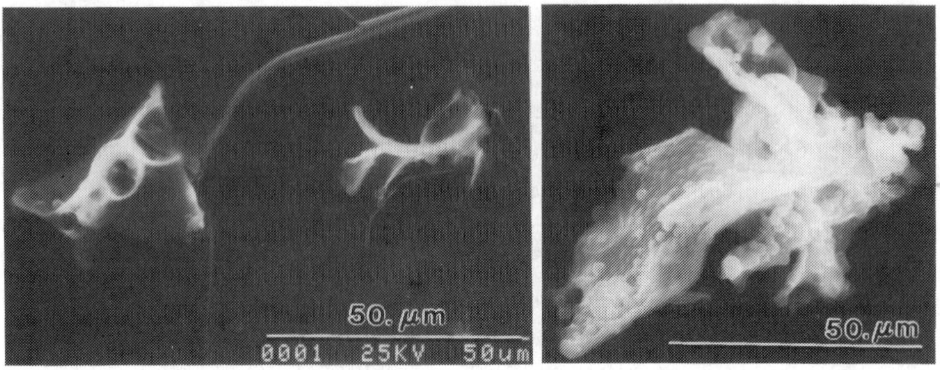

Figure 5. Powders calcined at (a) 900°C and at (b) 600°C showing the star-like morphology. Powders are made of small particles that fused together.

The star-shaped morphology of the mixed oxide powders prepared from the new precursor is rather unique (Fig. 5). Powders calcined at lower temperatures revealed that the "fins" stretching out from the central "node" are actually made up of small particles (of approx. 2 µm) which sintered together during calcination. SEM investigation showed that most of the particles are in the size range of 10 to 50 µm. The powders prepared with lower DP PVA (type B and C), resulted in chunkier, more rounded particles as the DP decreased (Fig. 6). Powders from the smallest DP PVA (type C) precursors exhibited a morphology that is similar to the powders prepared through the Pechini method [11].

Figure 6. The SEM micrograph of the powders prepared with type A PVA. The particles are chunkier and more rounded than powders from higher DP PVA precursors.

Despite this variance in particle morphology, the size analysis of the powders calcined from precursors with different types of PVA showed that the size distributions are very similar to each other (Fig. 7). The median of the distribution (i.e. 50% finer) was about 7 μm for all the powders. Calcination temperatures in the range of 600° to 1000°C also did not have a significant effect on the particle size distribution of these powders.

On the other hand, the specific surface area (SSA) of the same powders showed a strong calcination temperature dependence (Fig. 8). Powders calcined at 550°C for 1 h had a BET SSA of 12.5 m^2/g, whereas powders calcined at 900°C for the same time had a SSA of only 2 m^2/g.

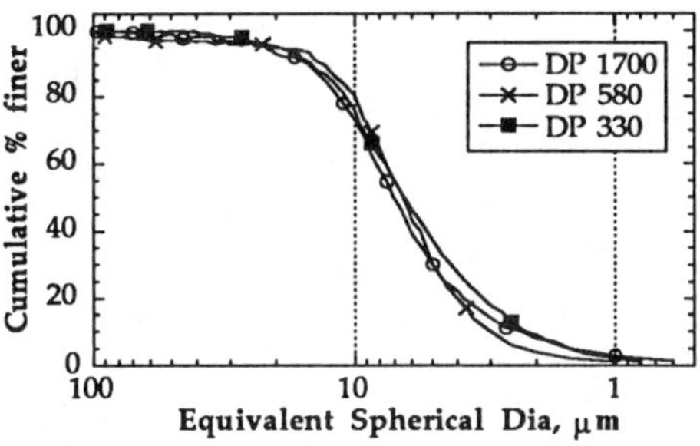

Figure 7. Particle size distribution as a function of DP of PVA.

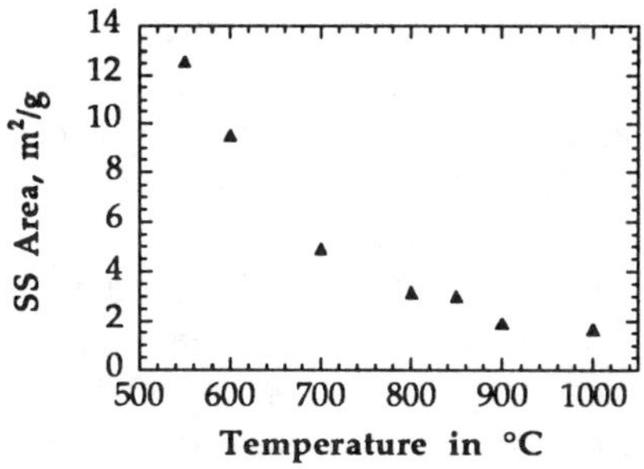

Figure 8. Specific surface area as a function of calcination temperature.

DISCUSSION

The solution polymerization technique employing polyvinyl alcohol as the carrier organic phase is a viable, inexpensive technique for producing mixed oxide powders. It has a ceramic to polymer yield of ~ 1.8, as compared to 0.15 for the Pechini method [12].

The results indicate that chelation of the metal ions by the carboxylic end groups, as in the Pechini resin, is not the only route to obtain stable precursors. Previous results from Pechini precursors with very high metal ion to polymer end group ratios have shown that the resin was able to support a higher amount of cations than it can chelate [11]. In this study, the precursor had a positively charged valence from cations to negatively charged functional end group ratio of 4:1. Despite this large charge imbalance, no precipitation from solution was observed. This could also be due to the very high aqueous solubility of nitrate salts of calcium and aluminum. The results from precursors with PVA indicated that polymers with non-chelating functional groups can be used as the carrier in the system. Further investigation is underway to study mixed oxide systems whose precursor salts are less soluble, e.g. distrontium silicate.

X-ray results showed that the powders thus produced are single phase calcium aluminate when calcined above 900°C. Below 900°C they are x-ray amorphous when calcined for 1 h. FTIR investigations and x-ray analyses of powders annealed at 750 and 800°C for extended times indicated that there is a small amount of carbonates, possibly calcium carbonate, forming in the system. It was also observed in ^{27}Al NMR studies, that Al was involved in more than one type of configuration, as shown by the peaks at 10 ppm, 40 ppm, and 75 ppm. These peaks can be assigned to six, five and four-fold coordination, respectively [13]. Al in $CaAl_2O_4$ and $CaAl_4O_7$ is in four-fold coordination in Al-O tetrahedra [14]. The existence of a small amount of six-fold coordinated aluminum and the absorption band in FTIR associated with carbonates is indicative of the formation of some intermediate phase(s) in the precursors below 900°C. The last weight loss below 900°C in TG analysis was most likely due to the decomposition of this intermediate phase. TGA results further revealed that most of the weight loss due to the burn-out of the organics in the system occurs below 600°C.

SEM studies with the powders calcined at different temperatures indicated that when calcined at about 600°C the powders are clusters of smaller round particles. For such equiaxed, smaller particles, one must probably stay at lower calcination temperatures, and break the agglomerates by milling. The effective mean particle size of the calcined powders is about 10-20 µm, as observed by SEM. The discrepancy between particle size measured by sedimentation and that observed in SEM, is probably due to the very shape of these particles. Powders with a morphology of fins stretching out of a central node will have a very small settling rate. Thus, any technique employing particle settling rate and assuming spherical morphology will give results for these powders that are skewed towards smaller particle sizes.

A comparison between the powders prepared in this study and powders prepared from a Pechini precursor showed that the carrier polymer has a significant influence on the morphology of the particles. Powders prepared from a Pechini precursor which uses smaller organic molecules result in chunkier particles with a small aspect ratio. The powders produced in this study are star-shaped with rather high aspect ratio. Furthermore, it was shown that precursors from PVA with smaller DP result in chunkier particles as well. Thus, it appears that using large chain molecules as the polymeric carrier, the solution polymerization technique can be utilized to produce mixed oxide powders with controlled morphology and aspect ratio. Powders with such morphologies as these may have very interesting characteristics as reinforcement phases in ceramic matrix composites.

CONCLUSIONS

A solution polymerization technique employing PVA as the organic carrier can be used as a viable, cost-effective method to produce single phase oxide powders with controlled morphology. As the size of the carrier polymer increases, the aspect ratio in the particles derived from the precursor increases. Spectroscopic analysis indicated that there may be a carbonate phase, and possibly other phases as well, forming as an intermediate before the $CaAl_2O_4$ forms. These intermediates decompose below 900°C, and heat treatment at 900°C gives single phase, crystalline calcium aluminate.

ACKNOWLEDGMENTS

The authors are grateful to Dr. Yuehui Xiao and Mr. Priyadarshi Desai for NMR and FTIR analyses, respectively. The use of the facilities at the Center for ACBM is greatfully acknowledged.

REFERENCES

1. N. G. Eror and H. U. Anderson, "Polymeric Synthesis of Ceramic Materials," Mater. Res. Soc. Symp. Proc., Vol 73, Materials Research Society, Pittburgh, PA (1986), p. 571-77.
2. M. Pechini, "Method of Preparing Lead and Alkaline-Earth Titanates and Niobates and Coating Method Using the Same to Form a Capacitor," U.S. Pat. No. 3 330 697, July 11, 1967.

3. D. Budd and D. A. Payne, "Preparation of Strontium Titanate Ceramics and Internal Boundary Layer Capacitors by the Pechini Method," Mater. Res. Soc. Symp. Proc., Vol. 32 (1984), p.
4. I. Nettleship, J. L. Shull, Jr., and W. M. Kriven, " Chemical Preparation and Phase Stability of Ca_2SiO_4 and Sr_2SiO_4 Powders," J. Euro. Ceram. Soc., Vol. 11, 291-98 (1993).
5. L-W. Tai and P. A. Lessing, "Modified Resin-Intermediate Processing of Perovskite Powders: Part I. Optimization of Polymeric Precursors," J. Mater. Res., Vol. 7 [2], 502-510 (1992).
6. L-W. Tai and P. A. Lessing, "Modified Resin-Intermediate Processing of Perovskite Powders: Part II. Processing for Fine, Nonagglomerated Sr-Doped Lanthanum Chromite Powders," J. Mater. Res., Vol. 7 [2], 511-519 (1992).
7. S. C. Zhang, G. L. Messing, W. Huebner and M. M. Coleman, "Synthesis of $YBa_2Cu_3O_{7-x}$ Fibers from an Organic Acid," J. Mater. Res., **5** [9], 1806-1812 (1990).
8. P. A. Lessing, "Mixed-Cation Oxide Powders via Polymeric Precursors," Am. Ceram. Soc. Bull., 68 [5] 1002-1007 (1989).
9. L-W. Tai, H. U. Anderson, and P. A. Lessing, "Mixed-Cation Oxide Powders via Resin Intermediates Derived from a Water Soluble Polymer," J. Am. Ceram. Soc., Vol. 75 [12], 3490-3494 (1992).
10. H. U. Anderson, M. J. Pennell and J. P. Guha, "Polymeric Synthesis of Lead Magnesium Niobate Powders," in Advances in Ceramics, Vol. 21: Ceramic Powders Science, Am Ceram. Soc., Westville, OH (1987), p. 91.
11. M. A. Gulgun, O. O. Popoola and W. M. Kriven, "Chemical Synthesis and Characterization of Calcium Aluminate Powders," J. Am. Ceram. Soc., Vol. 77 [2] 531-539 (1994).
12. B. R. Johnson, unpublished work, personal communication.
13. X. Cong and R. J. Kirkpatrick, "Hydration of Calcium Aluminate Cements: A solid -State ^{27}Al NMR Study," J. Am. Ceram. Soc., 76 [2] 409-416 (1993).
14. V. I. Yakerson, V. D. Nissenbaum, E. Z. Golosman, and V. M. Mastikhin, "High Resolution NMR study of Calcium Aluminate Catalysts," translated from Kinetika i Kataliz, 27, [6] 1419-26 (1986).

PRECERAMIC POLYCYCLOMETHYLSILAZANES: CHEMISTRY OF PROCESSING

Yigal D. Blum, Sylvia M. Johnson, and Gregory A. McDermott
SRI International
333 Ravenswood Ave.
Menlo Park, CA 94025

ABSTRACT

The chemistry involved in preceramic polymer synthesis, fabrication, and conversion to ceramic products is presented using polycyclomethylsilazane as an example. Reactions that are carried out during synthesis, curing, and pyrolysis are discussed. Chemical approaches to manipulate viscoelastic properties, polymer curing, ceramic yields, and final ceramic compositions are demonstrated.

INTRODUCTION

New approaches to the synthesis of low-cost, nonoxide, ceramic powders and processing of structural ceramics must be developed to develop a significant market for structural ceramics which is competitive with metallic alloys. Preceramic polymers can help or enable fabrication of nonoxide and oxide ceramic products (e.g., fibers, coatings, ceramic matrix composites, and net-shape structures), and can help to improve material properties. Many of the potential applications of preceramic polymers are concerned with lowering the cost of processes and improving the properties of the products.

The theoretical concept behind preceramic polymer technology is simple and derived from mature technology for fabricating carbon-carbon composites.[1] Preceramic polymers have inorganic skeletons but behave like organic polymers. Shape-forming of polymers and polymer-powder mixtures can be based on conventional plastics technology. After shape-forming, polymers are cured (crosslinked), and converted to the desired ceramic product by pyrolysis (heat treatment). Recent reviews have summarized progress in the field of polymeric precursors to silicon carbide, silicon nitride, and boron nitride.[2-4]

Moving from concept to practice is tedious. The polymers need to be designed to meet three sets of requirements. The first set are rheological properties: meltable versus soluble polymers, suitable molecular weight, control of viscosity, high tenacity (for fiber applications), good wetting and surface bonding (for coating applications), rapid post-fabrication curability, and shelf stability. Second are pyrolytic: high ceramic yield, control of stoichiometry after pyrolysis, low organic content, and easily cleaved organic groups. Third are economic constraints: low cost raw chemicals, simple synthesis, easy fabrication and processing to ceramics, and the ability to adapt the process to current manufacturing practices.

Research in this emerging field requires a multidisciplinary effort combining chemical synthesis, polymer characterization, fabrication processes, pyrolysis studies, and the characterization and evaluation of the derived ceramic products. The entire process from raw chemicals to final ceramic products is based on chemical reactions that start at temperatures as low as -70°C and are completed at temperatures exceeding 1500°C. These chemical steps determine the ability to synthesize the precursors and the composition and properties of the final product. In this paper we discuss the chemical aspects of developing practical preceramic polymers and converting them to the desired ceramic products. Chemistry must be considered whenever a material or a new application is developed. We have chosen polycyclomethylsilazane (PCMS), one of the polymers developed at SRI, to illustrate the role of chemistry in the development and use of preceramic polymers..

RESULTS AND DISCUSSION

Catalytic Approach for Polymer Synthesis, Fabrication, and Curing

Researchers at SRI International have developed new catalytic routes for polysilazane synthesis based on the ability of transition metal catalysts to activate Si-H bonds and dehydrocouple them with N-H bonds, as schematically illustrated in the following reaction:[5]

$$R_2SiH_2 + H_2NR' \xrightarrow{\text{trans. met. cat.}} -[R_2Si\text{-}NR']_n- + H_2$$

A variety of silane, amine (ammonia), and oligosilazane (synthesized by ammoniolysis) reactants and catalysts can be used to form polysilazanes with different structures, chemical contents, and thus, physical and pyrolytic behavior. This catalytic approach is extremely useful for (a) synthesis of polymers with controlled viscosity and chemical content, (b) modification of polymers, (c) efficient post-fabrication curing, and (d) control of pyrolysis properties.

The main catalyst used in our on-going development is $Ru_3(CO)_{12}$, a commercially available, air-stable, cluster compound chosen for its efficiency and practicality. The amount of catalyst commonly used to obtain polymers like PCMS is less than 200 mol ppm. Some polymerization reactions (e.g., aminated polycyclomethylsilazane, APCMS) proceeded with less than 10 mol ppm of $Ru_3(CO)_{12}$ at 60° to 90°C and were complete after 10 to 20 hours. The initial rates of the catalyst can exceed 1000 turnover numbers (catalytic cycles) per hour. Over 10,000 catalytic cycles were found to have occurred in several reactions without any signs of deactivation or decomposition of the catalyst. The long-term activity of the catalysis illustrated by the fact when dissolved in neat polysilazane the catalyst remained active over a period of two years.

The catalyst is homogeneous and remains in the polymer as a "living catalyst" that is later used to modify the polymer composition, manipulate rheological properties, promote rapid curing, and allow curing after shape-forming. It is also used to increase the nitrogen content of the final ceramic product.

Polycyclomethylsilazane Synthesis

Polycyclomethylsilazane belongs to a family of useful preceramic polymers developed at SRI. The polymer is produced from inexpensive methyldichlorosilane – a by-product of the silicone-oil and -rubber industry. Consequently, it is a low-cost precursor to ceramics which contain Si-N bonds. Figure 1 illustrates the synthesis of PCMS by a two-step reaction. First, a mixture of cyclooligomeric silazanes (CMS) is formed by reacting methyldichlorosilane with ammonia. The mixture of cyclomers is a low-viscosity liquid consisting mainly of volatile molecules that evaporate above 200°C. Cyclooligomeric silazanes are convenient reagents for synthesizing a variety of polysilazanes and polymetallosilazanes by the SRI[6] and other[7] methods. The polymer is then synthesized by adding the catalyst to the liquid cyclomers. The polymer can have a wide range of viscosities (1 to 1000 poise) and other attractive rheological characteristics. The polymer is converted to ceramic with yields of 60 to 85 wt%. The final ceramic compositions can be varied by changing the polymer formulation, additives, curing process, and pyrolysis conditions.

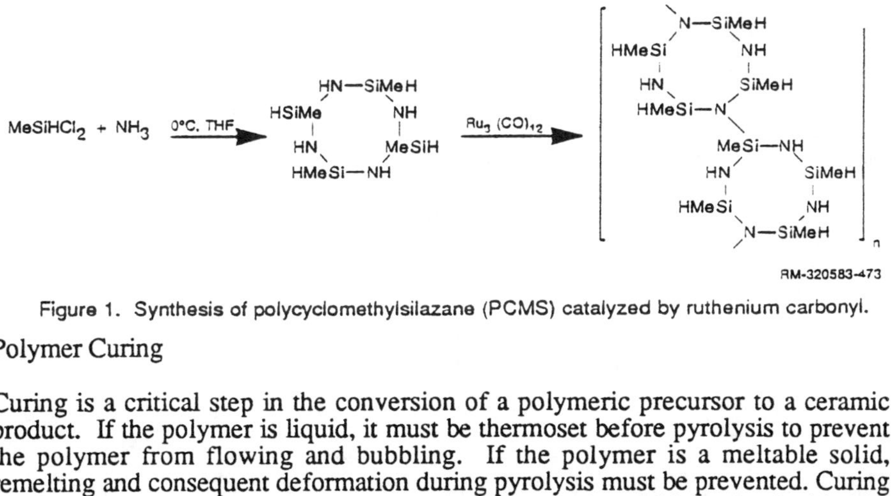

Figure 1. Synthesis of polycyclomethylsilazane (PCMS) catalyzed by ruthenium carbonyl.

Polymer Curing

Curing is a critical step in the conversion of a polymeric precursor to a ceramic product. If the polymer is liquid, it must be thermoset before pyrolysis to prevent the polymer from flowing and bubbling. If the polymer is a meltable solid, remelting and consequent deformation during pyrolysis must be prevented. Curing can also increase the ceramic yield. In certain cases the yield can be doubled by using an appropriate curing treatment before pyrolysis. The higher ceramic yield is the result of forming three-dimensional chemical bonds (crosslinking) that prevent the metathetic reactions responsible for cleaving small and volatile oligomeric fractions.[2] Crosslinking also prevents the volatilization of the low-molecular weight fraction found in the original polymer.

Polycyclomethylsilazane is a self-curing polymer. It can be cured by heating the polymer which contains the dehydrocoupling catalyst. A typical curing temperature is 150°C. Depending on the application (e.g., coating, binder) the curing time varies from 4 to 24-h in inert atmosphere. As indicated below, the dwell period of the curing can be drastically shortened when the polymer is cured in ammonia gas. Ammonia is an external source of N-H groups that accelerate the dehydrocoupling reaction. Typical times for curing under ammonia are 2 to 4-h at 150°C.

Other compounds containing multiple N-H groups can be used as additives for rapid curing. A few examples are CH_3NH_2, $CH_3CH_2CH_2NH_2$, $H_2N\text{-}NH_2$, and $H_2NCH_2CH_2NH_2$ (ethylene diamine; EDA). Ethylene diamine has been used as a reactive addition to increase the curing rates and ceramic yields. The two $-NH_2$ groups of EDA crosslink the polymer chains by a dehydrocoupling reaction with the Si-H groups. A series of reactions was carried out between PCMS and various amounts of EDA at 150°C in a nitrogen atmosphere to determine the curing rate and the effect of crosslinking on the post-pyrolysis ceramic yield. As shown in Table I, an optimal amount of EDA additive (approximately 10 wt%) accelerates the rate of gelation by a factor of about 200. The PCMS/EDA solution changed its viscosity from 1 poise to over 10,000 poise (typical gelling viscosity) in a period of 8 min.

The addition of curing agents allows the polymers to be used in applications where low viscosity is required. Infiltration of the polymer into structures with submicron pores, or into fiber weaves to produce composites by resin transfer molding, and injection molding formulations require low viscosity polymers, particularly if solvent is undesirable. Low viscosity is associated with low molecular weight, and usually means very low ceramic yields after the heat treatment. Use of the additives allows processing with a low viscosity polymer followed by further polymerization/curing *in-situ*. The polymer is cured rapidly and there is a high ceramic yield on conversion.

Table I: Effect of EDA and Ammonia on Gelation Time and Ceramic Yield

Amount of EDA (wt% of PCMS)	Gelation time (min)	Ceramic Yield[b] (wt%)
0	1560	60
0[a]	540	85
5	28	82
10	8	78
20	11	not measured

[a]Curing performed in bubbling ammonia through the liquid polymer;
[b]Pyrolysis conducted in ammonia.

Bubbling of ammonia through the liquid polymer can also accelerate the gelation time, but only by a factor of ~3. However, ammonia crosslinking is an important mechanism in curing the polymer because it can be used to cure polymers after they are used in a ceramic system. The ammonia curing route is commonly used at SRI as a first step in the pyrolysis of the fabricated polysilazanes.

Polymer Pyrolysis

General pyrolysis chemistry: Typical pyrolysis of SRI polysilazanes (as well as most other silicon-based preceramic polymers) consists of multiple chemical reactions that occur at various temperatures. Below 450°C the reactions are associated with (1) evaporation of volatile low-molecular-weight oligomers, split-out fragments, and volatile additives, (2) curing reactions, and (3) gas evolution from curing. Above 450°C the polymer is converted to an amorphous ceramic product. At this stage, the organic substituents are either removed as volatile organic compounds (i.e., methane, ethane) or converted to carbonaceous phases such as amorphous carbide or free carbon species.

Table II lists the vaporization species generated during the curing/pyrolysis process of PCMS and similar polymeric precursors to silicon nitride. Polymers that provide high ceramic yields are those in which (1) the evaporation of silicon-containing compounds (silanes and oligosilazanes) at 150° to 400°C is eliminated and (2) the weight fraction of the organic substituents is minimized (by reducing the size and number of the organic functional groups). Elimination of volatiles is feasible when (1) no low-molecular-weight precursors are present, (2) the polymer structure hinders the metathetical rearrangement of the polymer backbone, or (3) the cured polymer has very tight, three-dimensional, chemical bonding from crosslinking.

Table II: Chemical Reactions During Pyrolysis of PCMS and Related Polysilazanes

Temperature Range	Typical Reactions	Typical Volatiles
30 – 150°C	(1) Curing by deamination and dehydrocoupling (2) Solvent volatilization	(1) H_2, NH_3, RNH_2 (2) THF, cyclohexane, toluene, etc.
150 – 250°C	(1) Curing by deamination and dehydrocoupling (2) Si-N metathesis reactions (3) Si-H/Si-N metathesis reactions	(1) H_2, NH_3, RNH_2 (2) $[RSiNR']_n$ (e.g., $[CH_3SiNH]_n$) (3) $RSiH_3$ (only in specific cases)
250°C – 400°C	(1) Si-N metathesis reactions (2) Si-H/Si-N metathesis reactions	(1) $[RSiNR']_n$ (e.g., $[CH_3SiNH]_n$)
450°C – 700°C	(1) Si-R bond breaking (2) Deamination	(1) RH such as CH_4, CH_3CH_3, $CH_2{=}CH_2$, etc.(2) NH_3
700 – 1100°C	(1) Deamination (2) Dehydrocoupling	(1) NH_3 (2) H_2

Pyrolysis Environment: The pyrolysis environment plays a major role in determining the final composition of the ceramic. Polycyclomethylsilazane pyrolyzed under nitrogen or argon forms an amorphous, silicon carbonitride material. Pyrolysis in ammonia or air results in the formation of Si_3N_4 or SiO_2 respectively. Pyrolysis in the presence of a controlled amount of oxygen or water results in the formation of silicon oxynitride. Table III illustrates the elemental composition of PCMS pyrolyzed in nitrogen (with and without the addition of EDA), and in ammonia. The effect of ammonia on the elimination of carbon and the simultaneous increase in the nitrogen ratio relative to the silicon content is obvious. The addition of EDA does not affect the final ceramic composition.

The use of ammonia as a pyrolysis gas is critical to the success of many practical applications of precursors to ceramic nitrides. For the PCMS system, the ammonia is used both to promote further crosslinking during the pyrolysis and thus to increase the ceramic yield, and to cleave the carbon-containing species, which appears to be a requirement for achieving high quality silicon nitride products.

Table III: Effect of Pyrolysis Atmosphere on Composition of Pyrolyzed Polymer

Polymer	Atmosphere	Elemental Analysis (mol ratio)			
		Si	N	C	H
PCMS	N_2	1.00	0.99	0.73	0.45
PCMS	NH_3	1.00	1.34	<0.02	0.37
PCMS/EDA[a]	N_2	1.0	0.93	0.75	0.41

[a]10 wt% of PCMS.

Ceramic Yields: As discussed above, efficient curing of PCMS is important to obtain high ceramic yields. Though the polymer is self-curable, curing can be enhanced by the addition of external compounds having multiple N-H or Si-H bonds. For practical and cost reasons, the use of an N-H source is preferable.

Ammonia can be used as a curing agent to enhance the ceramic yields of low-viscosity PCMS. For example, a low-viscosity polymer cured slowly and completely gelled after approximately 9 h, had a ceramic yield of 58 wt% when pyrolyzed under ammonia. After treating the PCMS with ammonia (bubbling ammonia gas into the liquid polymer at 150°C for 30 min), the ceramic yield increased to 75 wt% as shown in Figure 2, even though the polymer viscosity remained low and almost unchanged. Ammonia has the potential for crosslinking the polymer because its multiple N-H bonds can undergo rapid dehydrocoupling with polymers that contain Si-H bonds. However, when a polymer was pyrolyzed using a standard heating rate (such as 5°C/min) in the presence of ammonia, no increase of ceramic yield was observed. This evidence contradicted the above result of a significant increase in ceramic yield when the low viscosity polymer was treated with ammonia in a solution at 150°C. Consequently, the pyrolysis schedule has been modified and a dwell period at 150°C for 4 h incorporated in the heating schedule. As shown in Figure 3, the hold period at 150°C in ammonia has a significant effect on the pyrolysis yield, and ammonia curing is efficient with this temperature/time combination (compare lines 2 and 3). Only a small improvement in ceramic yield is obtained when the polymer is held under nitrogen at 150°C for the same period (compare lines 1 and 4, and lines 2 and 5).

As indicated in Table I, the presence of a curing additive significantly increases the ceramic yield. When EDA is added to the polymer in the presence of a catalyst the

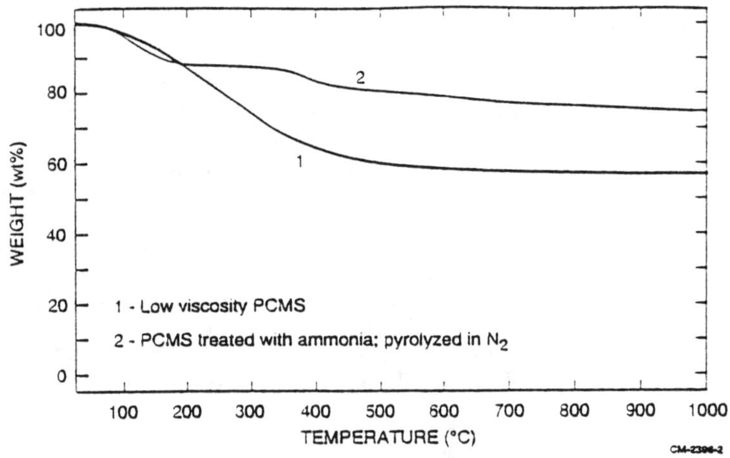

Figure 2. TGA results for low viscosity PCMS before and after ammonia bubbling through the liquid polymer at 150°C for 30 min.

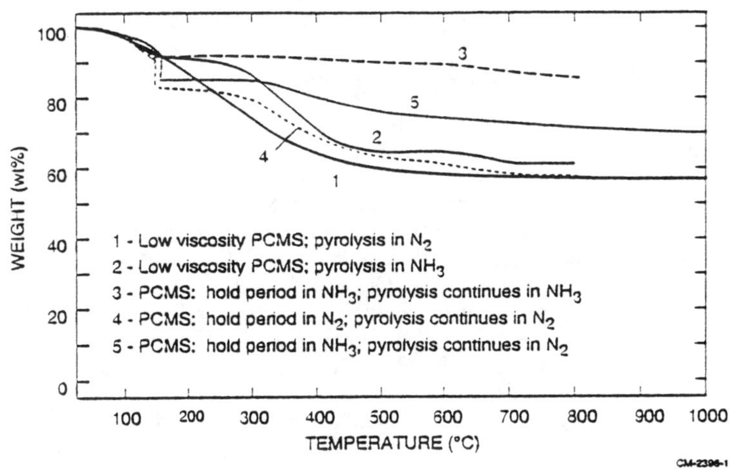

Figure 3. The effect of hold period at 150°C during the pyrolysis.

ceramic yields increase from about 60 to 80 wt%. A very small amount of EDA (5 to 10 wt%) is needed to obtain very high ceramic yields. TGA experiments revealed that EDA-crosslinked polymer provides a very high ceramic yield with either nitrogen (75 wt%) or ammonia (82 wt%), although pyrolysis in nitrogen resulted in a final product with a high carbon content. As shown in Table II, pyrolysis under ammonia resulted in significant reduction of this carbon.

CONCLUSIONS

Various polymers synthesized in our laboratory have been shown to be useful in fiber,[8] binder,[9] coating,[10] joining, composite, and sealing applications and as aids for microstructure control. Our experience reveals that the development of each application requires careful choice of precursor based on the specific processing and final property requirements. The results reported above demonstrate the need to understand and manipulate the chemistry to satisfy various requirements which can sometimes be in conflict (for example the need for low viscosity coupled with high ceramic yield). In-depth understanding of the chemistry involved at each stage of the entire process from the first synthesis of the chemical reagents to final ceramic product is required to develop optimized (or controllable) precursors and processes. This optimization is essential for making the preceramic polymer approach a viable commercial route to fabricating advanced ceramic products.

REFERENCES

1. K. J. Wynne and R. W. Rice, Ceramics via Polymer Pyrolysis, Annu. Rev. Mater. Sci. 14, 297-334 (1984).
2. Y. D. Blum, K. B. Schwartz, and R. M. Laine, Inorganic Polymer Pyrolysis. I. Silazane Preceramics and Their Pyrolytic Properties, J. Mat. Sci. 24, 1707-1718 (1989).
3. R. M. Laine and F. Babonneau, Preceramic Polymer Routes to Silicon Carbide, Chem. Mater., 5 (3), 260-279 (1993).
4. R. T. Paine and L. G. Sneddon, Borazine-Based Polymers Close in on Commercial Performance, Chemtech, July 1994, 29-37 (1994)
5. R. M. Laine, Y. D. Blum, R. D. Hamlin, and A. Chow, Organometallic Polymers as Precursors to Ceramic Materials: Silicon Nitride and Silicon Oxynitride, pp. 761-769, in *Ultrastructure Processing of Advanced Ceramics*, edited by D. D. Mackenzie and D. R. Ulrich, (J. Wiley and Sons) (1988).
6. Y. D. Blum, K. B. Schwartz, E. J. Crawford, and R. D. Hamlin, New Structures of Preceramic Polysilazanes Synthesized by Transition Metal Catalysis, *Better Ceramics Through Chemistry III*, MRS spring meeting proceedings, Vol. 121, edited by C. J. Brinker, D. E. Clark, and D. R. Ulrich, pp. 565-570 (1988).
7. a. D. Seyferth and G. H. Wiseman, U.S. Patent No. 4,482,669 (November 1984); b. M. Arai, T. Isoda, and O. Funayama, U.S. Patent No. 4,659,850 (April 1987).
8. Y. D. Blum and E. J. Crawford, Polysilazanes and Related Compositions, Processes, and Uses, US Patent 4,952,715 (August 1990).
9. S. M. Johnson, Y. D. Blum, G. A. McDermott, and M. I. Gusman, Interfaces of Polymeric Precursors with Ceramics/Metals, Scripta Metallurgica et Materialia, 31 (8), 1025-1030, (1994).
10. Y. D. Blum, R. M. Platz, and E. J. Crawford, Glass Strengthening by Polymer-Derived Ceramic Coatings, J. Am. Ceram. Soc., 73, 170 (1990).

THERMODYNAMIC ANALYSIS OF THE HYDROTHERMAL SYNTHESIS OF THE $Mn_{0.5}Zn_{0.5}Fe_2O_4$FERRITE

Tsuneharu Ogasawara and Ana Maria Rangel de Figueiredo Teixeira
Metallurgical and Materials Engineering Program (P.E.M.M.) of the Coordination of the Graduate Programs of Engineering (C.O.P.P.E.) of the Federal University of Rio de Janeiro (U.F.R.J.)
P.O.Box 68505 - ZIP CODE 21945-970
Rio de Janeiro-RJ, Brazil

ABSTRACT

Data on free energies of the formation of the $Mn_{0.5}Zn_{0.5}Fe_2O_4$ ferrite have been estimated using Mg-Ca silicate's model. Diagrams of Hydrogen Potential against pH and $-\log_{10}a_{Fe}$ against pH have then been developed for iron molal activities, a_{Fe} , ranging from 1 to 10^{-18} and temperatures in the range of 25°C to 300°C under hydrothermal conditions. The predictions of the diagrams match very well with the available experimental observations. Now it is possible to predict the behavior of that system and to explain more easily the experimental results.

INTRODUCTION

Mn-Zn ferrite and Ni-Zn ferrite outstand among soft ferrites [1-3], because the addition of some amount of Zn to Manganese ferrite and Nickel ferrite produces higher magnetic permeability and higher saturation magnetization [2]. Hydrothermal synthesis as a promising route to produce these types of ferrites has already been succesfully studied experimentally by Dias et al. [4,5] and Mohallem et al.[6] during 4 to 30 hours in the temperature range of 110°C to 190°C, while Komarneni et al.[7] manufactured their Mn-Zn ferrite through hydrothermal treatment during 4 to 20 hours in the temperature range of 135°C to 175°C. But, unfortunately, there was none thermodynamic analysis of the considered hydrothermal process. So, the present study has been carried out to supply a first thermodynamic insight to the problem.

THERMODYNAMIC DATA

Most of the thermodynamic data needed to calculate and to construct the diagrams has been supplied by HSC Chemistry for Windows 2.0, 1994, Outokumpu Oy, Pori, Finland), the original sources of these data being references [8-13]. Unfortunately, there wasn't any thermodynamic datum for Mn-Zn ferrites. Then, thermodynamic data estimation work has been carried out using well-known procedures to estimate entropies and constant pressure molar heat capacities; entropies of formation and heat capacities of formation of the different types of Mn-Zn ferrites have been calculated. Free energies of the formation of the Mn-Zn ferrite have been calculated assuming three different hypothesis in terms of the interaction between $MnFe_2O_4$, $ZnFe_2O_4$ and Fe_3O_4: (a) $MnZnFe_4O_8$ is formed as an ideal solution of $MnFe_2O_4$ and $ZnFe_2O_4$ at 298.15K; (b) an ideal solution ,$MnZnFe_7O_{12}$, is formed at 298.15K from $MnFe_2O_4$, $ZnFe_2O_4$ and Fe_3O_4; (c) $MnFe_2O_4$ and $ZnFe_2O_4$ are able to form a compound $MnO\cdot ZnO\cdot 2Fe_2O_3$ in which the relative interaction strength between the source ferrites is like that one working between calcium silicate and magnesium silicate to produce $CaO\cdot MgO\cdot 2SiO_2$, as expressed by equation (4).

$$Mn_{0.5}Zn_{0.5}Fe_2O_4 = 0.5\ [MnO\cdot ZnO\cdot 2Fe_2O_3] \quad (1)$$

$$MnO\cdot Fe_2O_3 + ZnO\cdot Fe_2O_3 = MnO\cdot ZnO\cdot 2Fe_2O_3 \quad (2)$$

$$CaO\cdot SiO_2 + MgO\cdot SiO_2 = CaO\cdot MgO\cdot 2SiO_2 \quad (3)$$

$$\Delta G^o_{(2)} / [G^o MnFe_2O_4 + G^o ZnFe_2O_4] = \Delta G^o_{(3)} / [G^o CaSiO_3 + G^o MgSiO_3] = 0.00534 \quad \text{at } 25^oC \quad (4)$$

The enthalpies and entropies of the MnZn-ferrite at 298.15K, as well as, their constant pressure molar heat capacity coefficientes ($c_p{}^o = A + B.10^{-3} + C.10^5T^{-2} + D.10^{-6}T^2$, in Joules per mole) are presented in Table I for the three above defined assumptions.

Table I - Estimated Thermodynamic Data for 298.15K to 573.15K temperature range $c_p{}^o = A + B\cdot 10^{-3}T + C\cdot 10^5T^{-2} + D\cdot 10^{-6}T^2$ in $J.mole^{-1}K^{-1}$.

Species	$H^o_{298.15K}$ kJ.mole$^-$	$S^o_{298.15K}$ $Jmol^{-1}K^{-1}$	A	B	C	D
$MnZn\text{-}Fe_4O_8$	-2419.48	279.491	294.139	18.702	-4.054	100.165
$MnZn\text{-}Fe_7O_{12}$	-3541.92	427.605	448.387	-54.731	-19.435	283.412
$Mn_{0.5}\text{-}Zn_{0.5}\text{-}Fe_2O_4$	-1214.77	139.746	147.072	9.351	-2.029	50.082

CALCULATION AND CONSTRUCTION OF THE DIAGRAMS

The calculations have been entirely performed using the Potential-pH Diagrams Software from HSC Chemistry for Windows 2.0; diagrams for selected conditions of temperature and molal activities of the elements in the aqueous solution were constructed. From these several Eh-pH diagrams for varying molal activities of Fe, Mn and Zn, the appropriate a_{Fe} versus pH data were taken through the plane Eh = Eh(at pH=0) - 0.000198441 T pH in the following way: Eh(25°C) = 0.3 - 0.0591651 pH; Eh(100°C) = 0.3 - 0.0740482 pH; Eh(200°C) = 0.27 - 0.0938923 pH; Eh(300°C) = 0.29 - 0.1137364 pH. These planes appear in Figures 1, 3, 5 and 7 in the form of a dotted line.

PRESENTATION AND INTERPRETATION OF THE DIAGRAMS

Figures 1, 3, 5 and 7 present Eh-pH diagrams for $a_{Zn} = a_{Mn} = 0.25\ a_{Fe}$ at 25, 100, 200 and 300°C, respectively, while Figures 2, 4, 6 and 8 contain pa_{Fe}-pH diagrams (where $pa_{Fe} = -\log_{10}a_{Fe}$) for $a_{Zn} = a_{Mn} = 0.25\ a_{Fe}$, respectively for 25, 100, 200 and 300°C. It is clearly seen the need to keep the values of the Eh-pH couple far below those of the equilibrium between the liquid water and 1 atm pressure of $O_2(g)$.

DISCUSSION

The conditions under which Dias et al.[4,5] and Mohallem et al.[6] (4 to 30 hours at 110°C to 190°C, 14 > pH > 7) and Komarneni et al. [7] (4 to 20 hours at 135°C to 175°C, 14 > pH > 7) have well succeeded in the production of Mn-Zn ferrite through hydrothermal processing fall well inside those predicted by the diagrams in Figures 3 to 6. Figures 1, 3, 5 and 7 show that the Mn-Zn ferrite Eh-pH existence domain increases with temperature rise up to 200°C and decreases at higher temperature. In turn, Figures 2, 4, 6 and 8 show that in the 25°C to 300°C temperature range: (a) the Mn-Zn ferrite pa_{Fe}-pH existence domain increases with temperature rise; (b) the pH giving minimum solubility (maximum stability) to the Mn-Zn ferrite decreases with the temperature rise. Neverthless, difficulties in the kinetics of the crystallization step of the hydrothermal synthesis may result in the production of a mixture of oxides or hydroxides of Zn, Mn and Fe when chemical precipitation is performed at low temperatures (like 25°C); in this case hydrothermal crystallization at higher temperatures will be necessary.

CONCLUSIONS

(a) The predictions of the diagrams developed in the present work match well the available experimental data.
(b) The pH range in which Mn-Zn ferrite can be obtained increases with increasing temperature of the system.
(c) The pH giving the maximum stability to the Mn-Zn ferrite decreases with increasing temperature of the system.

ACKNOWLEDGEMENTS

The authors thanks to Mr. Eduardo Soares Ogasawara for the valuable help on the text and figures preparation and to CAPES, CNPq, COPPE, UFRJ and UFF for the financial support.

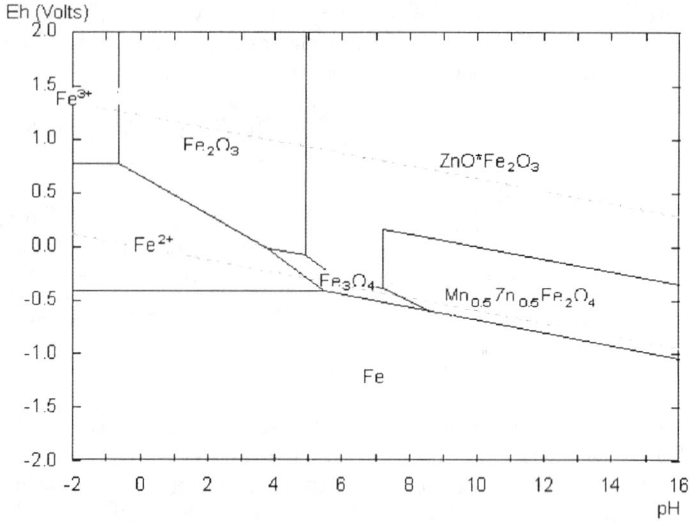

Figure 1 - Eh-pH diagram of the Mn-Zn-Fe-H_2O system at 25°C for 1 molal activity of Fe in the aqueous solution and $a_{Zn} = a_{Mn} = 0.25\ a_{Fe}$.

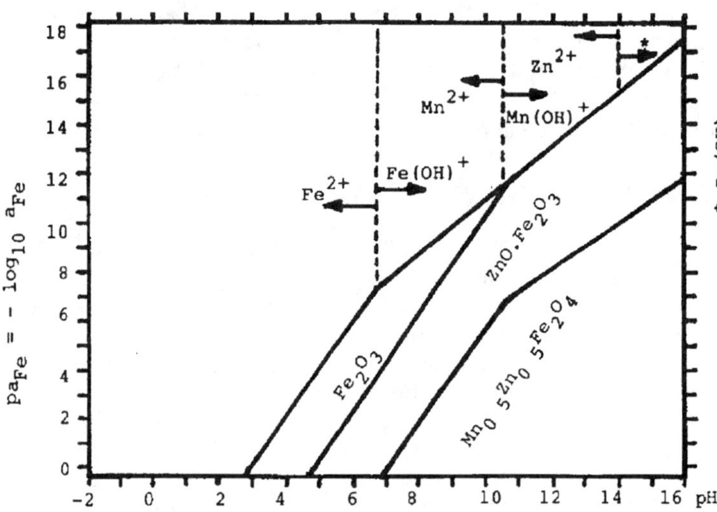

Figure 2 - The pa_{Fe}-pH diagram of the Mn-Zn-Fe-H_2O system at 25°C for $a_{Zn} = a_{Mn} = 0.25\ a_{Fe}$.

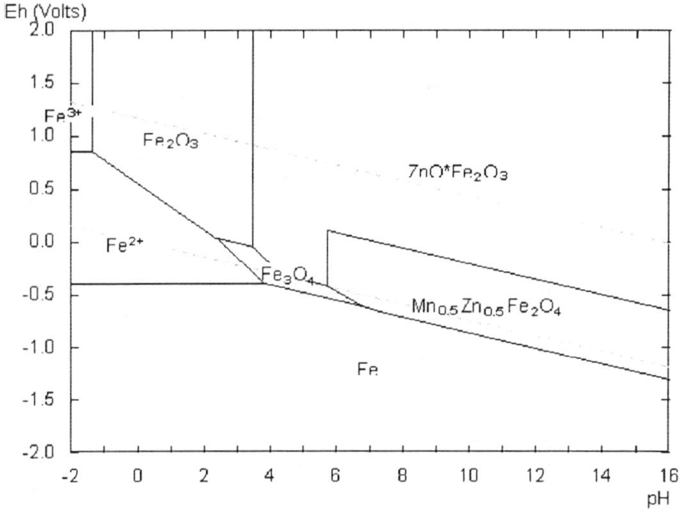

Figure 3 - Eh-pH diagram of the Mn-Zn-Fe-H_2O system at 100°C for 1 molal activity of Fe in the aqueous solution and $a_{Zn} = a_{Mn} = 0.25\ a_{Fe}$.

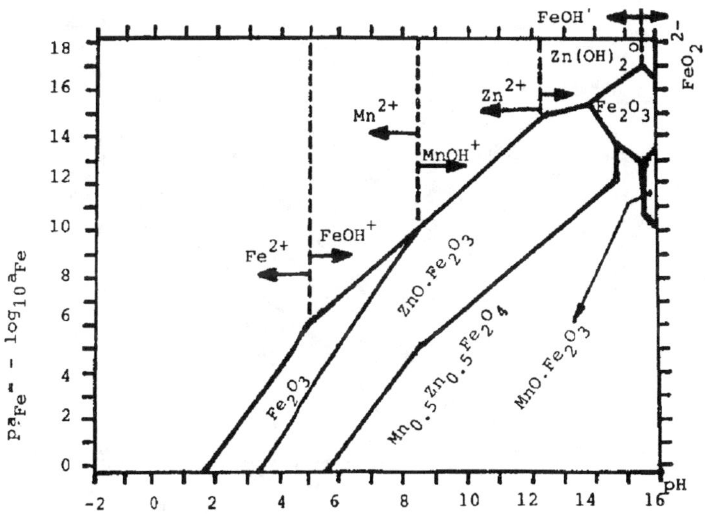

Figure 4 - The pa_{Fe}-pH diagram of the Mn-Zn-Fe-H_2O system at 100°C for $a_{Zn} = a_{Mn} = 0.25\ a_{Fe}$.

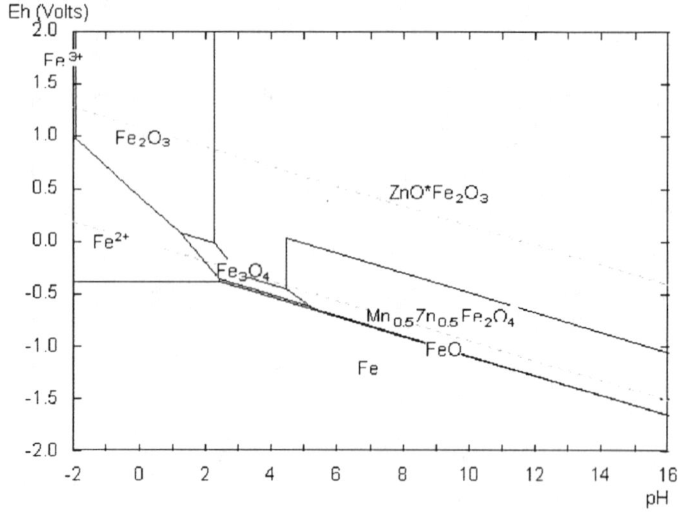

Figure 5 - Eh-pH diagram of the Mn-Zn-Fe-H_2O system at 200°C for 1 molal activity of Fe in the aqueous solution and $a_{Zn} = a_{Mn} = a_{Fe}$.

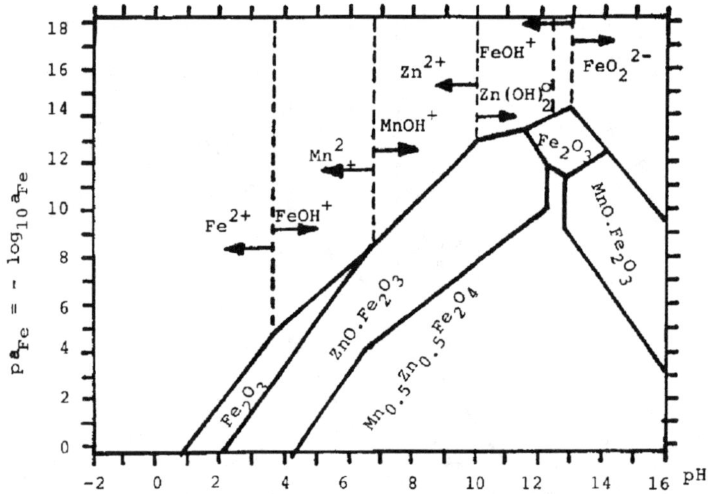

Figure 6 - The pa_{Fe}-pH diagram of the Mn-Zn-Fe-H_2O system at 200°C for $a_{Zn} = a_{Mn} = 0.25\ a_{Fe}$.

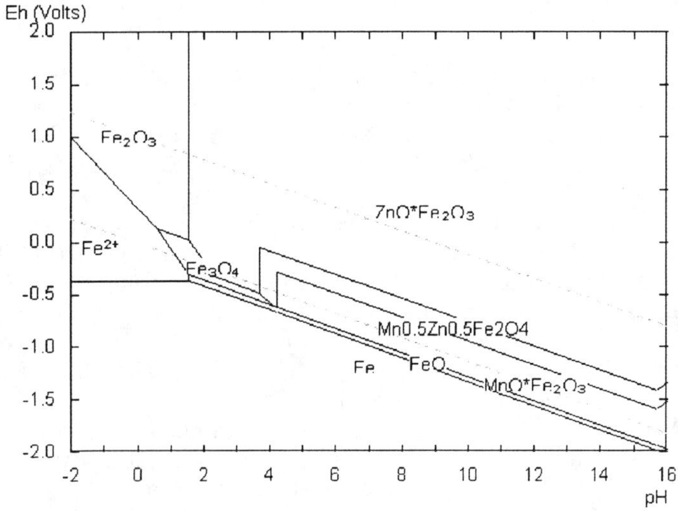

Figure 7 - Eh-pH diagram of the Mn-Zn-Fe-H_2O system at 300°C for 1 molal activity of Fe in the aqueous solution and $a_{Zn} = a_{Mn} = 0.25a_{Fe}$.

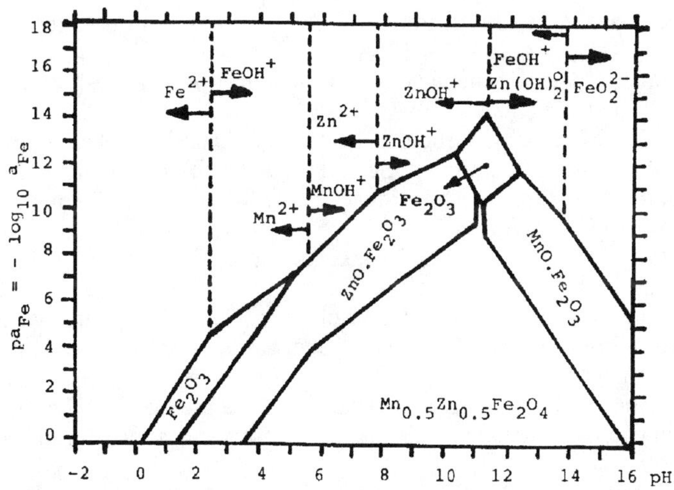

Figure 8 - The pa_{Fe}-pH diagram of the Mn-Zn-Fe-H_2O system at 300°C for $a_{Zn} = a_{Mn} = 0.25\ a_{Fe}$.

REFERENCES

[1] T.G. Reynolds and R.C. Buchanan, "Ferrite (Magnetic) Ceramics", Chapter 4, pp.227-264 in Ceramic Materials for Electronics, Relva C. Buchanan (editor), Marcel Dekker, Inc., New York,N.Y., 1986.

[2] L.L. Hench and J.K. West, "Principles of Electronic Ceramics", John Wiley & Sons, New York, N.Y., 1990.

[3] A. Morell and A. Hermosin, "Fast Sintering of Soft Mn-Zn and Ni-Zn Ferrite Pot Cores", Ceramic Bulletin, Vol. 59, No.6, pp.626-629, 1980.

[4] A. Dias, E.G. de Carvalho and V.T.L. Buono, "Morfological and Structural Characterisation of the Mn-Zn Ferrites Produced through Hydrothermal Process", Proceedings of the 35th Brazilian Ceramic Congress, Belo Horizonte-MG, 1991, Volume 1, pp.158-162, The Brazilian Ceramic Association, São Paulo-SP, Brazil.

[5] A.Dias, E.G. de Carvalho, M.S. Andrade and V.T.L. Buono, "Microstructure of the MnZn Mangetic Ceramics gotten by Hydrothermal Synthesis", Metallurgia & Materiais, (Metallurgy and Materials),Vol. 49, No.417, May 1993, pp. 338-342,ABM (Brazilian Metals and Materials Association), São Paulo-SP, Brazil.

[6] N.D.S. Mohallem, A. Dias,L.C.B.M. Pinto, A.A. Mendes Filho and E.G. Carvalho, "Preparation of Ceramic Powders Through Hydrothermal Process", Proceedings of the 35th Brazilian Ceramic Congress, Belo Horizonte-MG, 1991, Volume 1, pp.262-272, The Brazilian Ceramic Association, São Paulo-SP, Brazil.

[7] S. Komarneni, E. Fregeau and R. Roy, Journal of the American Ceramic Society, 71 (1), C26-C26, 1988.

[8] I. Barin, "Thermochemical Data of Pure Substances", VCH Verlags Gesellschaft, Weinheim 1993.

[9] S.M. Bailey, K.L. Churney, R.L. Nuttall, "The NBS Tables of Chemical Thermodynamic Properties, Selected Values for Inorganic and C1 and C2 Organic Substances in SI Units", J. of Physical and Chemical Reference Data, 11 (1982), Suppl.2.

[10] I. Barin, "Thermochemical Data of Pure Substances", VCH Verlags Gesellschaft, Weinheim, 1989.

[11] I. Barin, O.Knacke and O. Kubaschewski, "Thermodynamic Properties of Inorganic Substances", Springer-Verlag, Berlin and New York, N.Y., 1973, Supplement 1977.

[12] O.Knacke, O. Kubaschewski and K. Hesselmann, "Thermochemical Properties of Inorganic Substances", Springer-Verlag Berlin, Heidelberg, 1991.

[13] L.P.Ruzinov and B.S. Guljanickij, "Ravnovesnye Prevrasoenija Metallurgiceskin Reaktseij, Moskva, 1975.

THE EFFECT OF DIFFUSION FLAME CONFIGURATION IN AEROSOL SYNTHESIS OF NANOSIZE TITANIA AND SILICA POWDERS

Wenhua Zhu and Sotiris E. Pratsinis*
Department of Chemical Engineering
University of Cincinnati
Cincinnati, OH 45221-0171

Formation of titania and silica particles by oxidation of $TiCl_4$ and $SiCl_4$ vapors in a methane-air diffusion flame reactor was investigated. The effects of flame configuration (fuel and oxidant flow rate and position) on the product titania and silica particle size as well as on the phase composition of titania were studied. Specifically, the titania primary particle size decreased from 102 nm to 12.5 nm and the rutile fraction was reduced from 15% to 0.1 %, while the average size of silica particles was reduced from 29 nm to 10 nm by just exchanging the position of the fuel and oxidant streams in the burner.

INTRODUCTION

Titania and silica powders are important particulate commodities. The annual production volume of these powders amounts to over 2 million tons worldwide. Titania powders are extensively used in pigments, as catalyst supports, and more recently in synthesis of inorganic membranes and as photocatalyst in gas and water purification [1]. Silica particles have found applications as optical fibers, fillers, thickening and reinforcing agents, catalyst carriers etc. A large proportion of these product powders are manufactured by flame aerosol processes such as the so-called "chloride process" where oxidation of titanium or silicon tetrachloride ($TiCl_4$ or $SiCl_4$) vapor takes place in a flame reactor. In the past, studies on vapor phase synthesis of TiO_2 powders in electrically heated flow reactors provided valuable insight for this process. Akhtar et al. [2] showed that temperature, precursor concentration, and reactor residence time controlled the titania particle size and phase composition. However, titania powders are manufactured on a commercial scale mostly in flame reactors because they are economical and simple in construction and operation [1]. The synthesis of silica in hydrocarbon flames was studied both theoretically and experimentally by Ulrich et al. [3, 4]. They showed that the final silica aggregate and primary particle size were strongly

* **To whom communication should be addressed**

affected by flame temperature and residence time while nucleation and surface reaction did not affect particle dynamics.

Though a number of patents exist on reactor design and operation and several papers address specific systems, such as the synthesis of TiO_2, SiO_2, Al_2O_3 and superconductor powders, the understanding of flame synthesis of titania and silica is still limited. Powders made in these reactors are agglomerates of primary particles. Agglomerates can be beneficial in synthesis of catalysts and catalytic substrates but they can be detrimental in manufacture of powders for paints or ceramics. Since the particle synthesis process, from the formation of condensing species by chemical reactions to the growth of particles, occurs in the flame, the characteristics of the flame (e.g., the reactant mixing configuration, the flame height, temperature) can greatly affect the product particle characteristics. In the present studies, a diffusion flame reactor is selected for its widespread use in aerosol manufacture of commodity powders and its provision of a stable flame over a wide range of operating conditions [5]. A systematic investigation is carried out using four different flame configurations, depicting the effect of reactant mixing on the characteristics of the powder along with the comparative evaluation of sintering rate of different materials in identical flame conditions.

EXPERIMENTAL

Apparatus

A schematic of the experimental setup is shown in Figure 1. The burner (a) is comprised of a series of concentric quartz tubes. The diameter of the central tube is 4 mm and the spacing between the successive tubes (rings) is 1 mm. Particles formed in the flame are collected on glass fiber filters (Gelman Scientific; 47 mm) placed inside a stainless steel (Graseby /Anderson) open-faced filter holder (b). The filter is fixed at 120 mm above the tip of the burner in all experiments.

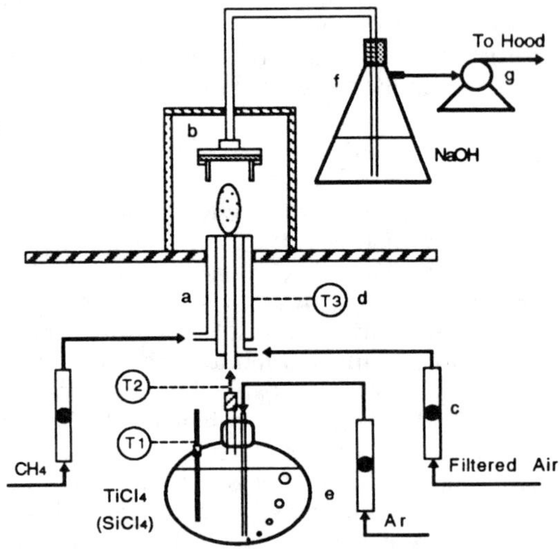

Figure 1 Experimental Setup

Methane (Wright Brothers, 99%) is used as fuel while air is used as oxidant. All reactant flow rates are metered by rotameters (c) while the temperatures are measured by thermocouples T2 and T3 (d) or thermometer T1. The precursor vapor ($TiCl_4$ or $SiCl_4$) is carried into the central tube of the reactor by passing argon (Wright Brothers, 99.8%) through a bubbler (e) containing liquid $TiCl_4$ or $SiCl_4$ (Aldrich, 99.9%). The bubbler containing $TiCl_4$ is kept at 50°C by an electrically heated mantel. The burner and the manifold are kept about 20°C above the bubbler temperature to prevent condensation of $TiCl_4$ in the lines. The temperature of the liquid $SiCl_4$ is maintained at 0°C by keeping the bubbler in ice water.

Sampling is facilitated by a vacuum pump (g) and corrosive byproducts such as Cl_2 and HCl are removed from the exhaust by passing them through a 1M NaOH aqueous solution (f) before being released to the fume hood.

Procedure and Measurements

Four flame configura -tions are investigated (Figure 2) in which Ar gas initially saturated with precursor vapor flows always through the central tube of the burner. The flow rates of CH_4 and air in each flame configuration are identical. From flame A to D, the position of the air stream moves from the central tube to the 3rd ring, meanwhile the position of methane moves from the outside ring to the center tube. It should be noted that in combustion terminology, flames A and B have an internal flame front between the supplied air and CH_4 streams as well as an external flame front between CH_4 and still air. In contrast, in flames C, D there is only one (external) flame front between the supplied CH_4 and air streams.

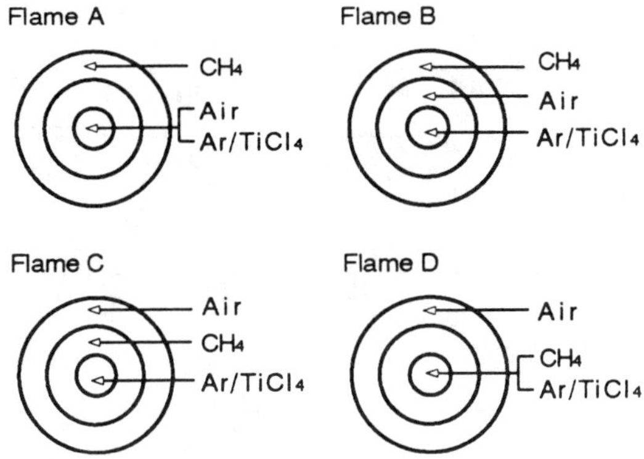

Figure 2 Mixing Configurations of Reactants

X-ray diffraction is used to determine the phase composition of the powders (D500, Siemens). The weight fractions of anatase and rutile phases in the samples are calculated from the relative intensities of the primary peaks corresponding to anatase and rutile as described by Spurr and Myers [6]. Transmission electron micrographs (TEM) are obtained on a Philips CM 20 microscope operating at 200 kV. The powder specific surface area is measured by nitrogen adsorption (Gemini 2360, Micromeritics) at 77K using the BET equation. Assuming spherical particles, the average grain size, d_p, is calculated from the specific surface area, A, by $d_p=6/\rho_p A$. The average grain size computed from the specific surface area is compared with the data measured directly from the TEM micrographs and it is about 5 ~ 15% greater than the microscopic measurements as the BET does not take into account the area lost at particle necks.

The maximum flame temperature at the tip of the flame is measured with a 0.015" Pt-Rh R-type thermocouple (Omega Engineering) and corrected for radiation losses calculated based on the measured flame temperature and the Reynolds number of the flow passing through the thermocouple junction [7]. Since particles generated in the flame quickly coat the thermocouple junction and cause measurement errors, the temperatures are measured in the absence of precursor, though the carrier gas Ar is flowing through the reactor.

RESULTS AND DISCUSSION

Table I lists the operation conditions and the corresponding flame heights and maximum temperatures taking into account the radiation losses. The flame height is defined as the distance from the burner mouth to the end of the luminous part of the flame.

Table I Reactant Flow rates and the Characteristics of the 4 Flames

CH_4: 0.31 l/min Ar: 0.25 l/min

Air	Flame A		Flame B		Flame C		Flame D	
l/min	Temp. (°C)	Height (mm)	Temp. (°C)	Height (mm)	Temp. (°C)	Height (mm)	Temp. (°C)	Height (mm)
2.5	1010	48	1099	60	1131	75	1138	85
3.8	1014	38	1096	55	1161	70	1166	75
5.5	1050	33	1091	48	1167	60	1176	65

Synthesis of TiO_2 Particles

The effect of flame configuration was investigated at various air flow rates and at constant CH_4 and $TiCl_4$ loading, 0.31 l/min and 5.8 x 10^{-4} mol/min, respectively. Figure 3 shows TEM pictures of titania particles synthesized in flames A, B, C and D at the above conditions and with 3.8 l/min air flow rate. Though the reactant composition is identical in all flames, the average primary particle size increases dramatically from flame A (15 nm) to D (102 nm)!

Figure 4 shows the average primary particle size of TiO_2 powders at air flow rate in the 4 different flames. Flame A produces extremely fine TiO_2 particles since the $TiCl_4$ stream is vigorously diluted with air (10 ~ 22 times dilution) prior to its oxidation in the flame. Pratsinis et al. [8] demonstrated that low precursor concentration and short residence times resulted in small primary particle size and loose agglomerates due to incomplete chemical reaction and low coagulation and sintering rates. Increasing the air flow rate dilutes the precursor stream reducing, thus, the primary particle size. Moreover, increasing the air flow rate reduces the flame height (Table I) since combustion starts earlier. Therefore, the collision rate of newly formed titania decreases, favoring formation of small particles.

As to flame B, the dilution of the $TiCl_4$ stream with air is taking place further downstream from the burner and near the flame. Therefore, the dilution is limited prior to its oxidation resulting in larger particles than flame A. In addition, the dilution of the precursor is not as homogeneous as that in flame A, as a result, the particles synthesized in flame B have a wider primary particle size distribution

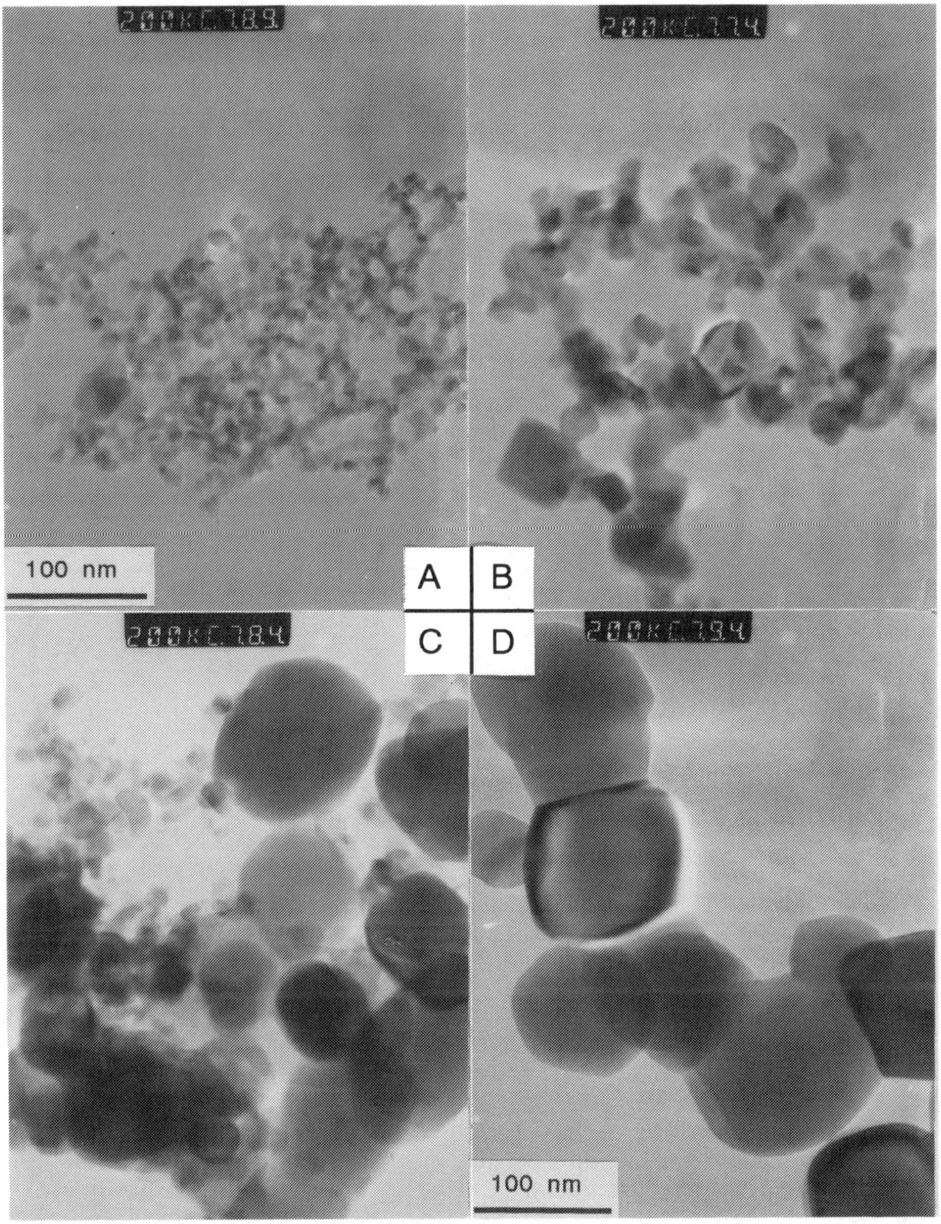

Figure 3 TEM Pictures of TiO_2 Particles Synthesized in the 4 Flame Configurations

than those made in flame A (Figure 3 a and b). Since the $TiCl_4$ stream comes in contact first with air, increasing the air flow rate increases the dilution rate reducing, thus, particle collisions and size in agreement with Formenti et al.[9].

Flames C and D have higher temperatures and narrower flame fronts than flames A and B. More importantly, the dilution of the $TiCl_4$-laden Ar stream with the CH_4 stream is not that significant since both streams have comparable flow rates 0.25 and 0.31 l/min, respectively. Furthermore, the flame height increases and the mixing of precursor vapor and methane takes place much earlier than that in flame A and B, the newly formed titania has a much longer time staying in the flame for sintering, resulting in rather large primary particles of titania (Figure 3, c and d). Flame D has an even narrower flame front than flame C since methane flows through the central tube of the burner and the flame is located at the first ring of the burner, the particles stay in the flame and sinter longer, as a matter of fact, the particle size remains at about 102 nm regardless of the air flow rate.

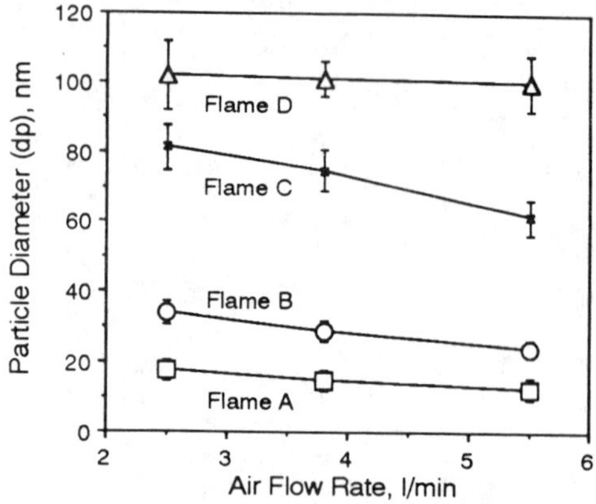

Figure 4 BET Equivalent Titania Primary Particle Size as a Function of Air Flow Rate in the 4 Flame Configurations

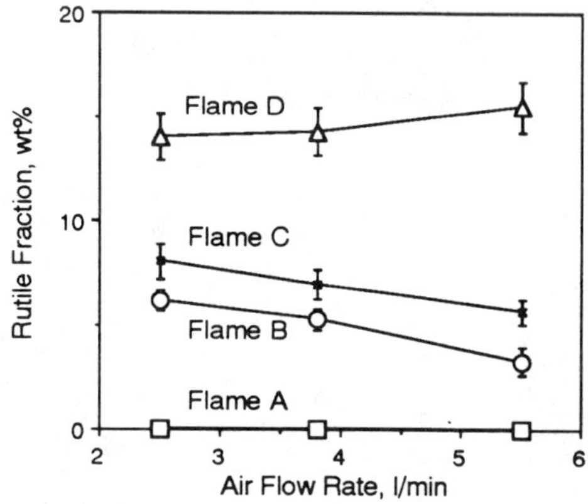

Figure 5 Rutile Fraction of Titania Particles as a Function of Air Flow Rate in the 4 Flame Configurations

Figure 5 shows that the rutile content of the titania powders increases from flame A to D. The rutile has a much closer atomic packing in its crystal pattern and is more stable than anatase. The transformation of anatase to rutile involves volume contraction and a cooperative movement of titanium and oxygen ions, which increases with increasing temperature [10, 11]. From the kinetics point of view, long residence times in the flame also enhance the anatase to rutile transformation. The flame temperature increases from flame A to D, resulting in the increase of the rutile fraction in the powders. In flame A, the air flow rate made no difference in the rutile content since even at the lowest employed air flow rate, virtually no rutile was formed due to the low temperature in the flame. Increasing the air flow rate does not affect the flame temperature much but reduces the flame height in flames B and C. As a result, the particle residence time is decreased reducing the rutile content of the powders made in flames B and C. This shows that flame height is probably a more accurate indication than maximum flame temperature is for the thermal history of the particles in these flames. Finally, in Flame D, the rutile content almost remains the same when air flow rate changes, in agreement with the constant particle size.

Synthesis of SiO_2 Particles

Figure 6 shows the evolution of silica particle size with respect to air flow rate in flames A and D. The flame condition is the same as listed in Table I and the precursor $SiCl_4$ loading rate is 10.3×10^{-4} mol/min, twice the loading rate of $TiCl_4$ in Figure 4. However, since the reaction yield of silica particle is about half of that of titania under the same flame conditions (Figure 7), the actual concentration of condensed species in the flame is comparable. Reaction yield is the rate of powder collected over the powder that would have been collected if all precursor halide had been converted to the corresponding oxide powder. It can be seen from figure 6 that though the silica particle size evolution with respect to flame configuration has the same tendency as that of titania particles, silica particles are much smaller than titania particles, especially in flame D. From flame A to D, the titania particle size increases dramatically (8 times!) while the silica particle diameter increases only 3 times. This is attributed to the different sintering mechanisms of titania and silica. Titania sinters by grain boundary diffusion while silica sinters by a viscous flow sintering mechanism. At the temperatures encountered in the present study, the sintering rate of titania is much faster than that of silica. In flame A, since both titania and silica are hardly sintered, they have comparable primary particle sizes. Titania particles sinter more effectively than silica in flame D since the later provides higher temperature and longer residence time resulting in the large difference in product particle sizes. At high air flow rate, the residence time is reduced, silica particles hardly sinter even in flame D, and as a result, particle sizes become comparable in flames A and D.

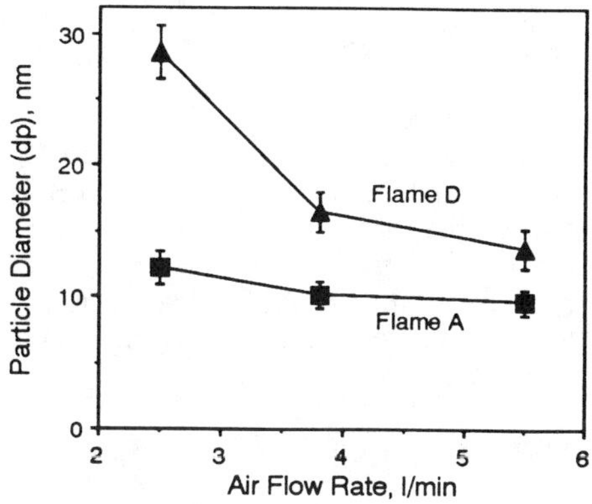

Figure 6 BET Equivalent Silica Particle Size as a Function of Air Flow Rate in Flames A and D

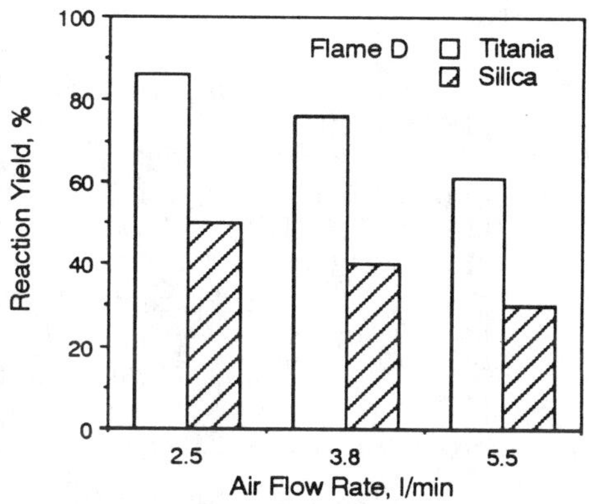

Figure 7 The Yield of Titania and Silica with Respect to Air Flow Rate in Flame D

Figure 7 shows the reaction yields of silica and titania in flame D with respect to air flow rate. The complete conversion of precursor to condensed phase is not reached due to the low flame temperature and short residence time under the present flame conditions. The reaction conversion of $TiCl_4$ is higher than $SiCl_4$ in accordance with the oxidation kinetics of $TiCl_4$ and $SiCl_4$ measured by Pratsinis et al. [12] and French et al [13] in furnace reactors.

CONCLUSIONS

Nanosized TiO_2 and SiO_2 are synthesized in a diffusion flame reactor. The effect of reactant mixing on product titania and silica particle size as well as on titania particle phase transformation were investigated. The better the mixing between fuel and precursor vapor, the coarser are the product particles and the higher is the rutile content. The reactant mixing affects titania particle size more significantly than that of silica due to the faster sintering rate of TiO_2. The low flame residence time and temperature limit the completion of the reaction of both precursors.

ACKNOWLEDGMENT

This research was supported by a grant from Advanced Fuel Research Inc. and the National Science Foundation, CTS-8957042.

REFERENCES

1. G.P. Fotou, S. Vemury and S.E. Pratsinis, *Chem. Eng. Sci.*, 49 (1994)4939.
2. M.K. Akhtar, S. Vemury and S.E. Pratsinis, " *AIChE J.*, 40 (1994)1183.
3. G.D. Ulrich, *Comb. Sci. Tech.*, 4 (1971)47.
4. G.D. Ulrich, B.A. Milnes, N.S. Subramanian, *Comb. Sci. Tech.*, 14 (1976) 243.
5. G.P. Fotou, S.E. Pratsinis, P.A. Baron, *Chem. Eng. Sci.*, 49 (1994)1651.
6. R.A. Spurr and H. Myers, *Analytical Chem.*, 29 (1957)760.
7. S.-L. Chung, Ph. D. Thesis, The Johns Hopkins University, Baltimore (1991).
8. S.E. Pratsinis, W. Zhu and S. Vemury, accepted by *Powder Technology*, (1995).
9. M. Formenti, F. Juillet, P. Meriaudeau, S.J. Teichner and P. Vergnon, *J. Colloid Inter. Sci.*, 39 (1972)79.
10. M.K. Akhtar, S.E. Pratsinis and S.V.R. Mastrangelo, *J. Am. Ceram. Soc.*, 75 (1992)3408.
11. M.K. Akhtar, S.E. Pratsinis, and S.V.R. Mastrangelo, *J. Mater. Res.*, 9 (1994)1241.
12. S.E. Pratsinis, H. Bai, P. Biswas, M. Frenklach, and S.V.R. Mastrangelo, *J. Amer. Ceram. Soc.*, 73 (1990)2158.
13. W.G. French, L.J. Pace and V.A. Foertmeyer, *J. Phys. Chem.*, 82 (1978)2191.

NITRIDING OF LARGE-SIZED ALUMINUM

Shigeru ITO, Kazuharu YOSHIDA, Shin-ichi SAITO and Kazuo AKASHI
Department of Industrial Chemistry, Faculty of Science and Technology
Science University of Tokyo, Noda Chiba 278, Japan

ABSTRACT

Aluminum chips with 2mm in size have been successfully nitrided with 99.3% of conversion, using yttrium oxide as a promoter. The product consisted of hollow spheres with 2mm in dia. and fine particles of aluminum nitride. This nitriding is divided into two steps. In the first step, nitriding begins from the surface of aluminum chip. Because the surface is activated by the addition of yttrium oxide, the reaction heat raises the temperature of aluminum enough to generate the aluminum vapor, followed by the second step; exothermic reaction of aluminum vapor and nitrogen gas. In addition, to accelerate the first step without additive, electrolytic etching was applied to aluminum wires with 2mm in dia. In this case, aluminum wires were also completely nitrided.

INTRODUCTION

Direct nitriding of aluminum is one of the industrial processes for the preparation of aluminum nitride. In this process, atomized powder of aluminum has been essentially used. This is because the nitriding is considered to be caused by the diffusion of nitrogen in the aluminum nitride layer on the aluminum melt. This leads to the use of atomized powder; that is, the smaller, the better On the other hand, whiskers are often observed in the product[1]. The whiskers seem to be formed by the vapor phase reaction of aluminum vapor and nitrogen gas. If main reaction of the nitriding is that of the vapor phase, the use of atomized powder is not exactly essential; that is, the large-sized aluminum will be able to be nitrided.

In this work, the nitriding of aluminum chips with 2mm in size (net wt. 8g) was examined. To generate the aluminum vapor, Y_2O_3 was selected as a promoter after the preliminary experiments[2]. In addition, electrolytically etched aluminum wires were also used as starting material to nitride without additive.

If large-sized aluminum is completely nitrided, the atomized process of aluminum will be removed. In addition, the product is expected to contain less amount of oxygen impurity, because surface area of large sized aluminum is much smaller than that of powder. And also, the large-sized aluminum is easy to handle, compared to the powder.

EXPERIMENTAL

Aluminum chips with 2mm in size (net weight 8g) were immersed in an ethanol solution of $Y(NO_3)_3 \cdot 6H_2O$. $Y(NO_3)_3 \cdot 6H_2O$ was uniformly coated on the surfaces of the chips by the vaporization of the solvent. $Y(NO_3)_3 \cdot 6H_2O$ immediately transforms to Y_2O_3 in the nitriding process at high temperature. Additive amounts of $Y(NO_3)_3 \cdot 6H_2O$ correspond to 0-3wt% as Y_2O_3 in the product. These chips were charged into a carbon reaction tube(18mm dia, 65mm in length), which was then put in the horizontal tubular furnace. The nitridings were performed at 1000-1300℃ for 2 hours, flowing nitrogen gas with 400 ml/min. The product was

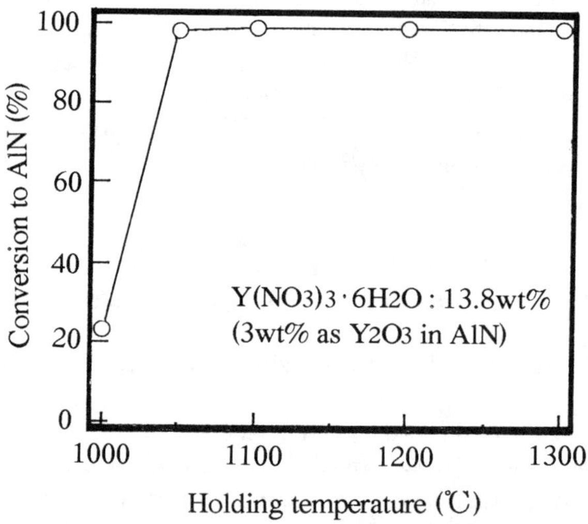

Fig.1 Conversion of aluminum chips to aluminum nitride. (Holding time : 2hours. Heating rate from room temperature : 15℃/min. Flow rate of nitrogen : 400ml/min.)

analyzed by kjeldahl method to determine the conversion to aluminum nitride.

RESULTS AND DISCUSSION

Nitriding with an Additive

Figure 1 shows the relationship between conversion to aluminum nitride and holding temperature. When the aluminum chips without additive were used, an aluminum block casted in the reaction tube was obtained. On the other hand, adding 3wt% of yttrium oxide, the aluminum chips with 2mm in size were completely nitrided above 1050℃. The conversion of the product obtained at 1300℃ reached to 99.3%. Single phase of aluminum nitride was confirmed by the x-ray diffraction pattern of the product. Figures 2 and 3 show the aluminum chips in the reaction tube and the appearance of the product, respectively. A number of spherical shells with 2mm in size were observed in the product. The spherical shell consisted of aluminum nitride and was covered with fine particles (ca. 1 μm) of aluminum nitride. Figure 4 shows the temperature in the reaction tube, which was measured by inserting a thermocouple into the aluminum chips. At 660℃, the inner temperature is lower than that of furnace temperature. This endothermic

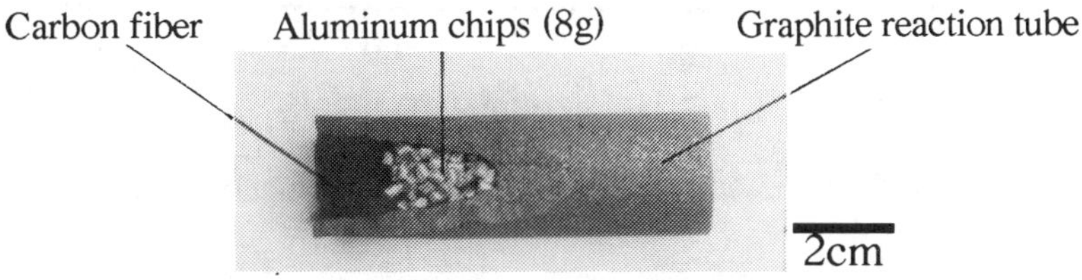

Fig.2 Aluminum chips in graphite tube.

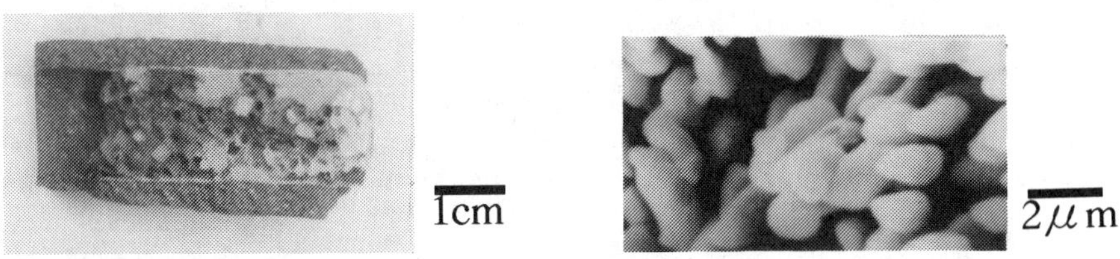

(a) Spherical shells and fine particles (b) Fine particles

Fig.3 Appearance of the product. (Holding temperature : 1300℃, Holding time : 2hours, Conversion : 99.3%)

reaction corresponds to melting of aluminum chips. However, above 800℃, inner temperature exceeds furnace temperature. This is due to the reaction heat of nitriding. When the temperature is reached to about 1300℃, inner temperature suddenly increases up to 1820℃. This temperature is enough to generate the

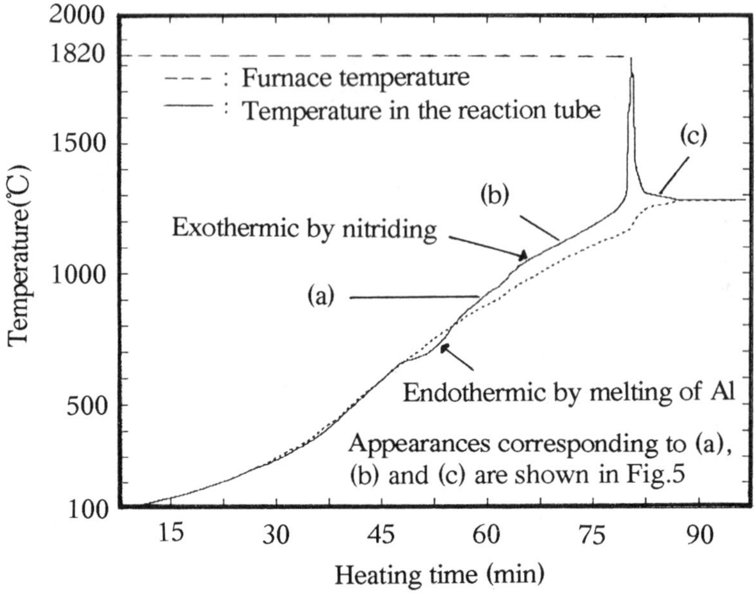

Fig.4 Furnace temperature and temperature in the reaction tube.

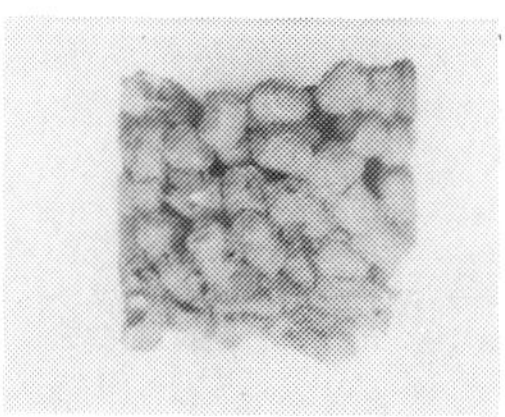

(a) 900℃ (Conversion to AlN : 1.9%, AlN layer : 6.4μm)

(b) 1150℃ (Conversion to AlN : 9.5%, AlN layer : 65μm)

1cm

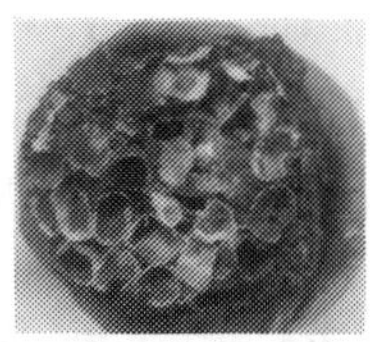

(c) 1300℃ (After rapid heat generation.) (Conversion to AlN : 99.0%, AlN layer : 120 μm)

Fig.5 Aluminum chips at every temperature.

aluminum vapor. Figure 5 shows the aluminum chips heated up to 900, 1150 and 1300℃, respectively. The aluminum chips heated up to 900 and 1150℃ maintained their shapes. On the other hand, after the rapid increase of the temperature, the aluminum chips transformed into the spherical shells.

Nitriding Model

From the results described above, following reaction model is suggested. This model is schematically shown in Fig. 6. This process is divided into two steps. In the first step, the nitriding proceeds from the surface of the chip, generating the reaction heat. In this step, the chip is covered with the aluminum nitride shell. As the temperature raises to 1050 ~ 1300℃, nitriding is accelerated and the generation of reaction heat is also accelerated. The temperature of the chip becomes to exceed the furnace temperature. When the temperature of chip raised by the reaction heat is enough to generate the aluminum vapor, the second step begins. Because the aluminum vapor is very active compared to the aluminum melt, the nitriding proceeds very rapidly to produce fine particles of aluminum nitride outside of the shell. The rapid nitriding generates a large amount of the reaction heat in a moment. The reaction heat causes new vaporization of aluminum. This vapor phase reaction continues until aluminum in the shell is wholly vaporized. As the result, the spherical shells covered with fine particles are formed.

Effect of Y_2O_3 Addition

In this model, the first step is very important; that is, yttrium oxide seems to promote the reaction of aluminum melt and nitrogen. It is well-known that the

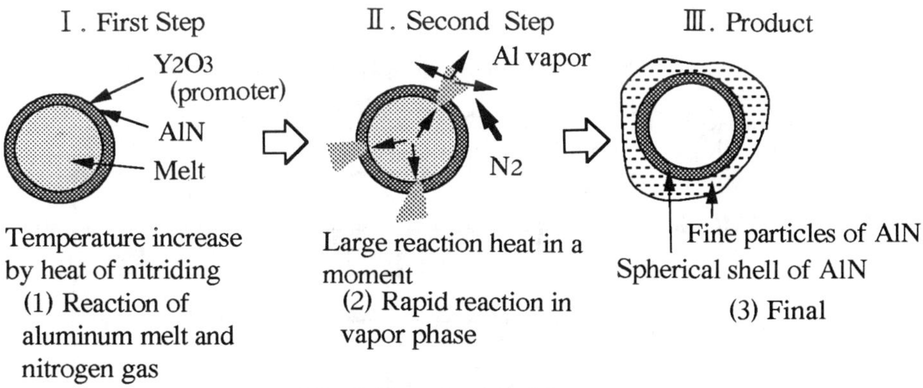

Fig.6 Reaction Model

thickness of aluminum nitride layer on the aluminum melt is about 18 μm at 1200℃[3], or about 30 μm at 1400℃[4]. In most cases, the diffusion of nitrogen in the aluminum nitride layer is too slow to complete the nitriding. However, in this study, the thickness of the shell grew up to be about 100-200 μm. This is the effect of yttrium oxide, which is also well-known as an additive for the sintering of aluminum nitride. The product contains 3 wt% of Y_2O_3, which is appropriate for the sintering. But, this product is not pure aluminum nitride. Table I shows the additive effect of Y_2O_3 on the nitriding of the aluminum chips. At 1300℃, 0.1 wt% of Y_2O_3 was even effective for the nitriding.

Nitriding without Additive

If the first step proceeds without additive, enough to supply the vaporization heat of the aluminum, the large-sized aluminum would be also nitrided. In order to accelerate the first step reaction, aluminum surface was electrolytically etched. In this case, aluminum wires with 2mm in dia. and 50 mm in length were used. The aluminum wires were electrolytically etched in 0.1N-HCl solution at 20-80℃ for 1 min, applying 2000A/m^2 of current density. The etched aluminum wires (net 4g) were then charged into the reaction tube. In order to maintain the etched surface, temperature was firstly kept at 600℃ for 2 hours. These wires were then heated at 1300℃ for 2 hours. Figure 7 shows the appearances of the product. Instead of the spherical shells, the aluminum nitride pipes formed. Fine grains of aluminum nitride also exists out side of the pipes. The maximum conversion was 99.7%.

Nitriding of Different Shaped Aluminums

The nitridings were also performed for some shaped aluminums; block, pipes and

Table I. Effect of Y_2O_3 on the nitriding of aluminum chips with 2mm in size. (Holding time : 2hours, Aluminum chips : 8g)

Holding temperature (℃)	Additive amount as Y_2O_3 in the product (wt%)				
	0.04	0.1	0.3	0.6	3
1050	1.2%	1.3%	1.1%	97.8%	98.0%
1100	0.8%	1.0%	1.7%	97.2%	98.1%
1200	1.0%	1.2%	96.7%	97.0%	98.5%
1300	1.6%	97.4%	96.0%	98.7%	99.2%

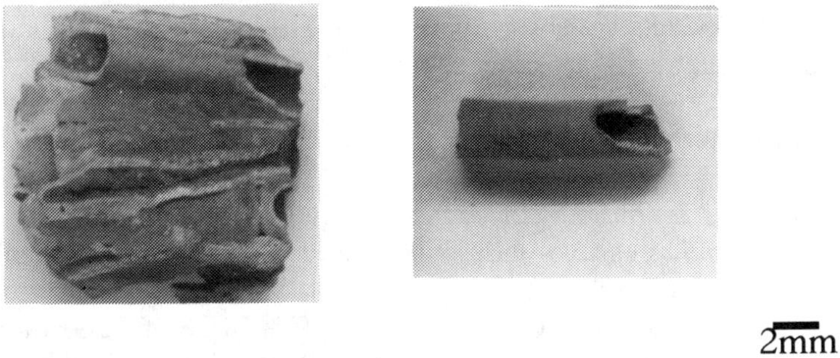

Fig.7 Products obtained from electrolytically etched Al wires.
Etching (Current density : 2000A/m^2 , 60sec, at 80℃)
Nitriding (2hours, at 1300℃)

Raw materials Products

(a) Block
(10mm × 10mm × 10mm)

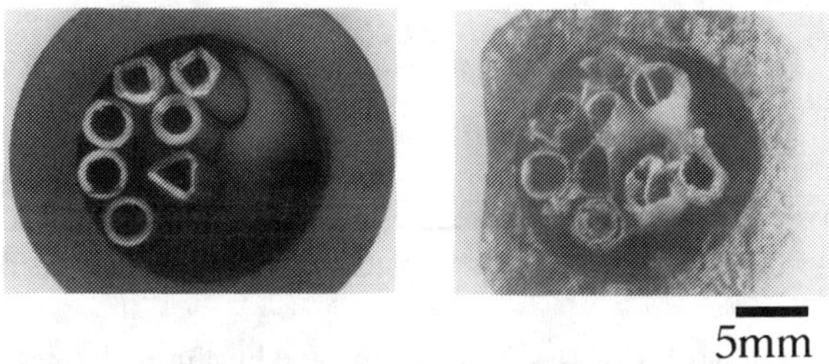

(b) Pipes (2.5mm in o.d. × 2mm in i.d. × ℓ 50mm)
Conversion to AlN ; 98.7%

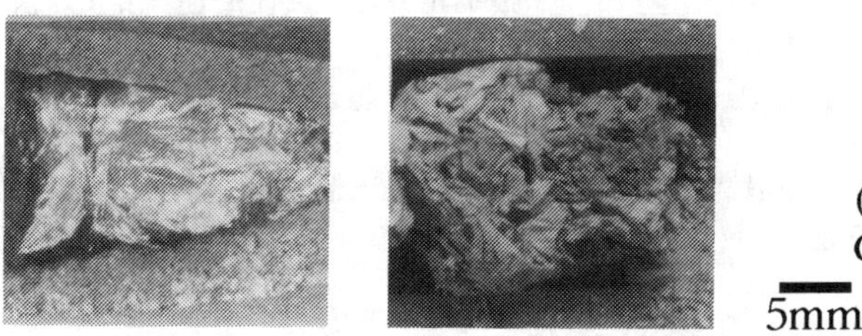

(c) Foil (thickness ; 15μm)
Conversion to AlN ; 97.4%

Fig.8 Nitriding of some shaped aluminums using Y_2O_3 as an additive.

foil. The products are shown in Fig. 8. The large-sized aluminums were almost completely nitrided in the other shapes.

CONCLUSIONS

In this work, the nitriding of large-sized aluminums was examined using aluminum chips with 2mm in size or wires with 2mm in dia. and 50mm in length. The results are summarized as follows,

1. Adding $Y(NO_3)_3{\cdot}6H_2O$ (0.04-1.28 g, 0.1-3wt% of Y_2O_3 in the product), the aluminum chips with 2mm in size (net wt. 8g) were nitrided at 1300℃ for 2 hours, giving 99.3% of conversion.
2. Nitriding model of large-sized aluminum was suggested. In the first step, the melt-gas reaction is accelerated by Y_2O_3 to generate aluminum vapor. In the second step, vapor phase reaction of aluminum vapor and nitrogen gas is completed in a moment. As the result, hollow, spherical shells covered with fine particles of aluminum nitride are formed.
3. The aluminum wires with 2mm in dia. and 50 mm in length were also completely nitrided by the pretreatment of aluminum wires; electrolytic etching. The conversion to aluminum nitride was 99.7%. In this case, nitriding was accomplished without any additive.

ACKNOWLEDGEMENT

This work was supported by a Grant-in-Aid for Scientific Research (C) No.05650844 from The Ministry of Education, Science and Culture.

REFERENCES

1. S. Ito, I. Ebato, H. Fukui, N. Koura and N. Yoneda, "Preparation of Aluminum Nitride Using Slightly Oxidized Al Powders," J. Ceram. Soc. Japan, Int. Edition, **100**, 622-626 (1992).
2. S. Ito, M. Sugiyama, N. Yoneda and Y. Sato, "Nitriding of Aluminum Powders Treated by Yttrium Nitrate," Proceedings of 1989 Annual Meeting of the Ceramic Society of Japan, p.291 (1989) (in Japanese).
3. H. Scholtz and P. Greil, "Nitridation Reactions of Molten Al-(Mg, Si) Alloys," J. Mat. Sci., **26**, 669-677 (1991).
4. K. Komeya, N. Matsukaze and T. Meguro, "Nitridation Behavior of Aluminum Melt," Proceedings of the Fall Meeting of the Ceramic Society of Japan, p.232 (1991) (in Japanese).

MANUFACTURING OF IRON CARBIDES IN FLUIDIZED-BED REACTORS

Seetharama C. Deevi, Sarojini Deevi, and Mohammad R. Hajaligol
Research and Development Center
Philip Morris, USA
4201 Commerce Road
Richmond, VA 23234

Abstract

Iron carbides with magnetic properties suitable for magnetic media were synthesized in a fluidized bed using reactive atmospheres. Fluidized-bed reactors of laboratory, bench, and pilot scale were employed for the synthesis of iron carbide, Fe_5C_2, by reducing and carburizing α–Fe_2O_3 in the temperature range of 300 to 550°C. A temperature of 400°C and a reaction time of 2 to 3 h were found to be optimal for the synthesis of iron carbide. A reducing /carburizing atmosphere such as carbon monoxide, resulted in the nucleation and growth of Fe_5C_2 directly from Fe_3O_4 with no intermediate phases such as FeO or Fe. Longer carburization times resulted in the deposition of carbon on the surfaces of Fe_5C_2 particles due to Boudouard reaction. The powders were characterized using X-ray diffraction, thermo-gravimetric analysis, and vibrating sample magnetometry.

INTRODUCTION

Iron carbides have been studied extensively due to their presence and importance in heat treated steels. Demand for high-quality magnetic tapes renewed interest in iron carbides as particulates for high density magnetic tapes due to their superior magnetic properties as compared to the conventional magnetic tapes based on γ–Fe_2O_3. Saturation magnetization of iron carbides is above 120 emu/g, whereas commercially available magnetic tapes contain maghematite, γ-Fe_2O_3, which has a saturation magnetization of 90 emu/g (and a coercive force of 400 Oe). Although the saturation magnetization of Fe is 130 to 150 emu/g with a high coercive force of 1000 Oe, the gradual oxidation of iron to iron oxides (of lower saturation magnetization) prevents the use of iron particles for high-density recording. The ferromagnetic nature of iron carbides, metallic conductivities and good wear resistance, coupled with their superior chemical stability as compared to metallic iron particles renewed interest in iron carbides as particulates for the next generation of magnetic tapes [1-8].

Three different iron carbides, related to three different crystal systems, are known. Of these, Fe_3C, is well known as cementite; Fe_5C_2, also known as X-carbide or per carbide, and Fe_7C_3,well known as an intermediate compound in the synthesis of diamond from iron and graphite under high pressure are also known. Hagg noted that the chemical composition of X - carbide was different from that of Fe_3C and approximated it to be Fe_2C. Since then, the carbide has been known as Hagg carbide. Senateur [5] identified Hagg carbide as Fe_5C_2 with a monoclinic structure based on the isomorphism with Mn_5C_2 and Pd_5B_2 systems. Senateur also [5] noted that the properties of Fe_3C and Fe_5C_2 are similar since their structures are almost identical except for a monoclinic distortion of 7.74°.

The structure of Fe_3C is orthorhombic belonging to space group Pnma. Six iron atoms form a coordination polyhedron in the shape of trigonal prism. Each prism shares an edge with a neighboring prism to form a trail of prisms. The structure could be represented as a succession of layers of prisms with triangular bases perpendicular to the b-axis [5]. The structure of Fe_5C_2 consists of a stack of layers identical in nature to the layers in cementitie, Fe_3C. The basic element of the structure is a right prism with triangular bases with carbon atom at the center. The lattice parameters of Fe_3C and Fe_5C_2 are expected to be essentially the same (due to the identical nature of the layers) in the direction parallel to the layers i.e., b-c for Fe_5C_2 and a-c for Fe_3C. Twinning of the layers in Fe_5C_2 leads to a lattice parameter almost double that of Fe_3C in the axis perpendicular to the plane of the layers , a for Fe_5C_2, and b for Fe_3C.

SYNTHESIS OF IRON CARBIDES

The literature indicates the synthesis of Fe_5C_2 from α-Fe_2O_3 and FeOOH in tubular furnaces using CO/H_2 mixture in the temperature range of 300 to 500°C. An undesirable by-product of Fe_5C_2 synthesis has been free carbon in the range of 5 to 20 wt. %. Attempts to limit carbon content resulted in the formation of Fe_3O_4 along with Fe_5C_2 [1,2,5]. The importance of process parameters in the synthesis of carbides was also described in several patents [9].

Manufacturing of iron carbides by fluidized-bed-reactor technology offers the ability to: (1) closely control the gas composition, (2) break up the agglomerates by fluidization of either iron oxide or a precursor of iron oxide, and (3) enhance the heat-transfer coefficient between the solid and the gaseous mixture at a desired temperature. Fluidized-bed manufacturing technology has been used during the last several years for a variety of oxidation/reduction reactions, coal gasification, drying of fine powders, and for surface protection of fine powders [10,11]. This manufacturing process offers advantages such as higher conversion of reactants to product in a shorter time, uniform product within a batch, and batch-to-batch consistency.

FLUIDIZED BED TECHNOLOGY

In the fluidized-bed process, solid particles become fluidized when a gas mixture flows through the powder leading to increase of inter-particle distance. At a certain velocity of gaseous mixture, enhancement of inter-particle distance leads to bulk movement of particles causing solids to behave like liquids and allowing the solids to conform to the shape of the container. The gaseous mixture, containing the required reactant gases, acts as a heat-transfer medium in fluidized-bed technology, and such an indirect heating approach reduces the final cost of the product significantly due to the optimum gas flow, reduced equipment size, and due to lower operating and capital costs.

Laboratory-, bench-, and pilot-scale fluidized-bed reactors of Procedyne Corporation (New Brunswick, New Jersey) were employed to synthesize iron carbides. Reactors of three different sizes were chosen to investigate whether fluidized-bed technology can be adopted to obtain carbides of iron with desirable phases and magnetic properties. A laboratory-scale reactor with a diameter of 10.16 cm, a bench-scale reactor with a diameter of 15.24 cm, and a pilot-scale reactor with a diameter of 35.56 cm were used with 1 kg of α-Fe_2O_3, 4 kg of α-Fe_2O_3, and 22.73 kg of α-Fe_2O_3, respectively. Carbides were synthesized using 50% carbon monoxide and 50% nitrogen mixture at a temperature of 400°C with a fluidization velocity of 22.86 cm/s. Nitrogen gas was used as a diluent to control the temperature rise associated with the exothermicity of the carburization reaction.

Bench-scale fluidized-bed experiments were carried out using a 15.24-cm-diameter and about 60-cm working height vessel (Figure 1). A screw-plate gas distributor was used for fluidization of the solids. The vessel was flanged to a 15.16-cm freeboard section containing two gas/solid filter tubes which cleaned the off-gases and retained the fine powders in the vessel. An automatic blow-back device was incorporated in the filter assembly to clean each tube alternatively of excess collection. The reactor was heated by immersing it in an electrically heated fluidized bath of alumina powder. The reactor was cooled after each experiment in a water-jacketed fluidized-alumina bath.

Commercially available iron oxide (Fisher Scientific Co., New Jersey), α-Fe_2O_3, was used for the synthesis of iron carbides . The average particle size was very low and required drying of the powder in an oven at 110°C for stable fluidization. Dried powders were sieved to remove the agglomerates prior to fluidization. A known amount of α-Fe_2O_3 was fed into the reactor for fluidization with nitrogen gas prior to heating. The reactor was slowly heated to a desired temperature while the bed was fluidized with nitrogen gas. A constant linear velocity of gases was maintained by adjusting the volumetric gas-flow rates to account for thermal expansion of the gases. At the desired temperature, nitrogen gas was replaced very slowly with a reducing gas such as hydrogen or with a reducing/carburizing gas such as carbon monoxide. The exothermicity of the carburization reaction increased the bed temperature, and constant monitoring of the temperature was found to be a necessary step. The fluidized-bed reactor was cooled to room temperature after carburization of the iron-oxide powder. The pyrophoricity of the carbide powder necessitated passivation of the surface by controlled oxidation of the surface with air.

RESULTS AND DISCUSSION

Conditions for the synthesis of iron carbides, specifically Fe_5C_2, were determined from controlled experiments carried out in a TGA using <1 g of α-Fe_2O_3 in the temperature range 250 to 550°C. A temperature range of 350 to 450°C was found to be optimal for the synthesis of Fe_5C_2, and is consistent with the temperature ranges reported earlier for the formation of Fe_5C_2 [1,9,11]. The initial reaction is confirmed by TGA and X-ray diffraction (XRD) to be the reduction of α-Fe_2O_3 to Fe_3O_4, followed by carburization of Fe_3O_4 by carbon monoxide, with increase of reaction time. The XRD pattern of the product obtained after 60 minutes of carburization in a 1:1 mixture of carbon monoxide and nitrogen in a laboratory-scale fluidized bed reactor at 400°C is shown in Figure 2. Carburization for times longer than 2 hours resulted in the formation of free carbon along with iron carbide, Fe_5C_2.

Attempts to obtain pure Fe_5C_2 resulted in a mixture of Fe_5C_2 and Fe_3C or Fe_5C_2 with a small amount of carbon, or a mixture of Fe_5C_2 and Fe_3O_4. For example, reduction of α-Fe_2O_3 (at 400°C in hydrogen for 1 h) followed by carburization at 325°C for 2 h with CO/H_2 mixture in the ratio of 1:7 ($CO:H_2$) gave a product containing a mixture of Fe_5C_2 with a small amount of Fe_3C. A mixture of carbides was observed with no Fe_3O_4 when the carburization experiments were carried out with reduction in hydrogen followed by carburization in a mixture of carbon monoxide and hydrogen gases in the ratio of 1:7 at 400°C. Products with similar phases were obtained when the carburization experiments were carried out in a 35.56-cm-diameter reactor with 22.7 kg of iron oxide under the same conditions as those used in the 10.16-cm reactor, with a 1:1 mixture of carbon monoxide and nitrogen.

The overall reaction of iron-carbide formation can be represented as:

$$5\, Fe_2O_3 + 32\, CO \text{ ----> } 2\, Fe_5C_2 + 19\, CO_2 \ ,$$

with an associated weight change of 24%. Carburization led to deposition of free carbon due to Boudouard reaction:

$$CO\,(g) + CO\,(g) \text{ ----> } CO_2\,(g) + C\,(s) \ .$$

The amount of carbon deposited on the surface of iron carbide particles can be determined by carrying out the oxidation of iron carbide in a TGA, since oxidation of iron carbide to iron oxide will lead to a weight gain of 31.7%, while the amount of free carbon will be indicated by a loss in weight due to the oxidation of carbon. Free-carbon contents of carbides synthesized using fluidized-bed reactors are in the range of 0.2 to 9 wt %.

MAGNETIC PROPERTIES

Saturation magnetization (σ_S) values of iron carbides are in the range of 77 to 125 emu/g depending on the size of the reactor, experimental conditions, and the product phases. The low σ_S values are associated with the products containing Fe_3O_4 phases since the σ_S of Fe_3O_4 is smaller than the σ_S of iron carbides. When carburization was carried out for longer times at a given temperature, the σ_S values did not change significantly, since the saturation magnetization values of Fe_5C_2 and Fe_3C are almost identical. The values obtained in the present study are in excellent agreement with the saturation magnetization values reported in the literature by several investigators [1,2,9]. Saturation magnetization of a carbide obtained with hydrogen as a reducing agent and methanol as a carburizing agent at 400°C for 90 min is 125 emu/g suggesting that carbides with appropriate magnetic properties can be prepared using fluidized-bed reactors.

Squareness ratio, which is a ratio of residual magnetization to saturation magnetization, of all the fluidized-bed carbides is in the range of 0.117 to 0.344, and is very much in agreement with the values reported earlier [9]. A representative hysteresis curve of iron carbide is shown in Figure 3. Coercive forces of carbides are in the range of 290 to 750 Oe. Coercive forces and the squareness ratios depend on a variety of factors such as precursor type, morphology of the precursor, acicularity of the particles, and the magnetic anisotropy of the particles.

Since carbides of iron can easily ignite upon sudden exposure to air, the powders were passivated with a very small amount of oxygen to prevent ignition of an entire batch of sample. This precaution, coupled with the fact that the sample was collected in air and stored in air, precluded magnetic measurements on fresh samples or samples collected in toluene. Tasaki et al. [14] noted that the saturation magnetization of the samples exposed in air was 25% lower than the ones collected in toluene. To understand the effect of exposure to air and the length of exposure, a second set of coercivity measurements made after a lapse of 13 months indicated a 20 to 25% drop from the previous measurements.

ACKNOWLEDGMENTS

This manuscript is prepared while one of the authors (S.C. Deevi) is on a sabbatical under the Philip Morris Continuing Educational Fellowship Program in Metals and Ceramics Division at Oak Ridge National Laboratory.

REFERENCES

1. S. Hirano and S. Tajima, "Synthesis and Magnetic Properties of Fe_5C_2 by Reaction of Iron Oxide and Carbon Monoxide," Journal of Materials Science, 25, 4457-4461 (1990).

2. S. Tajima and S. Hirano, "Synthesis and Magnetic Properties of Fe_7C_3 Particles with High Saturation Magnetization," Japanese Journal of Applied Physics, 29(4), 662-668 (1990).

3. J. P. Senateur, "Contribution to the Magnetic and Structural Study of Hagg Carbide," (Ph.D. Thesis, Paris, 1967), translation from Annales de Chimie (Paris), 2, 103-122 (1967).

4. F. J. Berry and M. R. Smith, "A Mossbauer Investigation of Iron-containing Catalysts Prepared at Low Temperatures and Active for Carbon Monoxide Hydrogenation," Journal of Chemical Society, Faraday Transactions, 1, 85(2), 467-477 (1989).

5. G. Le Caer et al., "Characterization by Mossbauer Spectroscopy of Iron Carbides Formed by Fischer-Tropsch Synthesis," Journal of Physical Chemistry, 86, 4799-4808 (1982).

6. D. A. Dowden, "The Reactions of Hydrocarbons on Multimetallic Catalysts," Catalysis, Vol. II (The Chemical Society, London, 1978), 1-27.

7. J. W. Niemantsverdriet et al., "Behavior of Metallic Iron Catalysts during Fischer-Tropsch Synthesis Studied with Mossbauer Spectroscopy, X-ray Diffraction, Carbon Content Determination, and Reaction Kinetic Measurements," Journal of Physical Chemistry, 84, 3363-3370 (1980).

8. L. N. Mulay et al., "Transmission and Backscatter Mossbauer Spectroscopy of Zeolite Fe (and Fe-Co) Synfuel Catalysts," Proceedings of Indian National Academy of Sciences, Physical Sciences, 457-459 (1982).

9. T. Arase, Y. Shibuya, and S. Daimon, "Particles Containing Iron Carbide," U.S. Patent 4,900,464 (1990).

10. A. Maroglou and A. W. Nienow, "Fluidized Bed Granulation Technology and Its Application to Tungsten Carbide," Powder Metallurgy, 29(4), 291-295 (1986).

11. J. Bhatia, "Fine Particle Processing in Fluid Beds," (Paper presented at the Fine Particle Society Meeting, Boston, Massachusetts, 1986).

12. I. S. Pap et al., "TG Study on the Reduction Kinetics of Hematite and Iron Oxide-Containing Wastes," Journal of Thermal Analysis, 33, 578-583 (1988).

13. P.R. Viswanath, B. Viswanathan, and M.V.C. Sastri, Transactions of the Japan Institute of Metals, 18, 148 (1977).

14. A. Tasaki, K. Tagawa, and E. Kita, "Recording Tapes using Iron Nitride Fine Powder," IEEE Transactions on Magnetics, Mag-17, 3026-3028 (1981).

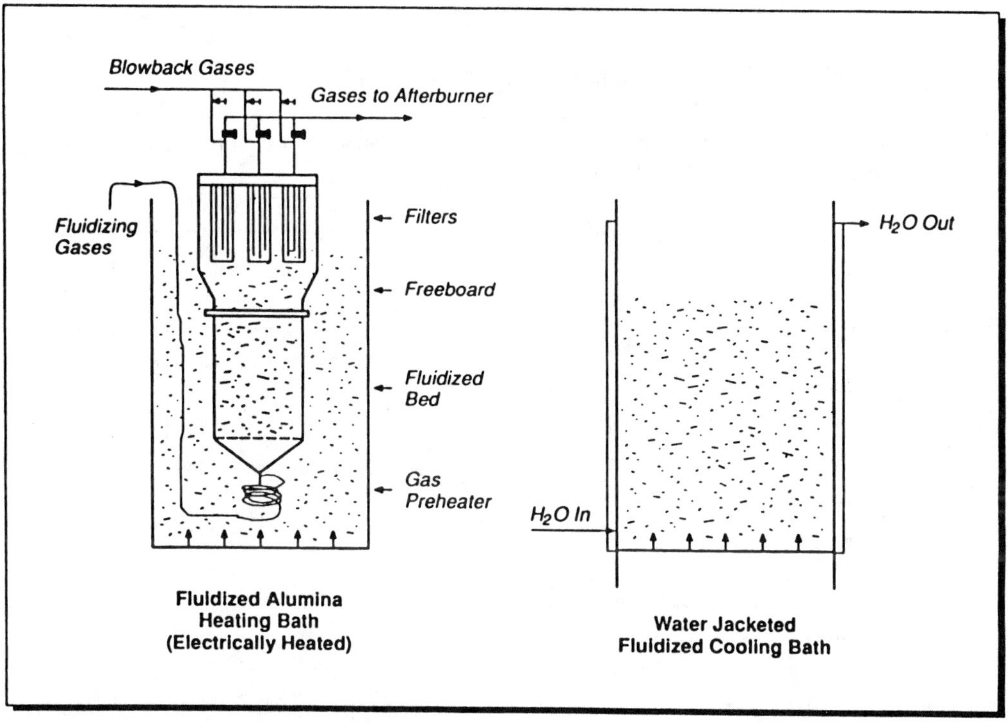

Figure 1 - Schematic of a bench-scale fluidized-bed reactor assembly employed in the synthesis of iron carbides.

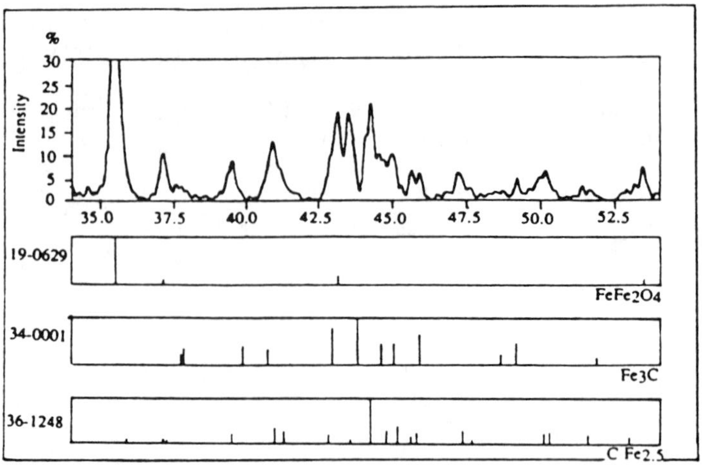

Figure 2 - X-ray diffraction pattern of the product obtained in a laboratory-scale fluidized-bed reactor during carburization of 1 kg of α-Fe_2O_3 in a 1:1 mixture of carbon monoxide and nitrogen at 400°C for 60 minutes. The numbers corresponding to the compounds are JCPDS file numbers.

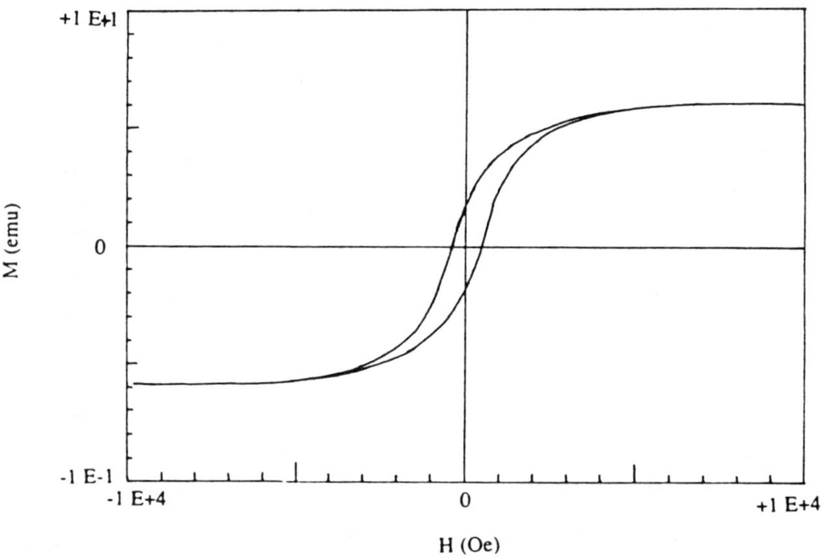

Figure 3 - Representative hysteresis loop of iron carbide synthesized using a pilot-scale fluidized bed reactor with a reaction time of 80 min at 400°C in a 1:1 mixture of carbon monoxide and nitrogen.

Synthesis of Ultra-fine Metallic Powders in a D C Plasma Reactor by Chemical Vapor Deposition

G. Y. Zhao and V. Hlavacek
Laboratory for Ceramic and Reaction Engineering
SUNY at Buffalo, Buffalo, NY 14260, USA

A 14 kW d.c. plasma unit has been developed for generating ultrafine metallic powders. Experiments to characterize the torch and the reactor showed that the plasma unit could be operated over a wide range of conditions without losing its discharge stability and a high temperature region could be produced for the synthesis of variety of ceramic and metallic powders. Ultrafine powders of Ta, Nb, Mo, and W were synthesized with high purity, uniform size distribution and good flowability which attracted industrial attention. Scale up of the plasma unit for industrial production of metallic powders is in process and some important problems in this respect are discussed in detail.

1. Introduction

Thermal plasma technology has evolved during the 1980's to a well - established interdisciplinary science with a wide range of important applications in materials processing. Very intensive work is being done on preparation and processing of materials with plasma technology. In a number of cases the processes developed lead to full - scale industrial production. The production of acetylene from light hydrocarbons using an electric arc process in Chemische Werke Huls AG in Germany [1] and the synthesis of titanium dioxide pigment through the oxidation of titanium tetrachloride in an oxygen plasma in Tioxide [2] are two typical examples.

Most of the processes involving plasma - chemical synthesis of ultra fine powders of high purity ceramic powders have been developed on a laboratory or a pilot plant scale. However, there is no real progress in recent years regarding the plasma preparation of ceramic powders. The main reason may be that the industrial demand of the fine ceramic powders did not surge as expected and the application of ceramic powders for the motor application was not brought about into reality. The ceramic motor development is seriously affected by the difficulty to sinter the ceramic powder with reasonable reliability and controllability, which is due to the inherent disadvantages and most limiting properties of ceramics: brittleness and poor thermal shock resistance. However, for refractory metals, there is no limitation of their properties such as brittleness. As a matter of fact, the industrial demand of fine refractory metal powders is increasing.

The aim of this research was to design and develop a 14 kW d.c. plasma sys-

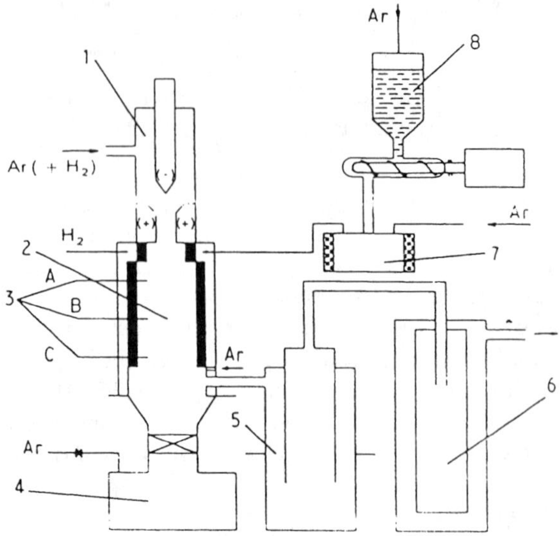

Figure 1: Experimental layout of the d.c. plasma system in the present study. (1) plasma torch; (2) reactor; (3) thermocouple; (4) collector; (5) cyclone separator; (6) bag filter; (7) evaporator; (8) feeding system.

tem for the preparation of ultrafine powders. Experiments have been conducted to generate powders of Ta, Nb, Mo and W by reduction of the corresponding chlorides. Preliminary research shows that this process can be scaled up for industrial applications if some key problems can be solved.

2. Experimental

Set - up

The design and fabrication of a d.c. plasma unit is a tedious job and should be handled very carefully. Figure 1 is a schematic of the plasma system developed in the present study. The system can be divided into five main parts: *Plasma Torch*, *Reactor*, *Feeding System*, *Collector*, and *Power Supply* (for details refer to [3]).

Plasma Torch Characteristics and Temperature Profile in the Reactor

We calibrated the entire system from ammeters, voltmeters, flow meters and thermocouples to temperature indicators, then started the following measurements. The plasma arc was generated by an arc discharge between a copper nozzle anode and a thoriated tungsten cathode. The plasma jet was expected to operate over a wide range of flow rate conditions at atmospheric pressure without losing its discharge stability. In order to confirm this point, measurements were then carried out for the current - voltage characteristics of the torch for various plasma gas flows. Then the results were used to illustrate the stability of the torch (for more details, refer to [3]). The plasma tail temperatures were measured by using a special thermocouple located at three sections of the reactor. Thermocouples could be moved freely along the radial direction, and the temperature was measured at different points under steady - state operating conditions. Temperature values for both radial and axial directions in the reactor were given by this kind of measurements. Figure 2 is an example of the radial temperature distribution profile for a section which is 62 mm from the exit of the plasma torch. Figure 3 is an example of the axial temperature distribution profile. Both of the examples were conducted at powers and flow rates indicated. From these measurements we can clearly see that there is a high temperature (above about 1200 K) region with a length of 20 cm and a diameter of 1.5 cm exists in the plasma reactor, which can be used for the reduction chemical reactions.

3. Synthesis of Metallic Ultra - Fine Powders

As a first step, we considered the following systems for generating ultrafine powders of Ta, Mo, Nb, and W, where the respective chlorides were reduced by hydrogen.

$$2TaCl_5 + 5H_2 = 2Ta + 10HCl \tag{1}$$

$$2NbCl_5 + 5H_2 = 2Nb + 10HCl \tag{2}$$

$$MoCl_6 + 3H_2 = Mo + 6HCl \tag{3}$$

$$WCl_6 + 3H_2 = W + 6HCl \tag{4}$$

The chlorides were fed through the screw feeder at a selected standard feeding rate to the evaporator, which was heated to 400^oC. The vapors of evaporated materials was carried away by the Ar carrier gas to the reaction chamber (into the plasma tail) where they were reduced, and the generated ultrafine powders were collected at various sections: collector, cyclone, and bag filter.

The main plasma - forming gas was the mixture of Ar and H_2. Part of the hydrogen bypassed the torch and entered the reactor directly for the reduction process to produce the ultrafine nuclei. The working gas temperature in the reactor could be adjusted by changing the supply current or the quantity of Ar and H_2 flow rates in the plasma - forming gases. The final operating conditions are summarized in Table 1.

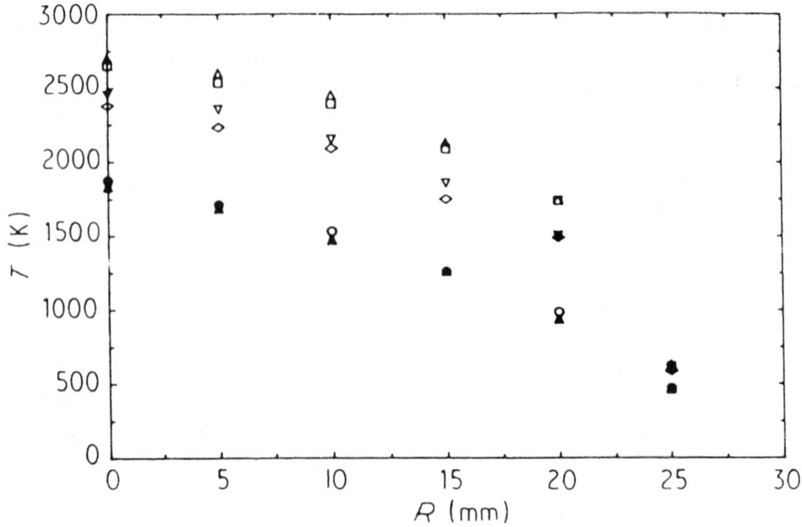

Figure 2: Radial temperature distribution in section A of the reactor. (1) P=10 kW: △ 2 m^3/h; □ 3 m^3/h, (2) P=7 kW: ▽ 2 m^3/h; ◇ 3 m^3/h, (2) P=4 kW: ∘ 2 m^3/h; △ 3 m^3/h.

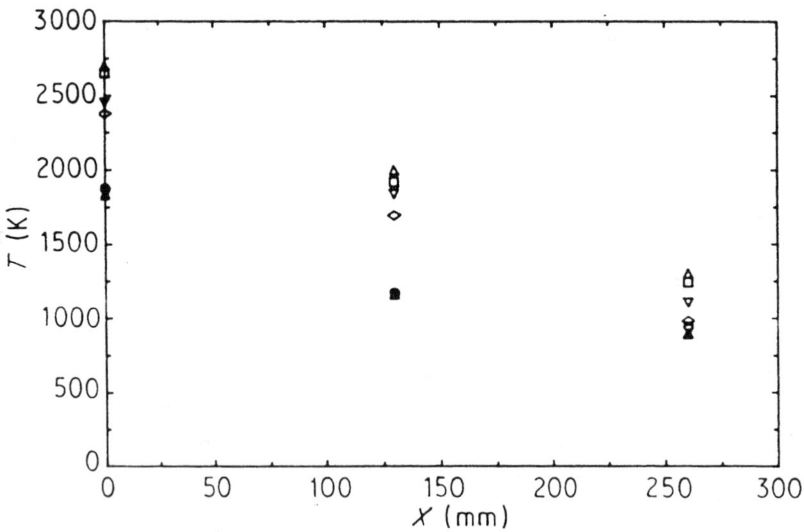

Figure 3: Axial temperature distribution in the centerline of the reactor. (1) P=10 kW: △ 2 m^3/h; □ 3 m^3/h, (2) P=7 kW: ▽ 2 m^3/h; ◇ 3 m^3/h, (3) P=4 kW: ∘ 2 m^3/h; △ 3 m^3/h.

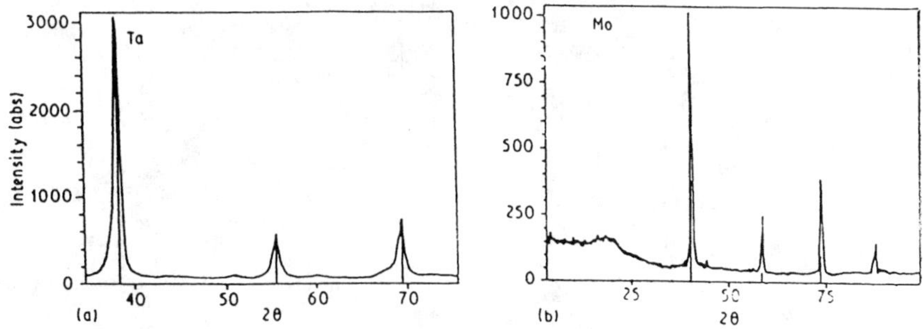

Figure 5: Some Samples of X-ray Diffraction. (a) Ta; (b) Mo.

ratio of hydrogen (H_2) to metal chlorides is very important to get nonpyrophoric Ta, Nb, Mo and W powders. We obtained these powders at the $H_2/$ metal chloride molar ratio between 5 and 12.

It is important that the system should be free of moisture and oxygen, as the powders generated are ultrafine and highly active. The possible sources of contamination are (i) the starting reactant, as it is or as an oxychloride; (ii) leakage in any part of the system; (iii) on exposure to the atmosphere (either oxidation or pyrophoric nature of the product); and (iv) materials of system and electrode burnout. The method of collection is also important. For example, if the sample is collected in atmosphere, 8% of oxygen in the powder is obtained within a few minutes, and if the collection is proceeded in N_2 or Ar atmosphere, oxygen contamination reaches 1%. The analysis of Ta powder was given below in p.p.m., using industrial grade $TaCl_5$ and pre - purified grade Ar and H_2 as starting materials.

$$Al[20], B[<6], Be[<6], Ca[<6], Cr[<200], Cu[200], Fe[600], Mg[<6],$$

$$Mo[200], Mn[20], Na[<6], Ni[200], Pb[<20], Si[200], Sn[<6], Ti[<6]$$

It appears that there is more Fe in the sample. Generally speaking, this depends totally on the purity of the original reactant. Once we used $TaCl_5$ from another source and the same grade of Ar and H_2, analysis showed that

$$Al[<6], B[<6], Be[<6], Ca[<6], Cr[<6], Cu[6], Fe[<6],$$

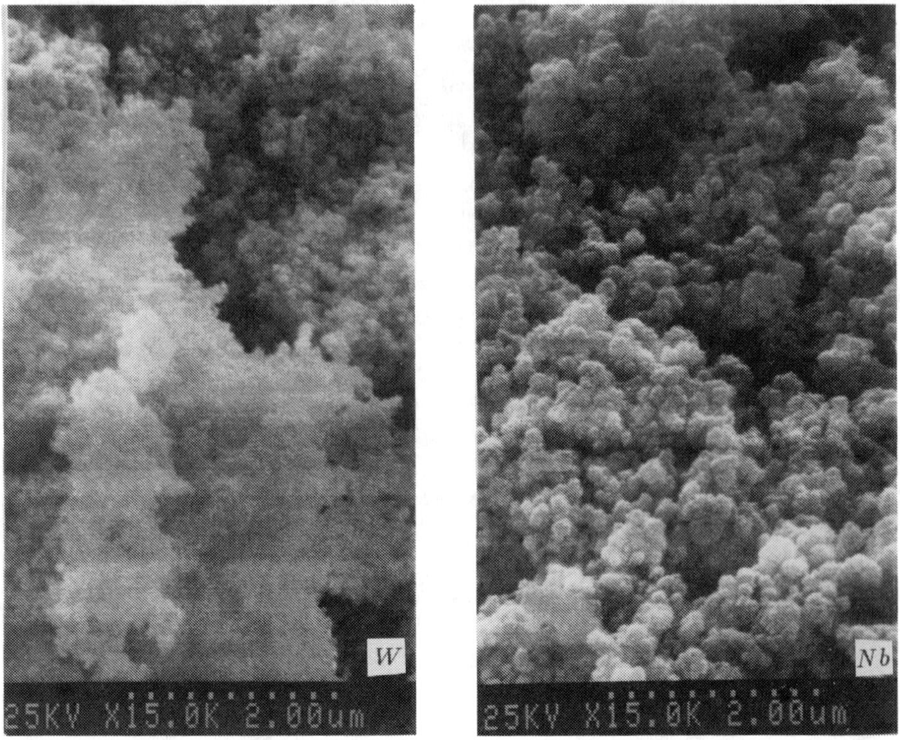

Figure 4: Scanning electron micrograph. (a) W; (b) Nb.

4. Results and Discussion

The reduction of $TaCl_5$, $NbCl_5$, $MoCl_5$, and WCl_6 as shown in equations (1 - 4) was carried out in the d.c. plasma reactor. The above mechanisms are the simple forms of the complex reactions occurring in the reduction processes. The $TaCl_5$ (m.pt 216^oC, b.pt 242^oC), $NbCl_5$ (m.pt 208.3^oC, b.pt 248.2^oC) $MoCl_5$ (m.pt 194^oC, b.pt 264^oC), and WCl_6 (m.pt 275^oC, b.pt 347^oC) are believed to be dissociated into several subchlorides[3], and the proportion of each being determined by the temperature to which the reactants were heated. These subchlorides finally react with hydrogen to produce Nb, Ta, Mo, and W vapors. Ultrafine powders of Ta, Nb, Mo, and W will be formed through homogeneous nucleation of the corresponding vapor, the subsequent growth of the nuclei and the deposit of the formed powders on the wall of the reactor, cyclone and bag filter (refer to figure 1). The whole process is called as synthesis of metallic powders in plasma reactor by chemical vapor deposition. The molar

Table 1: Experimental data (all flowrates are in l/min).

Starting materials	Feeding rate (g/min)	Ar plasma flow rate	Ar carrier flow rate	Hydrogen flow rate	Reaction temperature (^oC)
$TaCl_5$	2	35	10	1.360	900
$MoCl_5$	2	35	10	2.241	800
WCl_6	2	35	10	1.253	800
$NbCl_5$	2	35	10	1.405	700

$$Mg[<6], Mo[<400], Mn[<6], Na[<6], Ni[<6], Pb[<20], Si[100],$$
$$Sn[<6], Ti[<6], Va[<6], W[<20], Zn[<6], Zr[<6], Au[<10]$$

This indicates that in our case, it seems that the materials of cathode, anode and the reactor do no have essential effect on the impurity of the products.

The size and size distribution of particles are very important parameters for sintering. The SEMs and image analysis in Figure 4 show that the average particle size is around $0.10\mu m$ with the smallest particles being around $0.02\mu m$. All the particles appear to be below $0.5\mu m$. Particles are perfectly spherical or cubic, with spherical being the predominant shape. As shown in Figure 5, the peaks of X - ray diffraction pictures of Ta and Mo powders are in a very good concordance with lines from standard files. During the experiments, we also found that all the powders produced have very good flowability.

5. Conclusions of the Research

From the above research, the following points can be concluded: a 14k W d.c. plasma unit has been successfully designed and developed to synthesize ultra - fine powders by chemical vapor deposition method. Ultrafine refractory metals Ta, Nb, Mo, and W powders of around 0.10 micron with narrow size distribution and good flowability were produced and studied. With industrial grade chlorides, argon and hydrogen as starting materials, impurity of the metallic elements is about lower than 200 - 400 ppm.

6. Some Key Problems in Scale Up

Scale up of the plasma unit for industrial production of metallic powders is in process and some important problems in this respect are discussed in detail as follows.

Theoretical calculation conducted by the authors using chemical equilibrium approximation (being published in elsewhere) showed that the higher the hydrogen to chloride ratio in the reaction mixture gas, the lower the reactant temperature can be used in the plasma reactor. However, in the 14 kW d.c. plasma reactor, most part of the working gas is argon, just a small amount of hydrogen was directly heated by the d.c. discharge. Most part of the reactant gas hydrogen was heated by the working gas argon. In order to make full use of the energy from the d.c. discharge, a hydrogen plasma system or a plasma system with large ratio of hydrogen in plasma gas should be used.

In last section, we listed the possible sources of contamination. From our experience, the major impurities in the products were from the raw materials of the chlorides. Evaporation method seemed to be the best method to purify the chloride vapor reactants, because after the chlorides were evaporated experiments showed that a small amount of impurities were left in the evaporator. Even though our experiments showed that just a minor part of metallic impurities come from the burnout of the electrodes, when the d.c. power is increased, the impurities from electrodes will increase. An important method to eliminate the contamination from electrodes is to use the electrodes made of the product materials. This method is possible for the refractory metallic powder production. Leakage of air to the feeding system is one major source of oxygen contamination. Specially designed feeding system should be used to prevent the leakage. Another major source of oxygen contamination is from the exposure of the products to the air. However, the experiments here showed that powder collection at lower temperature could reduce the oxygen content. Finally, it should be stressed that the collection system of the 14 kW plasma reactor is good for collecting powder samples and the development from this system to an industrial one needs a dramatic reformation.

7. References

1. Peuckert, C., R. Muller, G. Prob, "Status of Plasma Technology and Development Work of Huls AG" 9th Int. Symp. Plasma Appl. Workshop Proc. 2 (1989) 95.
2. Orfeuil, M. et al., "Arc Plasma Processes - A Maturing Technology in Industry." UIE Arc Plasma Reviews. 1988. International Union for Electroheat, Paris, 1988.
3. Zhu, C. W., G. Y. Zhao, V. V. Revankar and V. Hlavacek, "Design and development of a D.C. plasma Reactor for Synthesis of Ultrafine Powders." *Journal of Materials Science, 27* (1992) 2211-2217.

Powder Processing—Dispersion/Rheology

MODIFICATION OF SILICON NITRIDE SLIP PROPERTIES BY POLY(ACRYLIC ACID)

Vincent A. Hackley and Subhas G. Malghan
National Institute of Standards and Technology, Ceramics Division
Gaithersburg, MD 20899

ABSTRACT

Acrylic-acid based polyelectrolytes are used for dispersion and rheology control of ceramic powder slips. This study focuses on the Si_3N_4/H_2O/poly(acrylic acid) (PAA) system, with the goal of improving our basic understanding of the mechanisms which may significantly affect slip properties during processing. A variety of experimental techniques were employed, including potentiometric titration, electroacoustic analysis, adsorption isotherms, and controlled-stress rheology. The slips exhibited complex behavior over a wide range of conditions in which solids concentration, PAA molecular weight and concentration, and pH were varied. Polymer charge and conformation were found to have significant impact on the flow properties. Pseudoplastic behavior was attributed to the presence of free polymer. The flow properties were also found to be highly pH dependent.

INTRODUCTION

Although polyelectrolytes such as poly(acrylic acid) (PAA) and poly(methacrylic acid) (PMA) have been routinely employed for some time as dispersants in the aqueous processing of ceramic powder slips, their role in these systems remains unclear particularly from a mechanistic standpoint. Previous studies have provided greater insight into the mechanisms of interaction between the solid phase and dissolved polyelectrolyte [1-4]. Many questions remain concerning among other topics, the role of free polymer and the effects of polymer conformation and size on slip dispersion and flow properties. In order to develop flexible strategies to cope with different processing requirements and materials it is necessary to understand the fundamental mechanisms of polyelectrolyte/particle interactions, and to specify the manner and degree to which different variables influence those interactions. The goal of the present work was to address this problem using several experimental methods to investigate the Si_3N_4/H_2O/PAA system.

EXPERIMENTAL

PAA of molecular weights 5000, 50,000, 90,000, and 150,000 were obtained from Polysciences, Inc. (Warrington, PA).* Silicon nitride powder used in these studies was high purity grade SNE10 manufactured by Ube Industries (Tokyo, Japan). This powder has a specific surface area measured by gas adsorption of 10.4 m^2g^{-1} and an average particle diameter in suspension of 0.4 μm. All additional chemicals used in this work were of reagent grade. Potentiometric titrations and dynamic mobility measurements were made on the Matec Applied Sciences (Hopkinton, MA) ESA 8000 single frequency electroacoustic analyzer. Polymer adsorption was measured by the depletion method using the Dohrmann (Santa Clara, CA) DC-80 low temperature total organic carbon analyzer. Rheology measurements were performed using the Rheotech International (UK) controlled stress rheometer, equipped with a concentric cylinder measurement system. Suspensions were prepared in high purity deionized water and dispersed with high intensity sonication at an output power level of 40 W. Samples were equilibrated for a period of two hours prior to analysis.

RESULTS AND DISCUSSION

PAA dissociates in water to produce an anionic polyelectrolyte as shown schematically in Figure 1. From potentiometric titration data, the degree of ionization α was determined as a function of pH (Figure 2). Using the data in Figure 2, the apparent dissociation constant (pK_{app}=pH - log $\{\alpha/1-\alpha\}$) for

---

*Certain trade names and company products are mentioned in the text or identified in illustrations in order to adequately specify the experimental procedure and equipment used. In no case does such identification imply recommendation or endorsement by National Institute of Standards and Technology, nor does it imply that the products are necessarily the best available for the purpose.

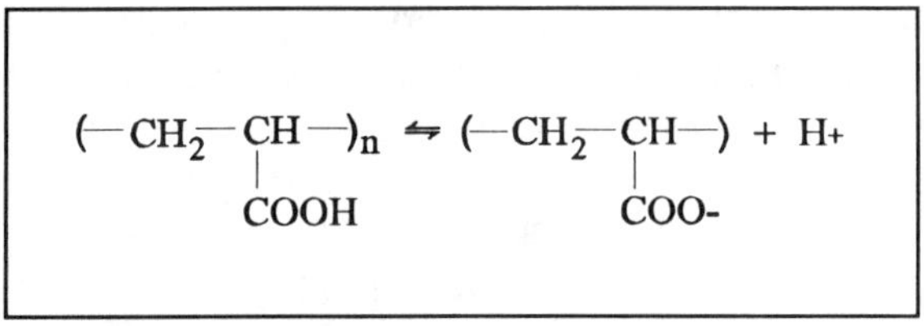

Figure 1. Dissociation of poly(acrylic acid).

PAA was calculated and then extrapolated in the linear region to α=0 from which the intrinsic dissociation constant K_0 was derived. The value of pK_0~4.8-5.1 found for the PAA used in this study falls in the same range as previously reported for acrylic polymers [2]. pK_{app} increases with pH due to the increasing free energy required to create a charged site on the increasingly ionized polymer chain. As a further consequence of ionization, the polymer chain undergoes a conformational transition between two forms [5]. The first conformation, which exists at low values of α (pH< 5), is compact and tightly coiled. At much higher α values (pH>9), an expanded conformation exists. Presumably the intermediate ionization states yield a transitional conformation falling somewhere between the two extremes.

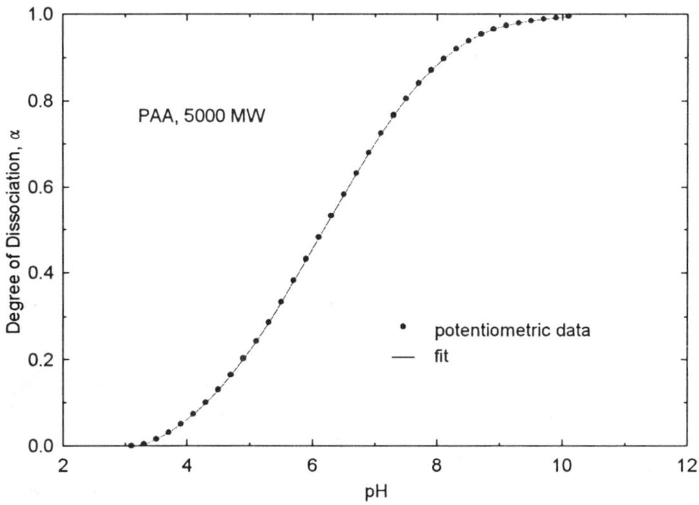

Figure 2. Ionization of poly(acrylic acid) as a function of pH.

Adsorption isotherms for PAA on Si_3N_4 indicate several trends. In general, the amount adsorbed increases with decreasing pH. At pH 9 nearly all of the PAA exists as free polymer, whereas at pH 3 high affinity adsorption leads to surface saturation. In the first case, at pH 9, mutual repulsion between the negatively charged particle surface and ionized PAA chain results in a weak interaction. Conversely, at pH 3 the observed adsorption is due in part to heterogeneous precipitation of the essentially uncharged and hydrophobic PAA onto the particle surface. Above pH 5 the plateau adsorption is relatively low and more or less independent of molecular weight; however, when the polymer is adsorbed at acidic pH values an irreversibly bound fraction is retained at pH 9. This fraction increases with molecular weight. Figure 3 shows the adsorption behavior for 50,000 MW PAA plotted versus the residual concentration of polymer in mg/L (ppm).

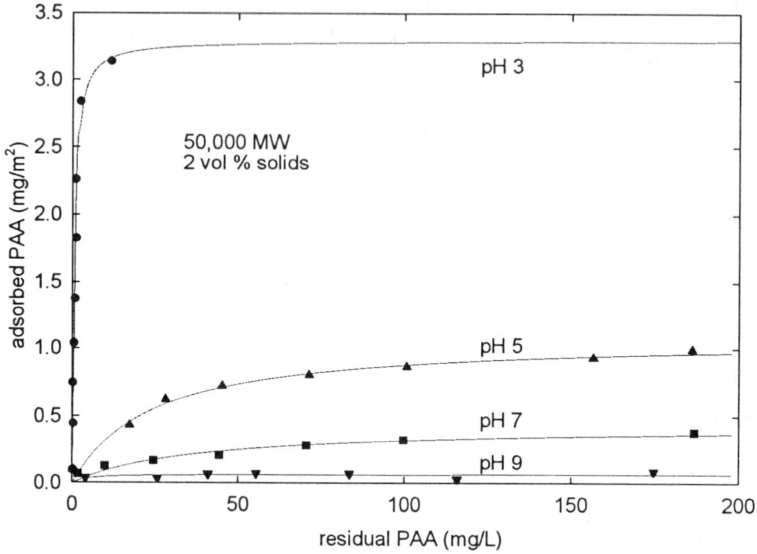

Figure 3. Adsorption isotherm for 50,000 MW PAA on Si_3N_4.

Electroacoustic measurements of the dynamic electrophoretic mobility were used to study the effects of PAA on the Si_3N_4 surface potential. This method probes the electrical double-layer at the hydrodynamic shear-plane some short distance from the surface [6]. The shear-plane potential is affected by processes such as chemisorption, electrolyte screening, and sometimes physical adsorption if the adsorbate is large enough to disturb the plane of shear. From measurements of mobility as a function of pH, PAA concentration, and molecular weight, the interaction mechanisms were deduced. It was found that PAA causes an acidic shift of the isoelectric point (iep) for Si_3N_4 (Figure 4), an indication of chemical adsorption. The iep shift is independent of molecular weight for the same weight fraction of polymer, further evidence for a chemical interaction mechanism involving the polymeric carboxyl group and surface amine group.

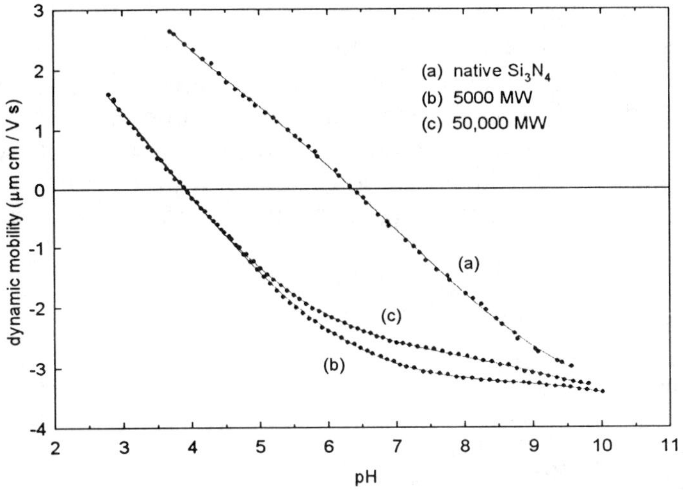

Figure 4. Dynamic mobility versus pH for 2 vol % Si_3N_4 and in the presence of 1000 mg/kg poly(acrylic acid). Titrated from pH 3 to 10.

The titration curves begin to diverge systematically as the pH is raised into the conformational transition region above pH 4, even though uptake measurements show no significant desorption initially. This indicates that a conformational change similar to the one associated with free PAA, also occurs in the bound polymer. Conformation greatly affects the charge density distribution in the direction perpendicular to the particle surface, and also the fraction of bound segments per molecule. The situation can be described in the following way. As the polymer chain expands due to mutual repulsion between neighboring ionized sites along the chain, the segment density distribution is broadened and shifted out from the particle surface. At pH values above the native iep (6.4) mutual repulsion between the similarly charged chain segments and particle surface further facilitates this process as chain segments detach from the surface and extend into solution. The conformational transition promotes the formation of loops, trains and tails, a condition necessary for electrosteric stabilization. The smaller 5000 MW PAA, although adsorbed to a lesser extent, appears to form a more intimate (flatter) configuration with the particle surface, and as a consequence tends to increase the negative shear-plane potential relative to the higher molecular weights in the transition region. The mobility curves reconverge near pH 10 which coincides with the fully expanded conformation, and corresponds closely with adsorption data.

The rheology of Si_3N_4 slips were characterized as a function of solids concentration, PAA concentration, molecular weight and pH. Equilibrium flow curves exhibited pseuudoplastic behavior in the presence of PAA (Figure 5). The data was represented using the Bingham flow equation, $\sigma = \gamma\, \eta_{pl} + \sigma_B$, where the plastic viscosity, η_{pl}, and apparent yield stress, σ_B, were determined from a plot of shear stress, σ, versus shear rate, γ, in the high shear region.

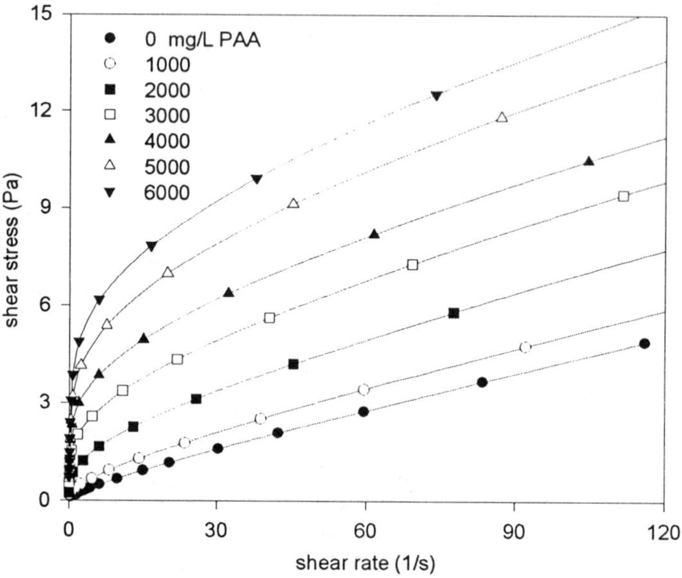

Figure 5. Equilibrium flow curves for 40 vol % Si_3N_4 slips at pH 10 as a function of PAA concentration (5000 MW).

It was found that both η_{pl} and σ_B increased monotonically with polymer concentration. Pure polymer solutions over the same concentration range showed no apparent pseudoplastic behavior, and only a slight increase in viscosity relative to water. Additionally, the yield stress did not follow the volume fraction dependence (ϕ^2) expected for simple particle interaction effects [7], but exhibited a higher exponent close to 5. Measurements were made as a function of pH at 3000 mg/L PAA concentration (see Table I). Above pH 9, the yield stress is independent of pH. Since nearly all PAA in this pH range exists as free polymer, the pseudoplasticity must derive primarily from the mutual interaction between expanded PAA segments in solution and allignment of polymer chains in the flow field. Possibly the mutually repulsive segments and highly charged particles form a loose gel-like matrix which acts to retard flow.

The pH dependence of the rheological behavior correlates closely with the conformational transition discussed previously. Between pH 9 and 8, the yield stress increases slightly and almost linearly for all molecular weights. This pH range corresponds with the onset of the transition region. The increase in yield stress may be the result of some polymer bridging as adsorption becomes more favorable. Between pH 8 and 5, the yield stress plateaus in the case of 5000 MW PAA, and decreases in the case of higher molecular weights. Below pH 5 the yield stress increases sharply. The decreasing yield stress region may result from the simultaneous removal of PAA from solution via adsorption (reduction in spatial dimension) and the contraction in volume occupied by each PAA molecule during the conformational transition. The sharp increase in σ_B and η_{pl} below pH 5 corresponds closely with the pK_0 value and the onset of polymer instability in solution which leads to heterogeneous precipitation (characterized by high affinity adsorption) and flocculation. Compare this behavior with suspensions without PAA, which show low viscosity and yield stress only above pH 8.5 and below pH 4. Intermediate pH values show a yield stress much larger relative to the PAA suspension. Finally, the flow is entirely reversible in the presence of PAA above pH 8, but increasingly exhibits hysteresis as pH is reduced below this value. This indicates that pseudoplasticity in alkaline pH arises primarily from polymer interactions in solution, whereas the acidic pH behavior is substantially different in nature and is probably due to structure formation resulting from particle aggregation in the unstable suspension.

Table I. Plastic viscosity, η_p (mPa·s), and yield stress, σ_B (Pa), versus pH for 3000 ppm 5000 MW poly(acrylic acid) in 30 vol % Si_3N_4.

	suspension pH										
	10.5	10.0	9.5	9.0	8.7	8.5	8.0	7.0	6.1	5.4	4.6
η_p	13.3	12.5	15.0	17.7	24.0	23.8	24.6	24.9	21.2	22.6	81.1
σ_B	1.2	1.1	1.5	1.7	6.2	7.9	11.3	11.6	13.6	19.3	74.0

CONCLUSIONS

Modification of slip properties by poly(acrylic acid) (PAA), is complex due to the combined effects of pH-dependent PAA and particle-surface chemistries. The presence of free polymer in alkaline pH slips corresponds with reversible pseudoplasticity and an apparent yield stress. A yield stress is required in the early stages of processing to prevent segregation of the slip components during casting. In the intermediate pH range plastic viscosity and yield stress vary partly as a result of polymer conformational changes and adsorption. At pH values below the pK_0, a low degree of ionization of PAA leads to incipient flocculation. This region is characterized by a sharp increase in yield stress, flow hysteresis, and high affinity adsorption.

REFERENCES

[1] J.E. Gebhardt and D.W. Fuerstenau, "Adsorption of polyacrylic acid at oxide/water interfaces," *Colloids Surfaces*, **7**, 221-31 (1983).

[2] J. Cesarano, I. A. Aksay and A. Bleier, "Stability of aqueous α-alumina suspensions with poly(methacrylic acid) polyelectrolyte," *J. Am. Ceram. Soc.*, **71**, 250-5 (1988).

[3] R.R. Vedula and H.G. Spencer, "Adsorption of poly(acrylic acid) on titania (anatase) and zirconia colloids," *Colloids Surfaces*, **58**, 99-110 (1991).

[4] Y. Hirata, A. Nishimoto and Y. Ishihara, "Effects of addition of polyacrylic ammonium on colloidal processing of α-alumina," *J. Ceram. Soc. Jap.*, **100**, 983-90 (1992).

[5] A.R. Mathieson and J.V. McLaren, "Potentiometric study of the conformational transition in poly(acrylic acid)," *J. Polymer Sci. A*, **3**, 2555-65 (1965).

[6] V.A. Hackley and S.G. Malghan, "The surface chemistry of silicon nitride powder in the presence of dissolved ions," *J. Mater. Sci.*, **29**, 4420-30 (1994).

[7] J.W. Goodwin, "Rheology of ceramic materials," *Ceram. Bull.*, **69**, 1694-98 (1990).

ADSORPTION OF ORGANIC ADDITIVES ON POWDER SURFACES

Amanda K. Berry, Leonard E. Bogan, and Sharon E. Agostine

Rohm and Haas Company
Spring House, PA 19477

The stability and rheological behavior of ceramic slips are influenced by the adsorption of additives on the powder surface. This paper characterizes the adsorption of several acrylic binders on alumina. We have found that acrylic binders with low pK_a adsorb faster and to a greater extent than binders with higher pK_a, and that adsorption can take many hours to reach equilibrium. The result of binder adsorption is an increase in slip viscosity, particularly at low shear rates. Dispersants with high acid number impede binder adsorption and result in slips with stable viscosity. Understanding the effects of additive adsorption on dispersion stability and rheological properties is crucial to obtaining stable, reproducible slips.

Introduction

The stability of a ceramic dispersion is dependent on the adsorption behavior of polymeric dispersants and binders on the particle surfaces.[1-4] Changes in slip rheology with aging can limit reproducibility and introduce variations in pososity, density or green strength in the final product.[3, 6, 7] Minimizing variations which occur early in the formulation stage of a process is critical to achieving uniformity in the end product.

We have shown that dispersion stability in the absence of binder can be controlled by selection of a proper dispersant.[5] Introduction of a binder to a stable ceramic dispersion can flocculate the particles and in most cases causes an increase in slip viscosity.[2] Although binder adsorption can change the stability of a slip, there is evidence that acrylic binder adsorption enhances the pressing properties of spray dried granules and improves green and sintered densities of pressed parts.[8] This study explores how binder and dispersant composition influence adsorption, and it characterizes the changes in slip rheology and stability which occur as a result of binder adsorption.

Experimental

(a) Raw Materials

Commerical alumina powder A-16SG (Aluminum Company of America) was used as supplied. Commercial dispersants Duramax™ D-3005 ceramic dispersant, an acrylic acid homopolymer; Duramax™ D-3002 ceramic dispersant, a methacrylic acid homopolymer; Duramax™ D-3019 ceramic dispersant, an acrylic acid copolymer (Rohm and Haas Specialty Materials) were used. Tamol SG-1 dispersant, a methacrylic acid copolymer (Rohm and Haas Company) was also used. Commercial binders Duramax™ B-1020 ceramic binder and Duramax™ B-1007 ceramic binder (Rohm and Haas Specialty Materials) and Rhoplex® B-60A and Rhoplex® E-2086 (Rohm and Haas Company) were used as supplied. Airvol 205 (Air Products) was used as supplied.

(b) Slip Preparation and Binder Adsorption

Slips were prepared by charging a 00 jar ball mill half filled with grinding media with alumina, dispersant (0.3 to 0.5 % based on alumina) and enough water to bring slurry solids to 50 wt.%. The mixture was milled for 30 minutes. After decanting the slip from the grinding media, binder (10% on alumina) was added and the slip was then placed on a roller for 15 minutes to incorporate the binder. The slurries were diluted (5:1) and centrifuged at 5000 rpm for 30 minutes. Gravimetric measurement of polymer solids of the supernatant gave the amount of unadsorbed binder, and the amount of binder adsorbed on alumina was then calculated by difference. Slips were aged under ambient conditions and binder adsorption was monitored over several days.

All acrylic binders in these studies had average particle size <150 nm. Control experiments were run to ensure that unadsorbed binder did not settle under the centrifugation conditions.

(c) Viscosity Measurement

Viscosities were measured with a Brookfield LV viscometer, using a #1 spindle at 0.1 and 1 sec^{-1}, allowing 10 minutes for the viscometer to equilibrate before reading the viscosity.

Results and Discussion

The primary interaction of an acrylic binder with an alumina surface is through ionized acid sites on the binder as shown in Figure 1.[9,10] Binder pK_a provides a measure of the degree of ionization of the acid sites. Lower pK_a values indicate more acid sites in the ionized form, and higher binder pK_a values indicate fewer ionized acid sites. Four acrylic binders were chosen which varied in the nature ionizability of the acid in the polymer backbone. Binder pK_as were determined by titration of the surface acid sites to be 7.0, 7.7, 8.2 and 8.3.

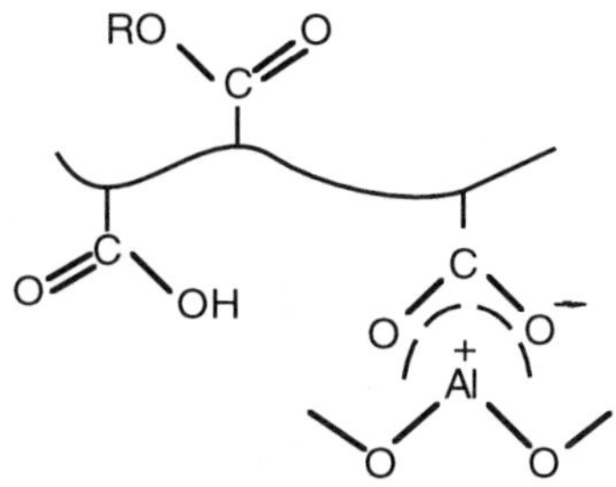

Figure 1. Interaction of acrylic binder with alumina occurs through ionized acid sites on the binder.

To compare adsorption of four different binders on alumina, we used a dispersant which at its saturation level left sites on the alumina surface available for binder adsorption. As shown in Figure 2, there is a correlation between binder adsorption and the extent of acid ionization. The lowest pK_a binder has the most readily ionizable acid sites and adsorption of this binder was rapid, with 25% of the binder adsorbed in within the first two hours of addition. Binders with the higher pK_a values have fewer acid sites in the ionized form and did not adsorb on the alumina surface.

It is common for both viscosity and rheology of slips to change with binder adsorption.[2,3,7] Studies involving poly(vinyl alcohol) binder have shown that there is a significant increase in viscosity which can be correlated with binder adsorption.[6] Figure 3 shows the viscosity increase that accompanies binder adsorption for an acrylic emulsion binder and a 60,000 MW poly(vinyl alcohol) binder.

The viscosity increase resulting from PVA adsorption is particularly evident under low shear conditions, and is due to a combination of the larger volume occupied by particles with adsorbed binder, polymer bridging between particles, and chain entanglement of the high molecular weight binder. Under high shear, polymer chains tend to align in the shear direction, thus reducing viscosity.

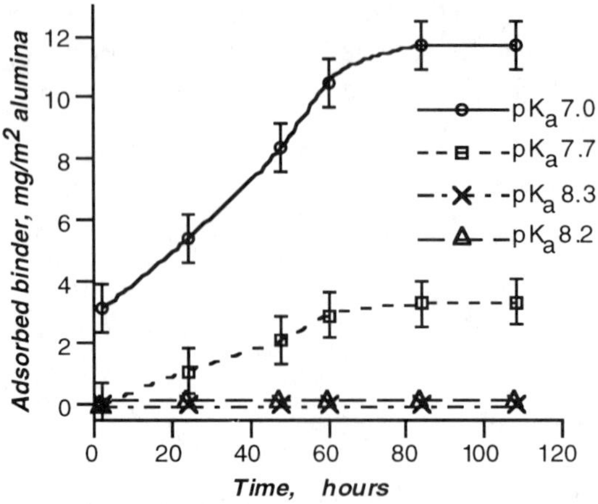

Figure 2. Adsorption on alumina (mg binder/m^2 alumina) of four acrylic binders with different pK_as. All slurries were at pH 9.3

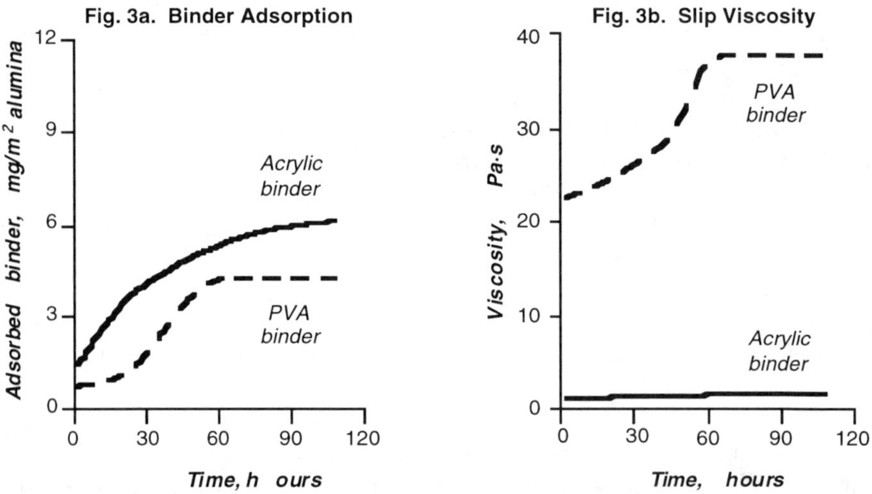

Figure 3. (a) Comparison of adsorption of acrylic binder (pK_a 7.0) with poly(vinyl alcohol) (Airvol® 205) on alumina dispersed with 13,000 MW poly(methacrylic acid). (b) Changes in slip viscosity which accompany binder adsorption.

The viscosity of a slip containing adsorbed acrylic binder is only slightly higher than a slip comparable with no adsorbed binder. Because acrylic emulsion binders are discrete particles, there is no extension of polymer chains into the aqueous phase of the slip to raise viscosity. Slip viscosity remains low because there is little change in the effective size of the particles with adsorbed binder.

The extent of binder adsorption on alumina can change significantly when a different dispersant is used. One way of differentiating dispersants is by their acid number, which is the number of acid sites per gram of the polymeric dispersant. A dispersant with a high acid number has a high affinity for the powder surface and readily saturates the powder surface, leaving few unoccupied sites for adsorption of other additives. In general, the higher the acid number of a dispersant, the greater its affinity for the alumina surface.

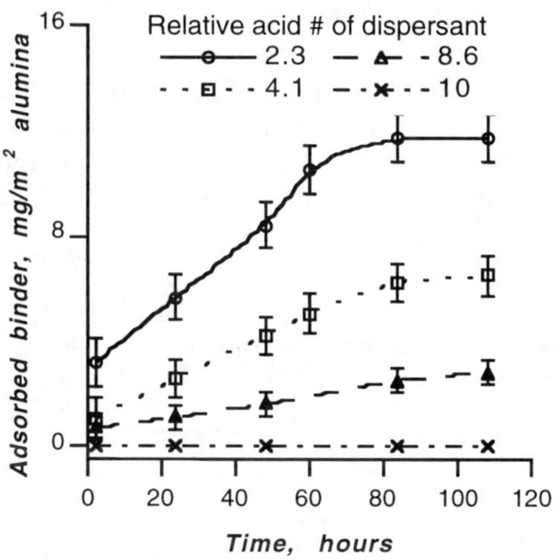

Figure 4. Adsorption of acrylic binder with pK_a 7.0 on alumina predispersed with four dispersants of varying acid number. Lower acid numbers correspond to less dispersant on powder surface at saturation. All slurries were at pH 9.3

Figure 4 compares adsorption of the most ionized or acidic (pK_a 7.0) binder on alumina predispersed with four dispersants, each with a different acid number. The dispersants are compared with respect to their relative rather than absolute acid numbers in Figure 4. The extent of binder adsorption can be rationalized in terms of dispersant acidity - there was no binder adsorbed on alumina dispersed with the highest relative acid number dispersant - the one with the greatest affinity for the alumina surface. The least acidic dispersant (relative acid number 2.3) has a much lower affinity for the alumina surface, leaving more sites available for adsorption of binder or other additives.

The viscosity of slips formulated with a high molecular weight dispersant can drift over time. We have correlated the change in viscosity during slip aging with dispersant adsorption.[5] Furthermore, we found that slips prepared with high molecular weight dispersants are slow to reach saturation, while slips quickly.

We now compare binder adsorption on alumina in freshly prepared and aged slips formulated with a high molecular weight poly(methacrylic acid) dispersant (relative acid number 8.6). Consistent with our previous result that this dispersant is slow to saturate the powder surface, we find that the level of binder adsorption changes as the slip is aged. Adsorption isotherms for an acrylic binder (pK_a 7.0) on alumina in freshly prepared and aged slips are shown in Figure 5. In a freshly prepared slip, we find that initial aliquots of binder adsorb readily on alumina and the level of binder in the aqueous phase is low. Aliquots of comparable levels of binder added to an aged slip adsorb to a lesser extent and are found predominantly in the aqueous phase of the slip.

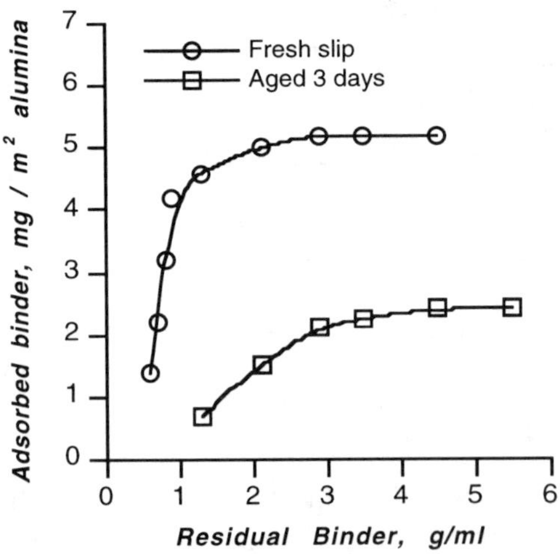

Figure 5. Adsorption isotherms for acrylic binder with pK_a 7.0 on alumina dispersed with 13,000 MW poly(methacrylic acid).

Thus, a binder which is added before dispersant adsorption has reached the saturation point will compete with the dispersant for sites on the powder surface. If the same binder is added to a slip after the dispersant has reached saturation, it adsorbs to a lesser extent because there are fewer sites available to it on the powder surface.

Conclusions

Adsorption of several acrylic binders on alumina has been characterized. Acrylic binders with low pK_a values adsorb faster and to a greater extent than binders with higher pK_as. Binder adsorption can take several hours to reach equilibrium. Slip viscosity increases with binder adsorption, although the increase is minimal for acrylic emulsion binders. The time at which a binder is added to a slip can affect its adsorption on the powder surface. A binder added to a slip before the dispersant has saturated the surface adsorbs more readily than if it were added several hours later. Binder adosrption can be controlled by choosing a dispersant which rapidly saturates the powder surface.

References

1. M. J. Garvey, T. F. Tadros and B. Vincent, "A Comparison of the Volume Occupied by Macromolecules in the Adsorbed State and in Bulk Solution," *J. Colloid Interface Sci.*, **49**, 57-68 (1974).

2. J. S. Reed, *Introduction to the Principles of Ceramic Processing*, Wiley, New York, 1988.

3. D. W. Fuerstenau, R. Herrera-Urbina and J. S. Hanson, "Adsorption of Processing Additives and the Dispersion of Ceramic Powders," pp. 333-51, *Ceramic Powder Science II*, A, G. L. Messing, E. R. Fuller, H. Hausner Eds., The American Ceramic Society, Westerville, Ohio, 1988.

4. B. V. Velamakanni and D. W. Fuerstenau, "The Influence of Polymeric Additives on the Rheology of Dense Slurries," pp. 211-223, *Flocculation in Biotechnology and Separation Systems*, Y. A. Attia, Ed., Elsevier Science Publishers B. V., Amsterdam, 1987.

5. S. E. Agostine, L. E. Bogan and S. A. Ibbitson, "Understanding and Preventing Slip Aging," Presentation at 96th Annual Meeting American

Ceramic Society, Indianapolis, Indiana, 1994.

6. C. S. Khadilkar and M. D. Sacks, "Effect of Poly(vinyl alcohol) an the Properties of Model Silica Suspensions," pp. 397-409, *Ceramic Powder Science II*, A, G. L. Messing, E. R. Fuller, H. Hausner Eds., The American Ceramic Society, Westerville, Ohio, 1988.

7. K. Nagata, "Rheological Behavior of Suspension and Properties of Green Sheet," *J. Ceramics Soc. Japan*, **100** (10), 1271-1275 (1992).

8. E. Carlstrom, "Surface Chemistry in Dry Pressing," pp. 245-78, *Surface and Colloid Chemistry in Advanced Ceramics Processing,* R. J. Pugh and L. Bergstrom, Eds., Marcel Dekker Inc., New York, 1994.

9. Y. N. Sun, M. D. Sacks and J. W. Williams, "Pyrolysis Behavior of Acrylic Polymers and Acrylic/Ceramic Mixtures," pp. 538-549, *Ceramic Powder Science II*, A, G. L. Messing, E. R. Fuller, H. Hausner Eds., The American Ceramic Society, Westerville, Ohio, 1988.

10. K. E. Howard, C. D. E. Lakeman and D. A. Payne, "Surface Chemistry of Various Poly(vinyl butyral) Polymers Adsorbed onto Alumina," *J. Am. Ceram. Soc.*, **73** [8] 2543-46 (1990).

IMPROVED AQUEOUS DISPERSION OF SILICON NITRIDE WITH AMINO-SILANES

Michael A. Buchta and Wei-Heng Shih
Department of Materials Engineering, Drexel University, Philadelphia, PA 19014

ABSTRACT

To improve whisker and powder dispersion of silicon nitride aqueous suspensions, a standard fiberglass surfactant, 3-aminopropyltriethoxysilane (APS) was added. APS coated aqueous suspensions were observed to have lower viscosities along with increased consolidation and dried green body densities compared to uncoated silicon nitride at pH values ≤ 8. APS adsorption behavior shows that silicon nitride was coated to saturation. Suspension pH measurements indicate that APS extracts one hydrogen ion for each APS molecule. Our results show that increased dispersion of silicon nitride can be achieved in acidic conditions.

INTRODUCTION

Silicon nitride has been a leading candidate for structural ceramic applications. Research completed recently[1,2] indicates that a colloidal approach is favored to process reliable silicon nitride bulk ceramics. In order to control the microstructures and properties of the final silicon nitride products, the dispersion of silicon nitride powders and whiskers[3] is essential. This project investigates a chemical adsorption method to improve both whisker and powder dispersion within aqueous suspensions. Silicon nitride's surface can be modified by the application of a standard fiberglass surfactant, 3-aminopropyltriethoxysilane (APS), $Si(OC_2H_5)_3(CH_2)_3NH_2$. APS readily reacts with the hydrolyzed silicon nitride surface in aqueous suspensions to form the aminoalkyl derivative of glass[4]. The hydrophilic nature of APS serves to increase both the wettability of the silicon nitride surface and the electrosteric repulsion between particles resulting in a higher degree of separation of whiskers or powders.

EXPERIMENTAL PROCEDURE

APS coated 10 vol% aqueous suspensions of silicon nitride whiskers (Ube Industries) or powder (Ube Industries, SN-E03) were prepared by 10 minutes of sonication (Ultrasonic Homogenizer 4710 Series, Cole Parmer Instrument Co.) of the silicon nitride/distilled water suspension followed by a 30 minute room temperature stirring period for complete hydrolyzation of the silicon nitride surface. 100μl of APS was added to the hydrolyzed suspension per 10g of silicon nitride. It can easily be shown through specific density and surface area calculations that only 18.8μl of APS is required to completely coat 10g of silicon nitride whiskers.[5] The pH is then adjusted accordingly by the addition of HCl or ammonium hydroxide.

Suspension viscosities were measured using a Rheometrics DSR200 rheometer. Suspensions were also centrifuged for 2 hours, after which a resulting cake density was calculated by comparing the initial suspension volume to the centrifuged cake volume. The presence of leftover APS after coating was qualitatively determined by the addition of salicylaldehyde (SA) to the centrifuged supernatant. The formation of a bright yellow salicylaldimine product between SA and APS will result[6]. Pressure filtration was performed for powder suspensions on a filtration cell in a static press. Pressure was slowly applied to the top of the cell as the water was allowed to leave the suspension through a porous stainless steel filter.

RESULTS

It was found that the pH values of the 10 vol% whisker suspensions are affecting the viscosity behavior of coated and uncoated suspensions differently. This behavior is clearly illustrated in Figure 1 where viscosity for both the uncoated and coated whisker suspensions is plotted at a constant shear rate of 300 1/s. For the uncoated whisker suspensions, viscosity simply decreases with increasing pH values. The viscosity behavior for coated whisker suspensions is more complex. The viscosity shows a local maximum at pH of 3 before increasing again at pH of 10. The local maximum is likely due to the proximity to the isoelectric point (pH of 3) of uncoated silicon nitride whiskers as measured by Hirata et al.[3] The results on viscosity indicate that the APS coating lowers the viscosity of whisker suspensions at pH values less than 8.

To study consolidation behavior, the suspensions of uncoated and coated silicon nitride whiskers were centrifuged and their resulting cake densities were measured. Figure 2 shows the cake density as a function of suspension pH. The centrifuged cake density and the whisker viscosity - pH behavior shown in Figs.1 and 2 correlate well in that as viscosity increases the consolidated density decreases. Therefore, for uncoated whiskers, the centrifuged density of suspensions increased with increasing pH values whereas for coated whiskers, the density tended to decrease with increasing pH values except around the isoelectric point of uncoated silicon nitride whiskers.

The viscosity and centrifuged density for silicon nitride powder suspensions were studied as well. The results are displayed in Figures 3 and 4. The uncoated powder was found to have a viscosity maximum in the pH range of 6-8 which corresponds to its centrifuged density minimum in that pH range. The coated powder displays a local maximum in viscosity close to the isoelectric

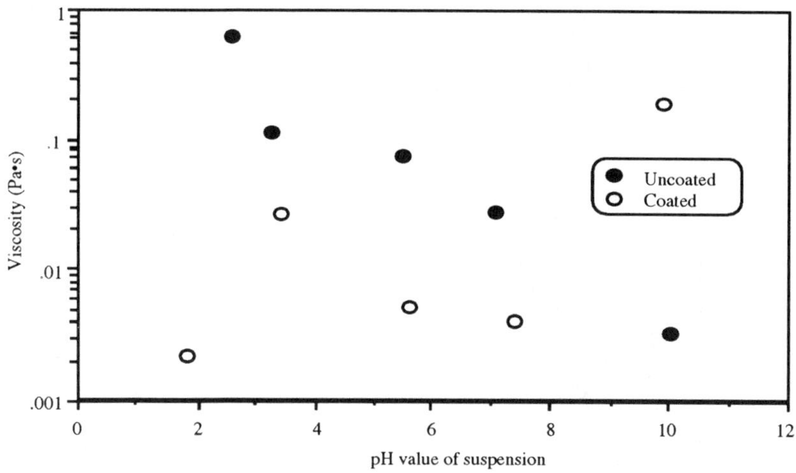

Fig.1 - Silicon nitride whisker suspension viscosity at a shear rate of 300 1/sec.as a function of pH.

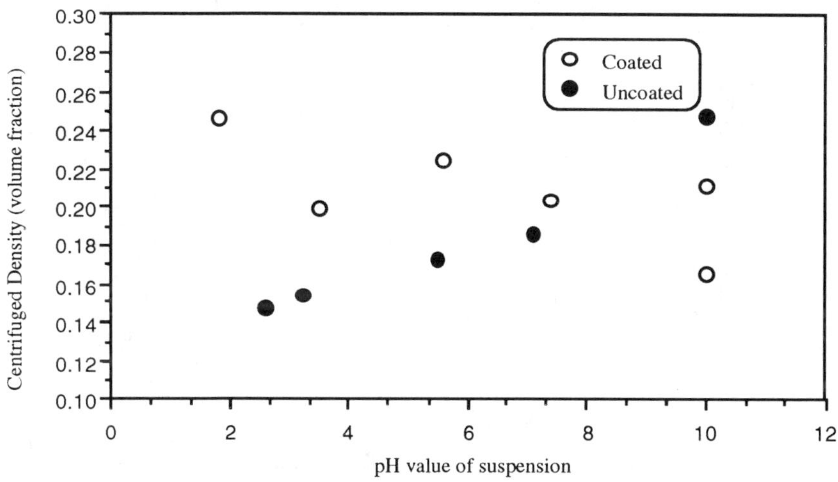

Fig.2 - Silicon nitride whisker centrifuged density as a function of suspension pH.

point [7] of uncoated silicon nitride powder (pH of 6.5). The coated behavior was found to be similar for silicon nitride powder and whisker suspensions. Comparatively, uncoated and coated whiskers had a higher viscosity and lower centrifuged density than silicon nitride powder suspensions due to their anisotropic geometry. The consolidation of 10 vol% silicon nitride powder suspensions was also studied using pressure filtration. Figure 5 displays the dried cake density after pressure filtration as a function of suspension pH. The consolidated cakes are

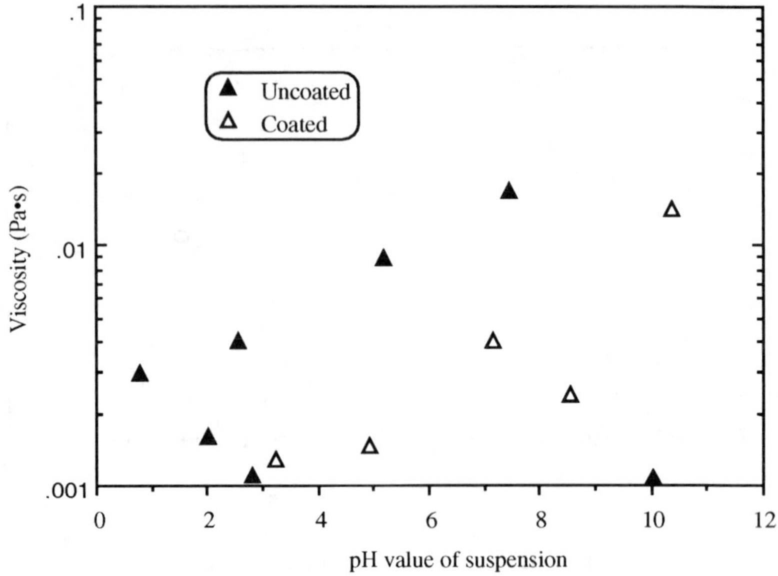

Fig. 3 - Silicon nitride powder suspension viscosity at a shear rate of 300 1/sec as a function of pH.

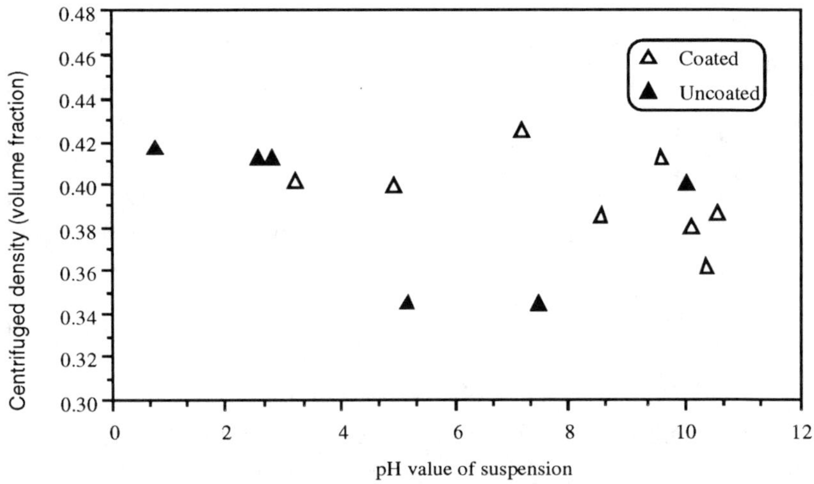

Fig.4 - Silicon nitride powder centrifuged density as a function of suspension pH.

cylindrical with an average thickness of 2.5mm and a diameter of 25.4mm. Fig. 5 shows that the APS coating aids in consolidation over a wide pH range for silicon

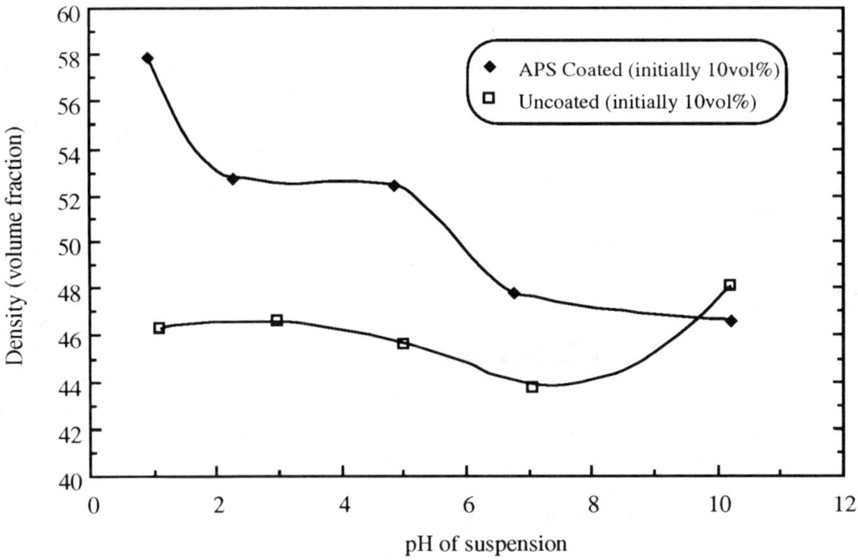

Fig.5 - Dried bulk density of silicon nitride powder after pressure filtration as a function of suspension pH.

nitride powder suspensions and a dried density of 58 vol% may be obtained. The consolidation behavior in Fig. 5 is similar, only higher, than that shown in Fig. 4.

DISCUSSION

The viscosity and consolidation results indicate that the APS surfactant is most effective at lower pH values whereas at high pH values it has an adverse effect. It is speculated that APS coated silicon nitride powder and whiskers both have an isoelectric point which shifted towards ≈10. This shift explains the APS coated silicon nitride behavior found in Figs. 1-5. The isoelectric points of uncoated silicon nitride whiskers[3] and powder[7] are 3 and 6.5 respectively. The APS coating reduces the flocculation behavior of whiskers and powders at those pH values, although a remnant effect near their corresponding isoelectric points is evident throughout our results. The remnant effect could originate from regions of the uncoated silicon nitride surface which can not react with the APS molecule. APS is solely expected to react with the silicon oxide surface during coating, not the nitrogen surface.

The centrifuged supernatant (from a 10g silicon nitride powder sample with 100μl APS and 29 ml water) was analyzed with the addition of SA. SA, APS and water were found to be immiscible since a bright yellow emulsion phase was formed. Qualitatively, we can conclude that APS was left over in the supernatant. To further our investigation of the adsorption behavior of APS and silicon nitride, suspension

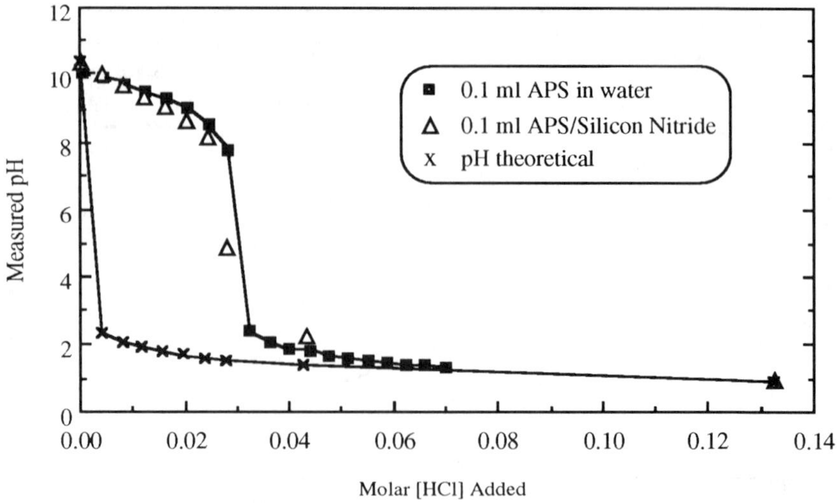

Fig.6 - Theoretical and experimental pH values for APS coating of silicon nitride powder.

pH was measured as a function of the amount of HCl added, and can be seen in Fig.6. If APS were not extracting hydrogen ions, then the suspension pH would follow the theoretical curve calculated from chemical equilibrium. This does not occur as the APS/water solution and APS/silicon nitride aqueous suspension exhibit a different behavior shown in Fig.7. There is a significant drop of pH values of solutions at [HCl]≈0.03 M. This drop correlates well with the concentration of APS in each solution, [APS]=0.028 M. Therefore, APS is concluded to extract one hydrogen ion for each molecule either chemisorbed on the surface or in the solution. A proposed reaction for the behavior of APS in water is:

$$Si(OC_2H_5)_3(CH_2)_3NH_2 + 4H_20 \rightarrow Si(OH)_3(CH_2)_3NH_3^+ + OH^- + 3C_2H_5OH \quad (1)$$

Reaction (1) explains the instant rise to pH to 10 upon the addition of APS to distilled water (pH of 7). The ability of APS to extract hydrogen ions is slightly stronger when no silicon nitride is present. This difference can be seen in Fig.6. as the pH for the APS/water solution is higher than that of APS/Si_3N_4 solution. To investigate this difference, pH values were measured for both 10g silicon nitride powder and 5g whisker suspensions (10 vol% each) as a function of the amount of APS added. The difference in the concentration of hydroxyl ions, $\Delta[OH^-]$ which is equal to APS/water solution $[OH^-]$ minus APS/silicon nitride suspension $[OH^-]$, as a function of the amount of APS added, is shown in Fig 7. Once the surface of silicon nitride is completely coated, one expects that there would be a constant $\Delta[OH^-]$ upon the addition further APS. It is observed that as the APS amount increases, the pH of the suspensions increase to a saturation level at high APS concentrations. It can be seen in Fig.7 that while for whiskers, $\Delta[OH^-]$ remained constant, silicon nitride powder went through a maximum and then saturated at 0.4 vol% APS. As expected, the whisker adsorption limit (≤0.001) was found to be

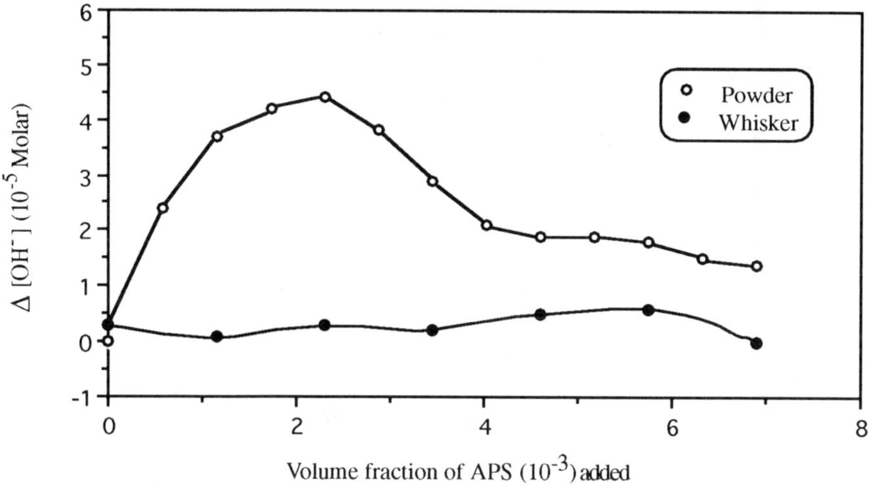

Fig.7 - Change in hydroxyl concentration between APS/silicon nitride suspensions and APS/water solutions as a function of APS amount.

smaller than the powder adsorption limit (≈0.004) because of its smaller specific surface area and sample size. We propose that the origin of the maximum found in Fig.7 arises from a small fraction of physisorbed APS molecules on the silicon nitride surface. A physisorbed APS molecule would have its amine group attracted to the surface in contrast to a chemisorbed APS molecule whose silicon end group chemically reacts with the silicon oxide surface and has its amine group pointing away from the surface. A physisorbed molecule would then not have the ability to extract hydrogen ions from the solution because it is hydrogen bonded to the oxide surface. A schematic drawing of physisorption and chemisorption of APS on the silicon nitride surface is shown in Figure 8. A maximum in Fig.7 would then occur if the number fraction of physisorbed molecules changes with the concentration of APS added to the solution. All samples studied can be concluded to have been coated to saturation since they contained an APS concentration of at least 1vol%.

CONCLUSIONS

The APS coating on silicon nitride in aqueous suspensions lowers the viscosity and increases both centrifuged consolidation and pressed green body bulk density in acidic conditions. All silicon nitride samples were coated to saturation by APS. APS increases the suspension pH level by extracting one hydrogen ion for each APS molecule. Our results show that increased dispersion of silicon nitride can be

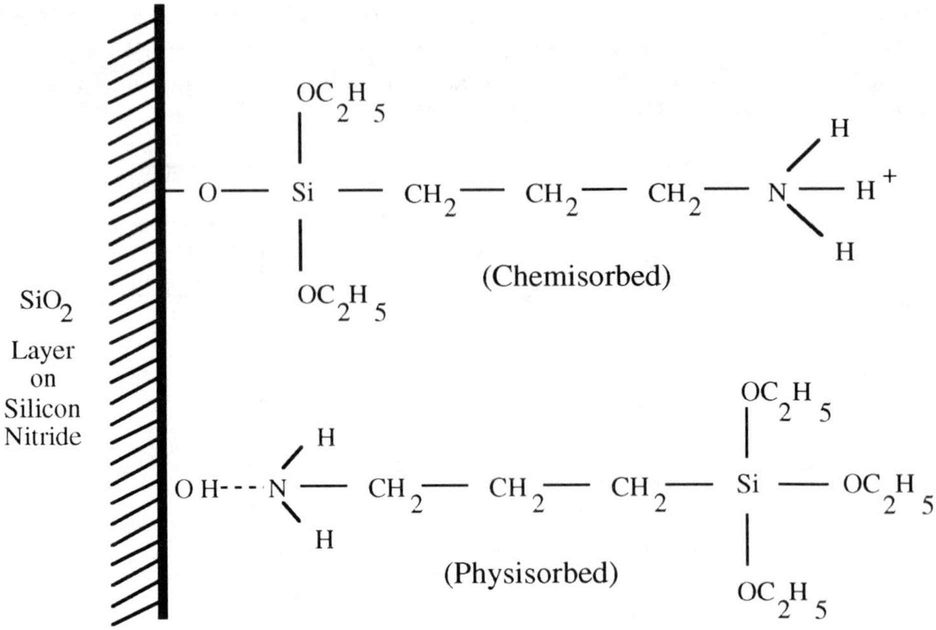

Fig.8 - Possible physisorption and chemisorption of APS on silicon nitride.

achieved in acidic conditions due to a shift in the effective iep of silicon nitride whiskers and powder towards ≈10.

REFERENCES

[1]F. F. Lange, "Powder Processing Science and Technology for Increased Reliability", *J. Amer. Ceram. Soc.*, **72** [1] 3-15 (1989).

[2]V. K. Pujari and D. M. Tracey, "Processing Methods for High Reliability Silicon Nitride Heat Engine Components", ASME paper 93-GT-319 (1993).

[3]Y. Hirata, S. Nakagama, and Y. Ishihara, "Dispersion and Consolidation of Si_3N_4 Whisker in Aqueous Suspension", *J. Mater. Res.*, **5** [3] 640-46 (1990).

[4]L. Jervis, N. Pettit, "Purification of Ribonuclease T1 on Porous Glass Affinity Adsorbents", *J. Chromatography*, **97**, 33-38 (1974).

[5]M. Buchta and W.-H. Shih, "Improved Dispersion of Silicon Nitride Whiskers", *Mat. Res. Soc. Symp.*, Vol.365, in press (1995).

[6]T. G. Waddell, D. E. Leyden, and M. T. DeBello, "The Nature of Organosilane to Silica-Surface Bonding". *J. Amer. Chm. Soc.*, **103** 5303-5307 (1981).

[7]W.-H. Shih, L.-L. Pwu, A. A. Tseng, "Boehmite Coating as a Consolidation and Forming Aid in Aqueous Silicon Nitride Processing", *J. Amer. Ceram. Soc.*, **78** [5] 1252-1260 (1995).

ORGANIC ADDITIVE SYSTEMS FOR SPRAY-DRYING AND DRY PRESSING SILICON NITRIDE

W.J. Walker, Jr. and J.S. Reed
Alfred University, Alfred, NY 14802

S.K. Verma
Union Carbide Corporation, South Charleston, WV 25303

ABSTRACT

Silicon nitride granules for dry pressing were prepared by spray-drying slurries containing polyethylene glycol as the primary binder combined with other organic additives. Differences in slurry viscosity, granule character, pressing behavior and green strength were found to depend on the choice of deflocculant.

INTRODUCTION

Organic additives play a number of roles in modern ceramic powder processing [1]. Surfactants improve particle wetting and dispersion. Deflocculants are necessary to stabilize a dispersed system and to prevent re-agglomeration. Binders give an unfired part the required strength for handling. Binder modifiers are used to fine-tune the properties of binders. Lubricants reduce particle/particle and particle/die wall friction.

Polyethylene glycol (PEG) binders have been the subject of recent studies [2-4]. Advantages include good lubricity, clean burnout and low order of toxicity. In this work, an amorphous PEG with MW=15-20 000* was investigated as a binder for silicon nitride dry pressing formulations. Structural applications of silicon nitride require that the number of defects with size greater than the particle size be minimized in order to obtain maximum reliability. In this paper, organic additive systems based on PEG Compound 20M and their influence on compaction and green strength are examined.

Granules can be viewed as controlled agglomerates containing powder, organic additives and porosity. The mechanical properties of the granules are dominated by the mechanical properties of the organic phase. Compaction behavior of granules containing PVA-based binders is sensitive to small changes in moisture content and to the presence of added plasticizers because the glass transition temperature T_g is close to room temperature. An advantage of PEG is that $T_g < -20^oC$, so added plasticizer is not required.

Compaction of granulated material occurs in three stages [1]. At low pressures, granules rearrange without deforming. Above some critical pressure (P_Y) which is related to their strength, granules start to deform and the compact density increases with log(pressure) as the interstitial

* CARBOWAX PEG Compound 20M, Union Carbide Corporation, Danbury, CT.

voids between granules are reduced in size. At high pressures, when these large voids are nearly gone, additional pressure increases the packing efficiency of the primary particles. It is useful to think of the compaction of a bed of granulated material in terms of its bimodal pore size distribution: large intergranular pores from granule packing in a matrix containing small intragranular pores from the packing of the primary particles [5]. Common microstructural defects from compaction include persistent large pores and poorly joined granule interfaces. The small pores are easily eliminated during sintering while large pores persist as strength limiting defects. Poorly joined interfaces can cause microlaminations in the compact as compaction pressure is released, resulting in cracks and porous regions in the sintered specimen which also act as strength limiting defects.

Table I: Slurry Formulations

Formulation	I	IA	II	III
Powder - Si_3N_4	100.0	100.0	100.0	100.0
Binder - PEG Compound 20M	3.0	3.0	3.0	3.0
Binder Modifiers				
PEG-1450*	0.5	0.5	0.5	0.5
PEO**	0.1	0.1	0.1	
Deflocculants				
TEPA***	0.5	0.5		0.5
NH_4-PAA+			0.5	0.5
Surfactant++	0.2	0.2	0.2	0.2
Water	48.5	54.8	48.5	48.5

EXPERIMENTAL

Silicon nitride powder with mean particle size 0.72 μm and specific surface area 7.3 m^2/g was used for this study. Slurries with formulations shown in Table I were ball milled for 8 h in 75 mm diameter polyethylene jars at 120 rpm, predispersed binder solutions being added after 6 h. Viscosity was measured with a concentric cylinder viscometer+++. Spray-drying was done in a mixed flow spray-drier with a two-fluid nozzle#. After conditioning in a desiccator with relative humidity of 51% which maintained an equilibrium moisture content of 0.4%, granules were compacted in a 2.54 cm diameter double action hardened steel die using a universal testing machine## to prepare compaction diagrams at low pressures and a laboratory press### to prepare compacts at high pressures. Density and pore size distributions of granules and compacts were determined using mercury intrusion porosimetry$. Green strength was determined using the diametral compression test.

* CARBOWAX, Union Carbide Corporation, Danbury, CT.

** POLYOX WSR N-10, Union Carbide Corporation, Danbury, CT.

*** Tetraethylene pentamine, Union Carbide Corporation, Danbury, CT.

+ Darvan 821A, R.T. Vanderbilt Co., Inc., Norwalk, CT.

++ Tergitol MDS-42, Union Carbide Corporation, Danbury, CT.

+++ Model RV3, Haake, Inc., Paramus, NJ.

Bowen #1, Niro Inc., Columbia, MD.

Instron Corporation, Canton, MA.

Laboratory Press Model M, Fred S. Carver, Inc., Menomonee Falls, WI.

$ Autopore II 9220, Micromeritics Instrument Corporation, Norcross, GA.

RESULTS AND DISCUSSION

Slurry Behavior.

Slurry viscosities as a function of shear rate are presented in Figure 1. Slurries I, IA and II are shear thinning over the entire range of shear rates measured, with Slurry I having the highest viscosity. Slurry III which had the lowest viscosity at low shear rates exhibited shear thickening above a critical shear rate of 150 s^{-1}. High shear rate viscosity values are more relevant to spray-drying since the shear rate during atomization in the type of nozzle used for this study was calculated to be 12 000 s^{-1}. The viscosity of Slurry I may be too high to be spray-dried using some types of atomizers. TEPA used alone appears not to be an effective deflocculant for Si_3N_4 for this reason. Slurry II which was deflocculated with NH_4-PAA was of low viscosity. Because of favorable results obtained using granules produced from Slurry I, two methods were used to reduce the viscosity of slurries containing TEPA into an acceptable range. Slurry IA contained the same organic additive system as Slurry I with additional water. Slurry III combined NH_4-PAA and TEPA as deflocculants, without an auxiliary binder.

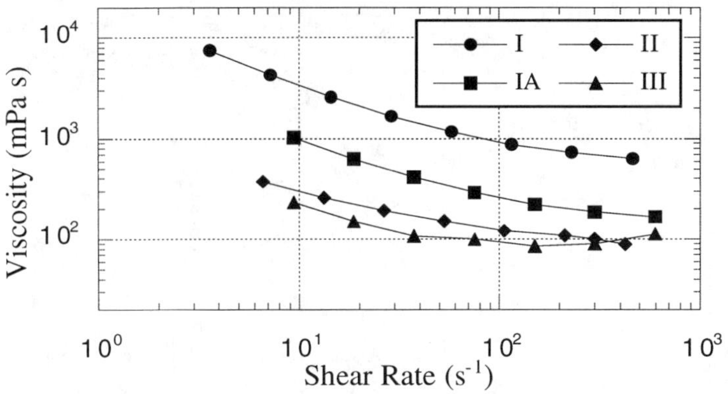

Figure 1. Slurry viscosity with respect to shear rate.

Spray-Dried Granules

Granule properties are presented in Table II. Despite the large differences in rheological behavior of the four slurries, the log normal granule size distributions were nearly the same. Slurry IA, which had lower solids loading produced granules that were of slightly smaller mean size. SEM micrographs of typical granules are shown in Figure 2. The granules produced from Slurry I were solid with a small dimple on their surfaces. Those prepared from Slurry II were hollow, with a single large open pore. Slurries IA and III produced hollow granules, some of which had a small surface opening, while others appeared solid until cut with a razor.

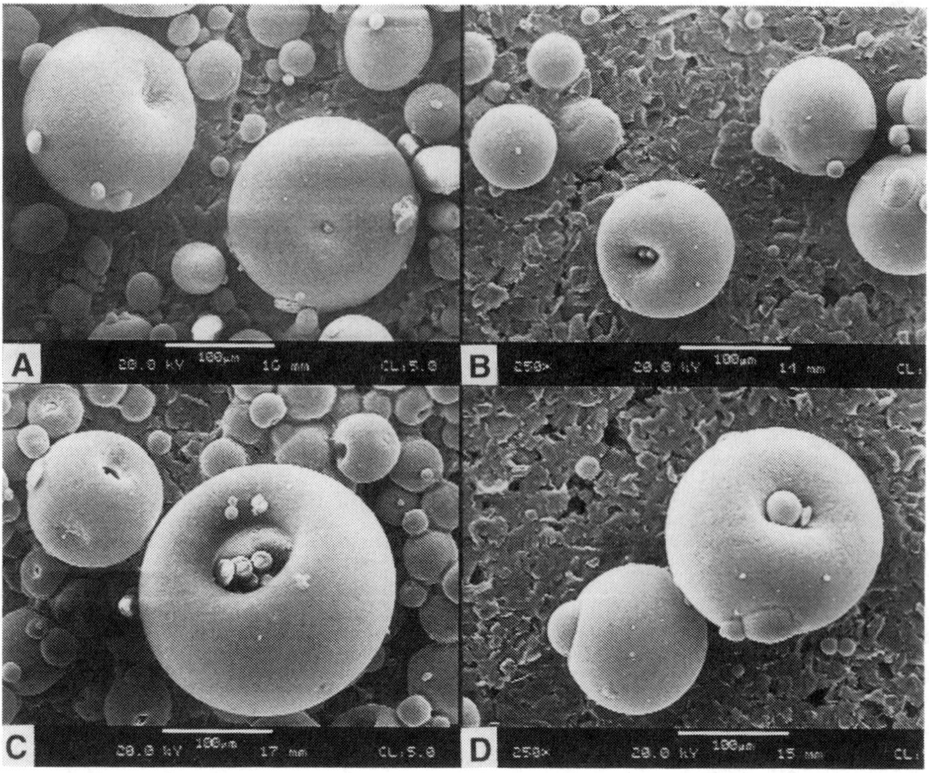

Figure 2. Scanning electron micrographs of spray-dried granules: (A) System I, (B) System IA, (C) System II and (D) System III.

Table II: Granule Properties

	I	IA	II	III
Median Size (µm)	86	82	86	86
Granule Density (%)	46.4	46.4	53.2	48.5
Tapped Density (%)	31.4	31.4	32.7	32.6
Packing Factor (%)	67.7	67.7	61.5	67.2

The granules prepared from slurries deflocculated only with TEPA (I and IA) were of lower density than those prepared from the slurry deflocculated with only NH_4-PAA (II). Combined deflocculants (III) resulted in granules with intermediate density. Systems I, IA and III showed granule packing efficiency of 67.2-67.7% which is slightly higher than the 64-65% normally expected for log normal distributions of spheres. The additional few percent packing efficiency could be caused by deviations from true log normal distribution and because the packing of these large granules is less influenced by the types of attractive forces that hinder the packing of fine powders. Somewhat lower granule packing efficiency was seen for System III. The difference may be due to the additional pore volume present from the macroscopic pores seen in Figure 2.

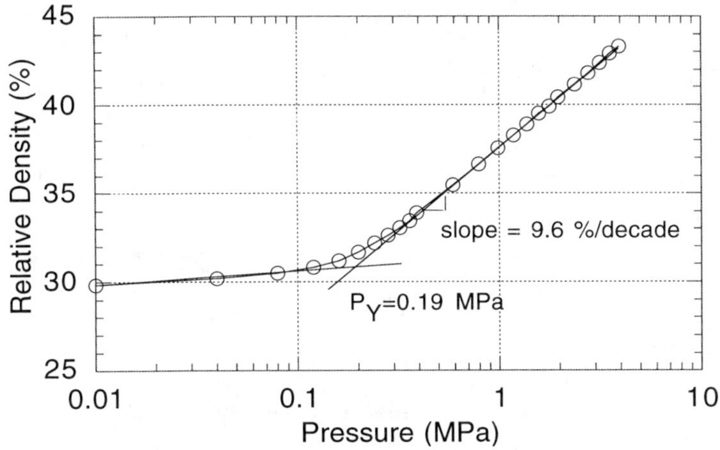

Figure 3. Compaction diagram for System I

Compaction and Green Strength

Figure 3 shows a typical compaction diagram. Granule yield pressure (P_Y) and pressing efficiency (slope) for each binder system are seen in Table III. Systems I and IA have the same value for P_Y. It is not surprising that the onset of deformation occurs at the same pressure, since they contain the same binder system. However, the slope is lower for System I. Since these granules are solid, more pressure is required for deformation, while the hollow granules from System IA can accommodate increasing pressure by collapsing.

Table III: Compaction Parameters

	I	IA	II	III
P_Y (kPa)	190	190	220	290
Slope (%/decade)	9.6	10.2	10.3	10.4

The higher values for P_Y of Systems II and III show that these granules are of higher strength. NH_4-PAA has a strengthening effect on the granules, either as a result of the higher granular density or through stronger interparticle bonding. System III contains both NH_4-PAA and TEPA which gives the highest strength granules, possible because they contain slightly more total organic content.

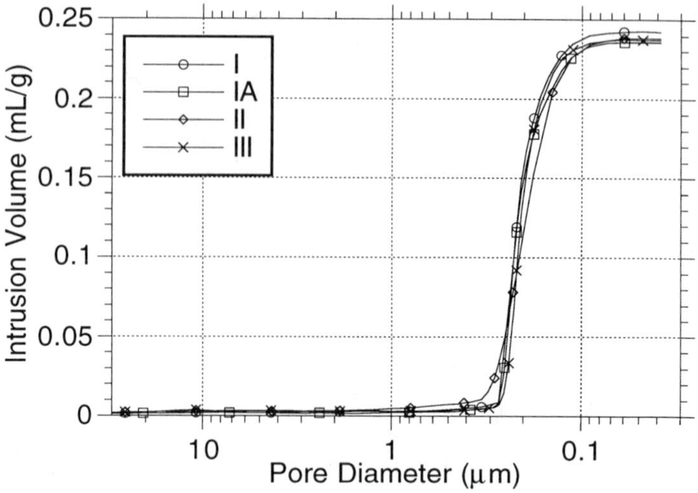

Figure 4. Pore size distribution of green specimens compacted to 51 MPa.

Compact density data determined from mercury porosimetry are given in Table IV. The structure of the compacts can be seen as a few large pores in a matrix with porosity determined by the packing of the primary particles. The lower limit for matrix density is the granule density. Pore size distributions for compacts pressed to 51 MPa are seen in Figure 4. Most of the intrusion volume is seen in the size range less than 0.3 μm, and is due to the packing of the primary particles. A small pore volume is seen with size larger than 0.3 μm, which is due to persistent intergranular porosity. A drawback of the mercury intrusion method is that the sizes of entry passageways are reported rather than actual pore sizes, causing error in the case of inkwell shaped pores. In order to minimize such errors thin specimens with large areas of fracture surface were used. Still, it is recognized that some large pores may not be detected. The matrix density reported in the table is calculated based on the porosity with size less than one half the mean particle size (0.36 μm). The difference between the matrix density and bulk density is the volume of large pores. Small differences represent a significant improvement in pressing behavior since these pores are too large to be eliminated during ordinary sintering.

Table IV: Compact Density and Green Strength

	I	IA	II	III
Density (%)				
pressed to 51 MPa				
Bulk	52.4	53.3	53.2	53.0
Matrix	52.9	53.7	54.1	53.4
Difference	0.5	0.4	0.9	0.4
pressed to 80 MPa				
Bulk	54.1	55.0	55.0	55.0
Matrix	54.3	55.5	55.9	55.4
Difference	0.2	0.5	0.9	0.4
Strength (kPa)				
pressed to 51 MPa	890	890	690	1050
pressed to 80 MPa	1080	1020	860	1140

At each compaction pressure, System I shows the lowest bulk density and the lowest matrix density. Systems IA, II and III show similar bulk density, but II has the highest matrix density. Figure 5 shows matrix density and matrix pore size plotted as a function of compaction. The matrix density and pore size of II are virtually unchanged as pressure increases, while the others show a large increase in matrix density which is accompanied by a decrease in pore size. The denser granules prepared by deflocculating the slurry with NH_4-PAA are less compactable than those deflocculated with TEPA. System III, which contains both deflocculants shows intermediate behavior. The lower deformability of the denser granules hinders compaction. These granules had a lower packing factor, leaving more voids to be eliminated. The result is a larger volume of persistent intergranular porosity.

The large pore volume remains unchanged between the two compaction pressures except in the case of I, where it is reduced. Paralleling the difference in pressing efficiency seen from the compaction diagrams this reduction of intergranular porosity at high pressure could be a result of granule structure. Contrasting the hollow granules which can collapse, the solid granules deform more uniformly until intergranular void space is more completely eliminated.

Green strength data are also seen in Table IV. System III had the highest green strength. Systems I and IA, which had identical binder system formulations had identical green strength when compacted to 51MPa, but a difference was seen at the higher compaction pressure. System II had a lower green strength, even though granule strength, P_Y was higher than for Systems I and IA. Figure 6 shows fracture surfaces of specimens compacted to 51MPa. The fracture path follows a large number of poorly joined granule interfaces in the case of System II, while in all other cases there is little or no obvious evidence of the granular precursor material. Clearly, granule joining is much weaker when TEPA is not present in the binder system.

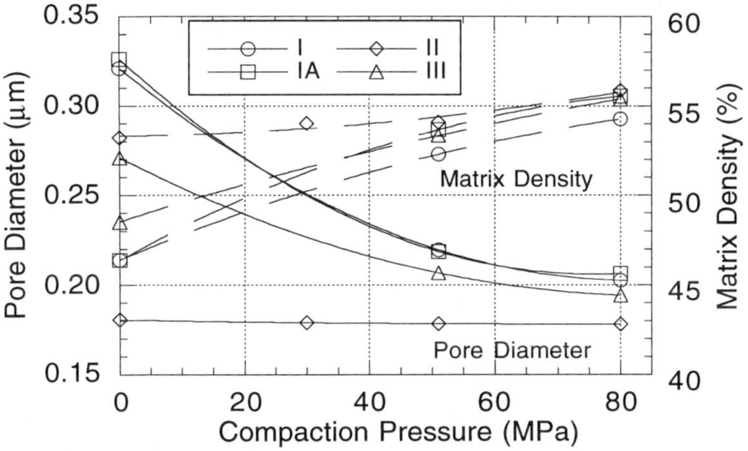

Figure 5. Changes in matrix density and pore size with compaction pressure.

A comparison in strengths between I and IA suggests that there is a difference in flaw size distribution at the higher pressure that is not present at the lower pressure. This is consistent with the differences observed in intergranular pore volume. Again, an explanation for this could be that the hollow granules (IA) can accommodate increased pressure by collapsing, thereby relaxing the structure and reducing the driving force for interfacial joining.

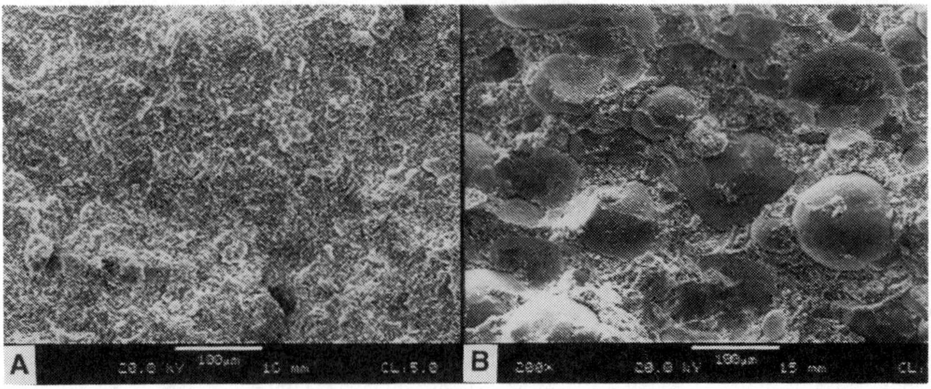

Figure 6. Scanning electron micrographs of green fracture surfaces: (A) System I and (B) System II. Systems IA and III had fracture surfaces similar to (A).

CONCLUSIONS

Four binder systems based on PEG Compound 20M binder were compared and significant differences in compaction behavior were observed. The following conclusions were drawn from this work:

1) NH_4-PAA is more effective than TEPA for reducing viscosity of silicon nitride slurries.

2) Granules spray-dried from slurries containing NH_4-PAA were denser and stronger than those processed with TEPA alone.

3) Green compacts containing TEPA were stronger, had better granule joining, and had a smaller volume of large pores than those not containing TEPA.

4) The structure of the granules (hollow vs. solid) was seen to influence compaction efficiency, compact density and, green strength.

REFERENCES

1. J.S. Reed, *Principles of Ceramic Processing*, 2nd. Edition, Wiley-Interscience, New York, 1995.
2. S.K. Verma, J.H. Robson, W.J. Walker, Jr., and J.S. Reed, "Amorphous Polyethylene Glycol in Ceramic Powder Processing," pp. 369-75 in Ceramic Powder Science IV, Ceramic Transactions, Vol. 22. The American Ceramic Society, Inc., 1991.
3. W.J. Walker, Jr., J.S. Reed and S.K. Verma, "Polyethylene Glycol Binders for Advanced Ceramics," *Ceram. Eng. Sci. Proc.* **14** [11-12] 58-79 (1993).
4. J. Zheng and J.S. Reed, "Ceramic Extrusion with PEG," *American Ceramic Society Bulletin,* **73** [12] 61-69 (1994).
5. J. Zheng and J.S. Reed, "The Different Roles of Forming and Sintering on Densification of Powder Compacts," *American Ceramic Society Bulletin,* **71** [9] 1410-16 (1992).

SURFACE MODIFICATION OF SILICON NITRIDE POWDERS BY METAL-ORGANIC SURFACTANTS

H. M. Wang and R. A. McCauley
P. O. Box 909, Center for Ceramic Research,
Rutgers University, Piscataway, NJ 08855-0909

ABSTRACT

As-received UBE-E10 silicon nitride powder was thermally and chemically treated for surface analysis. Surface characterization was performed by BET, XPS, TEM, and acoustophoresis measurement. The surface charge behavior of oxidized and leached powders was related to their surface oxygen contents. The surface morphologies and charge behaviors of powders changed after they were modified by metal-organics. The density of the coating layer was estimated and the surface area coverage of the coating was calculated. A mechanism for powder surface modification was also proposed.

INTRODUCTION

Silicon nitride has been studied for over two decades due to its promising high temperature properties. However, densification of silicon nitride is difficult because of its covalent bonding nature. Before achieving sufficient diffusion rates for atoms, silicon nitride decomposes above 1850 °C in 1 atm of nitrogen [1]. Generally, sintering additives are introduced to form liquid phases so that the sintering rate is enhanced before the thermal decomposition of silicon nitride can occur, especially for pressureless or gas pressure sintering processes.

Traditionally, the additives are introduced by mixing silicon nitride and metal oxide powders by ball milling. In order to achieve full densification, large amounts of metal oxides, usually over 5 to 10 wt.%, were used to overcome the heterogeneity of additive distribution. These metal oxides formed liquid phases at grain boundary and degraded the material performance when silicon nitride was used at elevated temperature. Possible solutions for the property degradation are to reduce the additive amounts, to crystallize the grain boundary phases, and to use more refractory metal oxide systems.

The purpose of this study was to solve the above problem by reducing the additive amounts. Our approach was to incorporate organic surfactants with desired metal cations from a colloidal process. The coatings were converted to metal oxides after organic burn-out, and were used for liquid phases during densification. The main advantages of this process are to introduce sintering additives more homogeneously and to reduce the additive amounts. The other advantage is the modification of surface charge behavior for coated powders, which allows easier in handling of slurries with multiple components.

EXPERIMENTAL PROCEDURE

As-received UBE E10 powder[1] was either heat treated or chemically leached in various conditions. The thermal treatments were performed from 300 to 1200 °C for 6 hours in a box furnace. Powder samples were chemically leached in 1 M HF solution for 5 days at room temperature. Aluminum and yttrium isopropoxides[2] were incorporated onto as-received powders by ball milling for one day in n-hexane. Organic was removed by heat treatment at 600 °C for 6 hours in air. In order to test the bonding between powders and coatings, as-received and coated powders were dispersed in water and were centrifuged at 3500 rpm for 30 minutes. One volume % of as-received or treated powders dispersed in 200 ml of 0.001 M KCl solutions was prepared for acoustophoresis measurement[3]. Surface area measurements for powders were done by a single-point BET[4]. Samples for TEM[5] observation were prepared on a copper mesh from dilute suspensions in ethanol. Powders for XPS[6] analysis were loaded on aluminum sample holders. All TEM and XPS samples were kept in a desiccator before any analysis.

RESULTS AND DISCUSSIONS

(A) Thermal Treatment and Chemical Leaching

XPS Analysis:

The surface compositions for as-received, thermally, and chemically-treated UBE E10 powders are listed in Table 1. It was assumed that only oxygen and nitrogen contents changed and silicon contents stayed the same on powder surfaces before and after treatments, except for those metal-organic modified powders. Therefore, the surface oxygen and nitrogen contents (in atomic %) could be normalized to that of silicon; i.e., (M/Si). The (N/Si) and (O/Si) for pure silicon nitride powders should be 1.33 and 0, respectively. Any process which introduces oxygen onto powder surface will increase the (O/Si) ratio and decrease the (N/Si) ratio. For as-

---

1. UBE Industry, Japan.
2. Aldrich Chemical Company, Inc., Milwaukee, WI.
3. Matec ESA-8000, Matec Applied Sciences, Hopkinton, MA.
4. Quantachrome, Quantasorb, Syosset, NY.
5. ISI ABT EM-0028, Akashi Beam Technology Corp., Tokyo, Japan.
6. XSAM-800, Kratos, Ramsey, NJ.

received UBE E10 powder (Table 1), the (N/Si) ratio was 1.16 and that for (O/Si) was 0.66. The (N/Si) and (O/Si) ratios for powders leached by 1 M HF were 1.32 and 0.08, respectively. This suggested that the oxygen on the powder surface was removed by HF so that the silicon nitride surface was cleaner than that of as-received powder. As the heat treatment temperatures increased from room temperature to 1200 °C, the (O/Si) ratio increased from 0.66 to 1.80, while the (N/Si) ratio decreased from 1.16 to 0.02. More oxygen was introduced onto powder surfaces during the oxidation process. The other observed phenomenon was the shift of Si_{2p} peak. The reported binding energy of Si_{2p} peak for silicon nitride film [2] and for silica [3]were 101.8 eV and 103.2 eV, respectively. If the Si_{2p} peaks were deconvoluted into two peaks located at 101.5 eV ($\pm$ 0.2 eV) and 103.1 eV ($\pm$ 0.2 eV), the shifts of peak positions from low to high binding energy were observed as more oxygen was introduced onto the powder surfaces.

Table 1. Surface Compositions for Treated Powders from XPS Analysis

Powder Treatment	Binding Energy (eV) Si_{2p}*		M/Si (N/Si)	M/Si (O/Si)
HF Leached	101.4 (95.1)	103.0 (4.9)	1.32	0.08
As-Received	101.4 (86.4)	103.0 (13.6)	1.16	0.66
300 °C/6 hrs	101.5 (86.5)	103.2 (13.5)	1.09	0.82
600 °C/6 hrs	101.5 (83.2)	103.3 (16.8)	0.97	0.92
900 °C/6 hrs	101.9 (74.3)	103.3 (26.6)	0.75	1.38
1200 °C/6 hrs	101.7 (14.0)	103.1 (86.0)	0.02	1.80

* Peak position (peak area, %)

Acoustophoresis Measurement:

Table 2 lists the isoelectric points, pH_{iep}, for powders with and without treatments. The pH_{iep} for the as-received UBE E10 powder was about 6. For 1 M HF leached powders, the isoelectric point shifted to a higher pH value, about 9.8. As powders were heat treated from 300 to 1200 °C for 6 hours, the pHiep values dropped from 5.8 to 2.44, which was very close to that of SiO_2, pH_{iep} = 2.0 [4]. These data implied that powder surfaces were modified to silica-like surfaces and the surface charge behaviors were altered after oxidation.

(B) Metal-Organic Modification

XPS Analysis:

The surface compositions of powders coated with various amounts and sequences of aluminum and yttrium isopropoxides are listed in Table 3. If the sampling depth by XPS probe was kept the same for every coated sample, the detected silicon content from powder surface was assumed to be constant. The (M/Si) ratio, where M = aluminum or yttrium, could be used to compare the metal contents which were incorporated onto powder surfaces. When 1.18 wt.% Al_2O_3 or 1.26 wt.% Y_2O_3

(converted from the added amounts of metal-organic) was introduced, the (Al/Si) and the (Y/Si) ratios were 0.29 and 0.11, respectively. When both metal-organics were incorporated at the same time, reasonable (M/Si) ratios were detected. However, coating sequence did affect the detected element amount due to the sampling depth (2 - 3 nm) of the equipment. Data in Table 3 show less amounts of metal were detected for those incorporated onto the powder surface first.

Table 2 Isoelectric Points for Treated Powders

Powder Samples	HF Leached	As-Received	Heat Treatment (6 hrs) 300 °C	600 °C	900 °C	1200°C
pH_{iep}	9.8	6.0	5.8	4.75	2.58	2.44

Table 3 Surface Composition for Metal-Organic Coated Powders

Metal-Organic Compounds*	Additive Amounts (%) Al2O3	Y2O3	M/Si Al/Si	Y/Si
Al	1.18	-----	0.29	-----
Y	-----	1.26	-----	0.11
Al + Y	1.87	1.26	0.01	0.21
Y+ Al	1.87	1.26	0.50	0.01
(Al, Y)	0.67	1.33	0.13	0.10

* Al + Y means the A1 layer coated first,then Y
(Al,Y) means A1 and Y were incorporated simultaneously.

Acoustophoresis Measurement:

The powder surfaces were modified to metal oxide-like surfaces when metal organics were coated onto their surfaces. Figure 1 shows that pH_{ieps} of powders changed as the incorporating amounts of metal oxides (calculated from the added amounts of metal organics) increased. A similar phenomenon was also observed by Liden et al.[5]. The change of pH_{iep} leveled off at added amounts over 1 wt.% for Al_2O_3 additive, while it reached the maximum after incorporating of 0.5 wt.% of Y_2O_3. This gave the information that either those additive amounts were high enough to change the charge behavior of powder surfaces or only certain amounts of powder surfaces were available for coating. From the literature [4], the pH_{ieps} for hydrolyzed alumina and yttria were about 9 - 10. The powder surfaces may not be modified completely by metal organics so the surface charge behavior did not match those of alumina or yttria.

BET Measurement:

The surface area increment for uncoated and coated powders with or without being centrifuged are shown in the Figure 2. The surface area of powders increased as the additive amounts increased. If the surface area increment, ΔS_A, was calculated from the difference between the surface areas for coated and uncoated powders, a linear relation between the fraction of additives and the surface area increment was found for both samples with and without centrifugal tests, which were applied to test the bonding between coatings and powders. Also, from the density measurements for powders coated with 1 wt.% alumina and uncoated powders, the

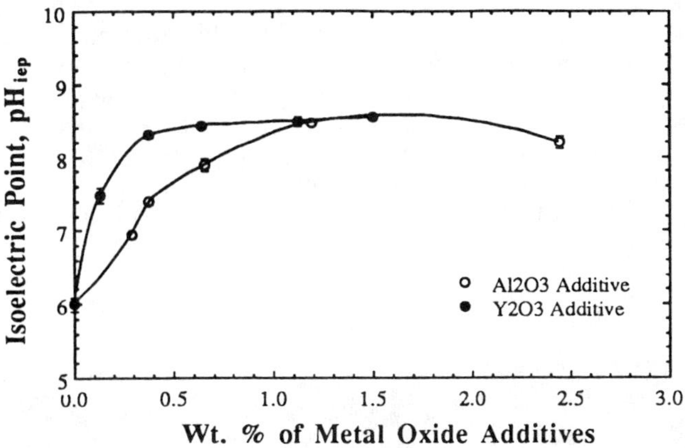

Figure 1. Variation of pHiep with Various Amounts of Metal Oxide Additives.

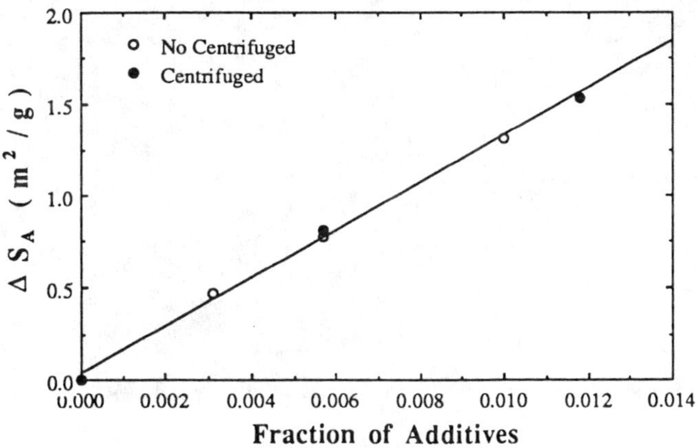

Figure 2. Surface Area Increment with Various Amounts of Aluminum Oxide Additives.

density of the alumina coating was estimated to be 2.72 g/cm^3. The calculated density was much lower than that of crystalline alumina, 3.95 g/cm^3. This might suggest that the coating layer was amorphous or porous.

TEM Observation:

High resolution TEM images for the coated and the uncoated powder are shown in Figure 3. Both powders have been heat treated at 600 °C for 6 hours in air. The uncoated powder showed a clean surface with a thin amorphous layer with a thickness of 2 - 3 nm. EDS analysis on the powder surface revealed that only silicon was present. This thin amorphous layer has been suggested to be a combination of silicon oxynitride and silica [6]. For the coated powder, a much rougher surface was observed. The surface bumps were about 5 - 6 nm in diameter and were analyzed to contain aluminum and yttrium from EDS. From the surface morphology, the coatings were not crystalline and didn't fully cover the powder

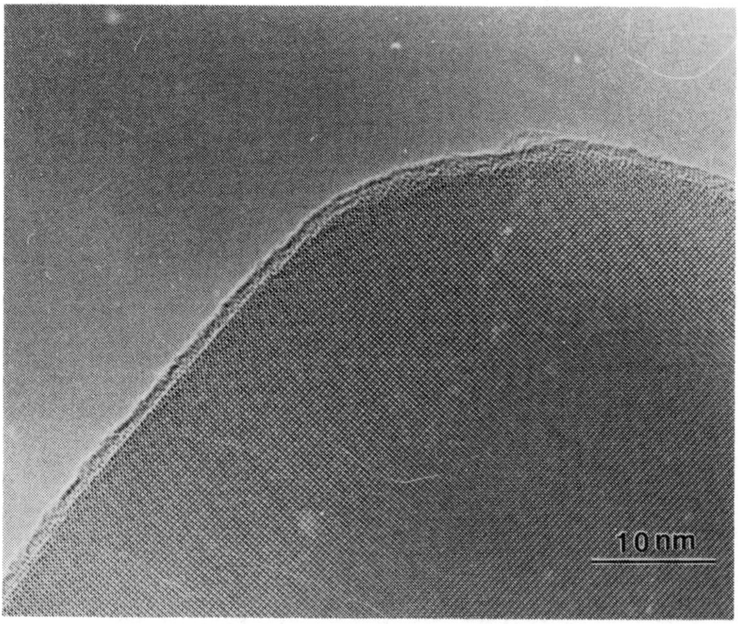

(a)

Figure 3. High Resolution TEM Images for (a) As-Received Powder after Heat Treated 600 °C for 6 Hours and,

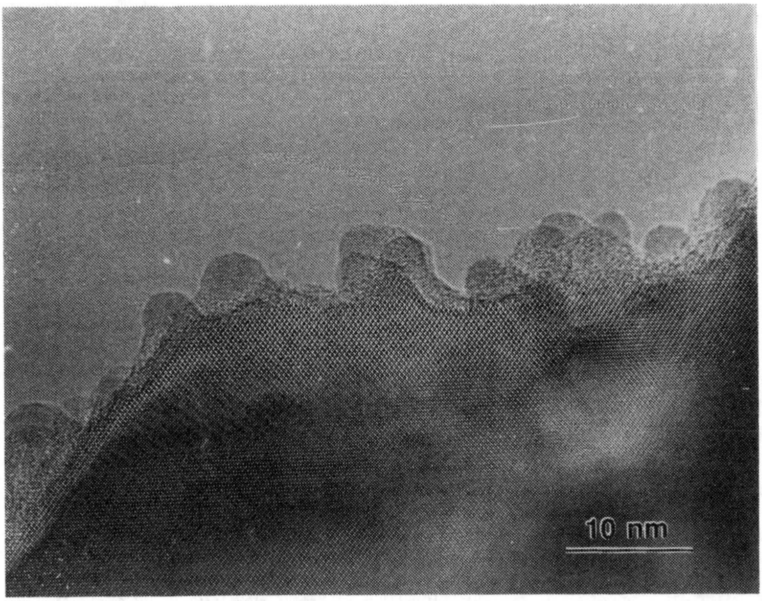

(b)

Figure 3. (Contd.) (b) Coated Powder with 1.33 wt.% of Y_2O_3 and 0.67 wt.% of Al_2O_3 Additives.

surface, which was consistent with the above measurements. A model was proposed and the surface area coverage was calculated to be 48% for 1.18 wt.% alumina coated powders [7].

(C) Proposed Coating Mechanism

From the above surface analysis and observation, the metal oxide coating layers were proven to be on the powder surfaces. Although the coated oxides did not fully cover the powder surfaces, the amounts were large enough to change the surface charge behavior of UBE E10 powders. The coating routes could be either one or both of following:
(i) M-$(OR)_3$, where M is aluminum or yttrium, reacts with silanol groups (Si-OH) on powder surfaces by chemical adsorption and then is polymerized by hydrolysis from the moisture near the powder surfaces, or
(ii) M-$(OR)_3$ hydrolyzes first with the moisture before attaching to powder surfaces and then is chemically adsorbed onto powder surfaces.
Either of the above reaction routes yields the same the result.

CONCLUSIONS

The surface oxygen contents can be characterized from XPS analysis and acoustophoresis measurement. As the oxygen contents on powder surfaces increased, the surface (O/Si) ratio increased and (N/Si) decreased. pHiep values also decreased and were close to 2, which is that of SiO_2. Incorporating metal-organic compounds onto powder surfaces did change the powder surface morphologies and surface charge behavior. The density of the surface layers modified by alumina were estimated to be 2.72 g/cm^3, and the surface area coverage was calculated to be 48 % for 1.18 wt.% alumina coating.

Acknowledgments

Financial supports from Center for Ceramic Research at Rutgers University and the New Jersey Commission of Science and Technology are gratefully acknowledged. Special thanks go to Dr. D. A. Hensely for his efforts in collecting XPS data.

REFERENCES

1. C. Greskovich and S. Prochazka, J. Am. Ceram. Soc., Vol. 64, No. 7, C 96 - C 97 (1981)
2. S. I. Raider, R. Flitsch, J. A. Aboaf, and W. A. Pliskin; "Surface Oxidation of Silicon Nitride Films," J. Electrochem. Soc., Vol.123, No.4, pp560-565 (1976)
3. M. Peuckert and P. Greil; "Oxygen Distribution in Silicon Nitride Powders," J. Mater. Sci., 22, pp 3717-3720 (1987)
4. G. A. Parks; Chem. Rev. 65 [2], pp 177-198 (1965)
5. E. Liden, L. Bergstrom, M. Persson, and R. Carlsson; "Surface Modification and Dispersion of Silicon Nitride and Silicon Carbide Powders," J. Euro. Ceram. Soc., Vol.7, pp 361-368 (1991)
6. P. N. Joshi, Master of Science thesis, Rutgers University (1992)
7. H. M. Wang, Ph.D. Dissertation, Rutgers University (1995)

ACRYLIC EMULSION BINDER FOR TAPE CASTING

Xun Tang*, Scott A. Ibbitson*, and Antonio T. Donato*
Rohm and Haas Company, Spring House, PA 19477

Hayes Byrd, Jr.*
AVX Corporation, Myrtle Beach, SC 29578

ABSTRACT

A crosslinkable acrylic emulsion binder has been demonstrated to significantly increase tensile strength of green tapes. A low Tg acrylic emulsion binder is blended with the crosslinkable binder to retain flexibility and elasticity of green tapes. The tensile strength of green tapes can be increased from 1.19 MPa to 2.34 MPa by incorporating a crosslinkable binder into the formulation while the tapes remain flexible and elastic with ~15% strain at break. Upon stretching, the green tapes do not yield before breaking. Increasing the drying temperature appears to slightly increase the tensile strength of the green tapes, and has a marginal effect on the strain at break of the tapes. This binder system does not liberate formaldehyde as a result of crosslinking.

INTRODUCTION

Tape casting is a very important forming step in manufacturing thin sheets of ceramic materials for multilayer capacitors, packages, and substrates. The technological advances in aqueous-based additives, especially aqueous-based acrylic emulsion binders, have allowed tape casting to expand from solvent-based processes to aqueous-based processes for water-insensitive materials. An ideal tape casting binder should provide adequate tensile strength, flexibility, and elasticity for green tapes. The casting process requires a certain degree of green tape strength and flexibility so that the tape can be removed from the carrier substrate and rolled (reeled) without breakage. The green strength also allows the tapes to withstand the subsequent cutting, printing, laminating, and final dicing processes. In addition to tensile strength and flexibility, a certain degree of elasticity is also important for dimensional control of green tapes.[1] Small applied stresses on green tapes are inevitable during tape processes. The stresses may cause dimensional changes, such as wrinkling and ridging during casting and stretching during subsequent processes. If the dimensional changes are not reversible, they will become permanent defects in the green tapes. These defects include misalignments and tape inhomogeneities that ultimately lead to problems in shrinkage control, cambering, and warping of the fired ceramics.[1]

Although the mechanical properties of green tapes prepared with nonaqueous and aqueous solution binders have been discussed in the literature,[2-4] it seems little work has been

---

*Member, American Ceramic Society

---

done to correlate the mechanical properties of green tapes to aqueous emulsion binder systems. In this study, we introduce an aqueous emulsion binder system that provides high-tensile strength, flexibility, and improved elasticity for green tapes.

EXPERIMENTAL PROCEDURE

Materials

The ceramic powder used in this study was Alcoa A-16 SG alumina. The surface area was 9.7 m^2/g (nitrogen BET) and the mean particle size was 0.5 μm (Coulter LS-130). The dispersant used was Duramax™ D-3005 dispersant, an aqueous low-molecular weight ammonium polyacrylate salt solution. The crosslinkable acrylic binder was Duramax™ B-1050 binder, and the low Tg binder (Tg: -40 °C) was Duramax™ B-1035 binder. The conventional acrylic binder was a blend of two acrylic polymers with Tg of -15 °C and 14 °C (ratio = 0.5 : 9.5), respectively. All the binders were aqueous emulsions and were used as received. The curing agent for the crosslinkable binder was Jeffamine® D-400, a liquid polyoxyalkylenediamine with a molecular weight of 400. Deionized water was used in all formulations.

Slurry Preparation

The slurry was prepared by charging a size of 0 jar mill with 252 g of water, 10.32 g of dispersant (0.3% on alumina), 1200 g of alumina, and ~2200 g of grinding cylinders. The slurry solids loading was 82 wt. %. The slurry was milled at 80-100 rpm for 3 hours.

Table I shows a typical formulation with a 4 to 6 ratio of the crosslinkable binder to the low Tg binder. The binders and the curing agent were added to the alumina slurry. The mixture was stirred for 30 minutes in a 250 ml jar by an overhead stirrer to ensure uniform mixing. The slurry was then filtered through a 100 mesh sieve.

Table I. Typical Formulation with Crosslinkable Binder

Materials	Function	Amount (g)
Alumina slurry (total solid: 82 wt. %)		121.9
Duramax™ B-1050 (total solid: 48 wt. %)	Crosslinkable binder	10.0
Jeffamine® D-400 (total solid: 50 wt. %)	Curing agent	0.21
Duramax™ B-1035 (total solid: 55 wt. %)	Low Tg binder	13.1

The total loading of the binders was 12 wt. % on alumina while the ratio of the crosslinkable binder to the low Tg binder varied according to the formulations. Concentration of the curing agent was 2.2 wt. % on the crosslinkable binder.

Tape Casting

Green tapes were cast with a doctor blade at a speed of ~12 cm/min onto Mylar tapes pre-coated with a thin layer of ammonium stearate. The wet thickness was 254 μm (10 mil). The wet tapes were dried at 150 °C in a forced air oven for 20 minutes. The dry thickness was 127-152 μm (5-6 mil) depending on the formulations.

Tensile Property Measurement of Green Tapes

The tensile properties of the green tapes were measured by stress-strain curve on Universal Testing Machine (Model 1122 Instron), according to a technique described by Cannon et al.[4] Standard tensile bar specimens (dog-bone shape) were prepared with a 0.635 cm tensile width and a 1.91 cm gauge length (narrow tensile section). The specimens were cut from the green tapes using a razor template. Air-accentuated grips with constant pressure were used to eliminate tear at the grip section.[4] A crosshead speed of 0.25 cm/min was used in this study. Data were acquired by computer (Series IX Automated Materials Testing System 1.01 program). When the tensile bar was fractured on either end of the narrow tensile section or fractured on the center section with apparent tearing, those data were excluded from the calculations. Each data point was an average of at least 10 specimen measurements.

RESULTS AND DISCUSSION

Crosslinking of Binder

A crosslinked polymer is one in which linkages exist between the polymer chains.[5] For the crosslinkable binder investigated in this study, the crosslinking is achieved by a reaction similar to the well-known reaction between acetoacetic ester and amines.[6] As shown in Scheme 1, the crosslinkable groups (left side of the reaction equation) on the polymer backbones react with a polyoxyalkylenediamine to form a crosslinked polymer network. The reaction takes place when the binder is dried. At this point, ammonia is liberated, and the polyfunctional amine links the polymer chains to form a network. Unlike formaldehyde-containing or -generating crosslinkable polymers, this crosslinkable polymer does not release formaldehyde as a result of crosslinking. No formaldehyde was detected (detection limit: 25 ppm) when the polymer was held at 190 °C for 30 minutes. In addition, using polyoxyalkylenediamine, compared to using monomeric diamines, favors formation of a more elastic crosslinked polymer.

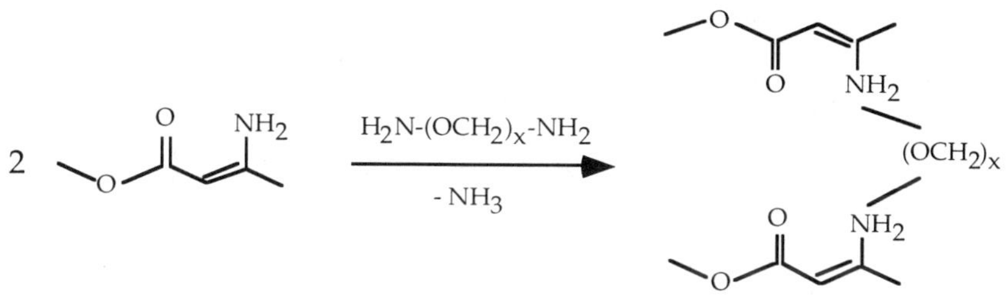

Scheme 1. Crosslinking reaction in the crosslinkable binder.

In general, crosslinked polymers tend to become stiff. To control flexibility of green tapes, a low Tg emulsion binder was blended with the crosslinkable binder. Unlike common plasticizers, the low Tg polymer not only "lubricates" the relatively stiff crosslinked polymer chains but also acts as an additional binder to enhance binding in ceramic formulations.

Effect of Crosslinking of Binder on Tensile Properties of Green Tapes

The effect of binder crosslinking on tensile properties of green tapes is shown in Table II. The green tapes were cast with the formulation given in Table I, containing a total of 12 wt. % blend binder in which the ratio of the crosslinkable binder to the low Tg binder was 4 : 6. When the curing agent of polyoxyalkylenediamine was absent, the binder was not crosslinked. The resulting green tape had a tensile strength of 1.19 MPa. When the curing agent was included in the formulation, however, the tensile strength of the resulting green tape was almost doubled to 2.34 MPa. Such a significant increase in tensile strength of the green tape was apparently due to the linkage between polymer chains via the polyoxyalkylenediamines.

Table II. Effects of Crosslinking on Tensile Properties of Green Tapes

Binder	Tensile Strength (MPa)	Strain at Break(%)
Uncrosslinked	1.19 ± 0.05	12.4 ± 0.6
Crosslinked	2.34 ± 0.11	14.7 ± 1.6

Green tapes cast with a blend of crosslinkable binder and the low Tg binder also showed better elastic properties than tapes that were cast with uncrosslinked conventional binders. Figure 1 shows a comparison of the stress-strain curves of two green tapes. The solid curve represents the green tape containing a total of 12 wt. % binder in which the ratio of the crosslinkable binder to the low Tg binder was 4 : 6. The dashed line represents green tape containing 12 wt. % of the conventional emulsion binder. Besides the significantly higher tensile strength, the green tape cast with the blend shows a higher strain at break (~15%) than the tape cast with conventional binder (~6%). Most importantly, green tape cast with the blend did not yield before breaking. In contrast, the green tape cast with conventional binder yielded at only ~6% of stretching and then continued to elongate until ~8%.

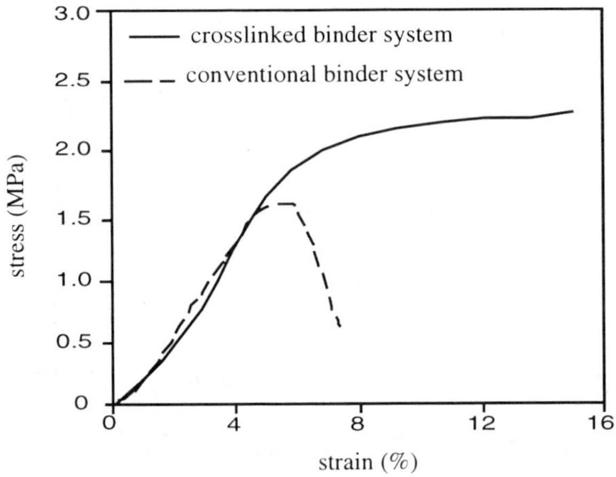

Fig. 1. Comparison of green tapes containing crosslinked binder and conventional binder.

It was observed during tensile tests that upon breakage, a gap existed between the two broken pieces of a green tape. For the green tape cast with the blend, the gap closed without an overlap between the two broken pieces when the Instron grip returned to its original position. In the case of the green tapes cast with the conventional binder, an overlap between the two broken pieces was always observed.

These results indicate improved elasticity of the green tapes cast with the blend compared to tapes cast with conventional binders. The improved elasticity provides dimensional reversibility for the green tapes. Upon stretching, the tape cast with the blend has a tendency to resist permanent dimensional changes. When the applied stress is released, the tape tends to go back to its original dimension. However, for the tapes cast with the conventional binder any stretching beyond the peak point will leave permanent dimensional damage to the green tapes even though the tapes may appear undamaged. Damage includes misalignments and tape inhomogeneities that ultimately lead to problems in shrinkage control, cambering, and warping of the fired ceramics.[1]

The improved elastic properties of green tapes may be a result of combined effects. Blending the crosslinkable polymer with another uncrosslinkable polymer may result in a lower degree of crosslinking in green tapes compared to the use of crosslinkable polymer alone. A low degree of crosslinking tends to increase elasticity. The length of the crosslinking agent might allow the linked polymers to have a certain degree of movement even after crosslinking. The low Tg polymer may "lubricate" the crosslinked polymer chains, making them more elastic. The low Tg polymer itself also may contribute to the elasticity of green tapes. All these combined effects favor the formation of elastic properties.

Effects of Degree of Crosslinking on Tensile Properties

The degree of binder crosslinking in green tapes was controlled by altering the ratio of the crosslinkable binder to the low Tg binder in the binder blend. The ratio of the curing agent to the crosslinkable binder was constant. The total binder loading was 12 wt. % based on alumina.

As shown in Figure 2, the tensile strength of the green tapes increased with the increase in the ratio of the crosslinkable binder to the low Tg binder, suggesting an increased rigidity of the polymer chains. However, strain at break of the green tapes showed a slightly different trend. As shown in Figure 3, the strain at break increased from 11.4% to 16.3% at a crosslinkable binder to low Tg binder ratio of 1:9. Increasing the ratio of the crosslinkable binder in the binder blend had little effect on strain at break, considering the error bars from the measurements.

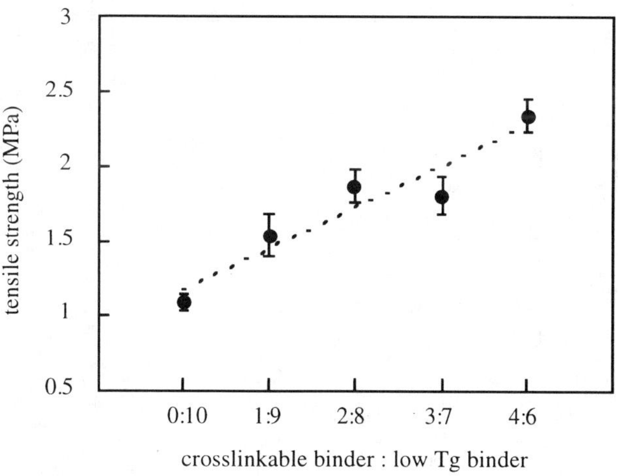

Fig. 2. Effect of crosslinking on the tensile strength of green tapes.

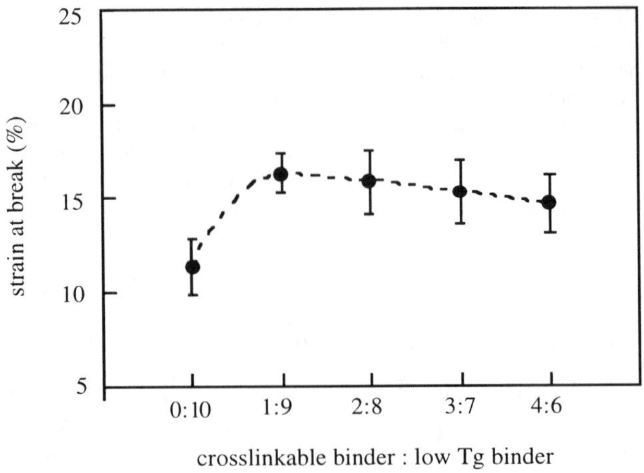

Fig. 3. Effect of crosslinking on strain at break of green tapes.

The combination of the increased tensile strength and the unchanged strain at break is very beneficial to tape processes. Increasing the tensile strength of green tapes reduces breakage of green tapes. Maintaining a proper strain at break allows for dimensional control of the tapes during processes.

Effects of Drying on Tensile Properties

The effects of drying temperature on the tensile properties of green tapes were investigated by drying the tapes at various temperatures for 20 minutes. The green tapes were cast with a blend binder containing 3 : 7 ratio of the crosslinkable binder to the low Tg binder. The results for tensile strength are shown in Figure 4 and for strain at break in Figure 5.

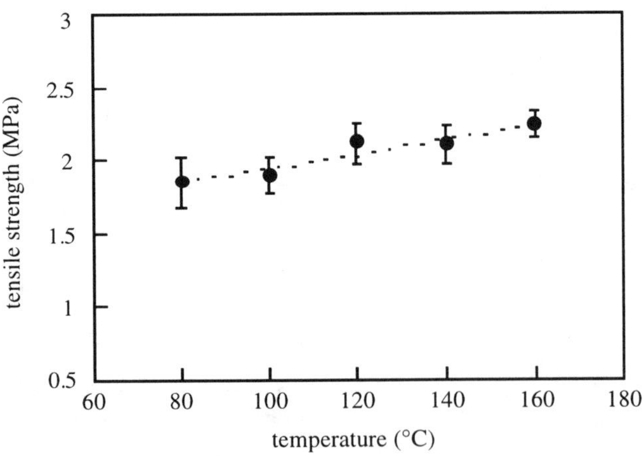

Fig. 4. Effect of the drying temperature on the tensile strength of green tapes.

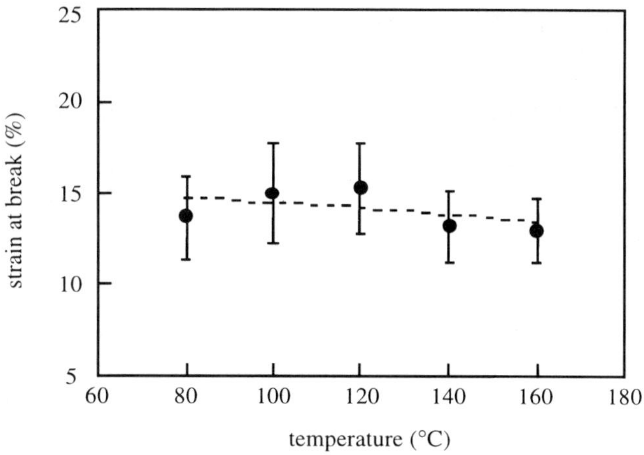

Fig. 5. Effect of the drying temperature on strain at break of green tapes.

While the drying temperature had essentially no effect on strain at break (given the error bars of measurements), the tensile strength showed a slight increase with increases in temperature. The increase of tensile strength was likely caused by removal of trace amounts of moisture, as has been observed by Cannon et al.[4] No effort was made to quantitate the moisture effect.

Since both the investigated temperature range (80-160 °C) and the heating time (20 minutes) represent the whole temperature gradient of the heating zone, as well as the longest time span in the manufacturing scale of the tape casting process, this result suggests a maximum range of tensile strength variation that may be expected by drying green tapes.

CONCLUSIONS

When blended with a low Tg binder, a crosslinkable acrylic emulsion binder can significantly increase tensile strength and elasticity of ceramic green tapes. The tensile strength increases with the increase of the crosslinking density in green tape, but the strain at break remains essentially unchanged. Heating temperature has less effect on the tensile strength than does the crosslinking density.

REFERENCES

[1]R.A. Gardner and R.W. Nufer, "Properties of Multilayer Ceramic Green Sheets," *Solid State Technology*, May 38-43 (1974).

[2]K. Nagata, "Effect of Functionalities of Binders on Rheological Behavior of Alumina Suspensions and Properties of Green Sheets," *J. Ceram. Soc. of Japan*, **101** (8) 845-849 (1993).

[3]A. Karas, T. Kumagai, and W.R. Cannon, "Casting Behavior and Tensile Strength of Cast $BaTiO_3$ Tape," *Advanced Ceram. Mat*, **3** (4) 374-77 (1988).

[4]S. Forte, J.R. Morris, Jr. and W.R. Cannon, "Strength of Tape Casting Tapes," *Am. Ceram. Soc. Bull.,* **64** (5) 724-725 (1985).

[5]H.R. Allcock and F.W. Lampe, "Synthesis and Reactions of Polymers", pp. 7-8 in *Contemporary Polymer Chemistry,* Prentice-Hall, Inc., New Jersey, 1981.

[6]C.R. Hauser and M.J. Weiss, "Cyclization of 2-Aminopyridine Derivatives to Form 1,8-Naphthyridines," *J. Org. Chem.*, **14** 453 (1949).

Duramax is a trademark of Rohm and Haas Company.
Jeffamine is a registered trademark of Texaco Comapny.

VISCOELASTICITY OF NANO-ALUMINA DISPERSIONS

B. Rand and R. Fries
School of Materials
The University of Leeds
Leeds, LS2 9JT, UK.

ABSTRACT

The flow and viscoelastic properties of electrostatically stabilized nano-alumina dispersions have been studied as a function of ionic strength and volume fraction of solids. At low ionic strength the suspensions were deflocculated and showed a transition from viscous to elastic behavior as the solid content increased associated with the onset of double layer interpenetration. The phase transition was progressively shifted to higher solids fractions with increasing ionic strength. At higher ionic strength, above the critical coagulation concentration, the suspensions formed attractive networks characterized by high elasticity. Two independent methods of estimating the effective radius of electrostatically stabilized 'soft' particles, a_{eff}, are presented based on phase angle data and a modified Dougherty-Krieger equation. The results suggest that a_{eff} is not constant for a given system but changes with both solids fraction and ionic strength.

INTRODUCTION

The rheological properties of dispersions of ceramic powders are important to processing operations, first because those operations frequently involve shear or flow processes and second because the consolidation into the ceramic 'green' body is determined by the structure that exists in a slip and how this is modified in the consolidation process (1). Rheological behavior is also strongly influenced by the slip structure and therefore rheological measurements are often used as a characterization technique in monitoring processing slips (2). Conventionally, continuous shear rheological measurements are made, but these have the disadvantage that the structure is changed during flow and no indication is given about the elasticity present in the system. Elasticity can arise from both electrostatic repulsion due to double layer overlap and attraction between particles that are assembled into a flocculated structure (3). Elasticity due to flocculation should only be manifest at stresses smaller than a characteristic yield stress, since the continuous interconnected network of flocs that tends to form in such systems can readily be broken down to a discontinuous set of flow units, 'flocs', under shear. Elasticity due to double layer overlap is strongly dependent on concentration and particle size. Elasticity can also arise from the presence of significant quantities of polymer in the slip, as in the case of tape casting systems where polymer is an essential constituent.

Where electrostatic stabilization is used to control the properties of slips, as in most aqueous systems, the rheological behavior is affected significantly by the ratio of the particle radius, a, to the electrical double layer thickness, $1/\kappa$, i.e. by the parameter κa. This determines whether the slip behaves as an assembly of 'hard' or 'soft' particles. Thus, smaller particles will be more likely to show 'soft' particle interactions when the electrical double layer is well extended into the solution. The general move towards the use of nanosized powders in order to decrease sintering times, allow control of grain growth and also to develop novel properties, places greater constraints on the processor due to the increased tendency for particles to agglomerate as the particle size is reduced. The use of colloidal consolidation techniques such as filter-pressing, centrifugal sedimentation and tape-casting will be required to provide the level of control needed to achieve high green densities and good homogeneity (4). For such systems electrostatic stabilization, which is often the most convenient, may pose difficulties due to the extension of the electrical double layer becoming commensurate with the particle radius. Where this takes place electroviscous effects and elasticity may be evident in the dispersions and may limit the solids loading that can be achieved both in the slip itself and in the green body produced from it.

The aim of this study was to investigate the extent to which the ratio of particle radius-to-double layer thickness, κa, influences the rheology of nano-dispersions of alumina with special emphasis on the development of elasticity. A subsequent study has extended this work to show how the interparticle forces that give rise to the viscoelastic behavior may affect the densification during filter-pressing (5).

EXPERIMENTAL

The powder required had to be of nano-size and of relatively well understood surface chemistry. It was decided that alumina would be the most appropriate material. Ultra-fine aluminas characteristically comprise metastable phases, thus the material chosen was a δ-alumina from Degussa, Al_2O_3 C, produced by flame hydrolysis of $AlCl_3$ (6). The fact that the phase transition to the α-modification inhibits sintering of the metastable forms was not a problem since the study was not designed to extend beyond the green consolidation stage. The BET surface area of the powder was 89 $m^2\ g^{-1}$ which corresponds to an average equivalent spherical particle radius of 10.7 nm. The size was confirmed by TEM examination. The powder tended to agglomerate strongly when exposed to the atmosphere due presumably to the effect of surface hydration. Therefore the powder was lightly calcined under vacuum to 300 °C and stored in a dry box. Powders from the box were immediately dispersed in the suspension fluid for characterization. All suspensions were ball-milled for at least 12 hours to break up agglomerates. The ionic strength was adjusted with KCl which was considered to be an ideal indifferent electrolyte since no sintering was to be attempted in this model investigation. Salt concentrations in the range from 0.001 to 2 mol dm^{-3} were investigated. Two different pH-values, namely 3.6 and 11.5, achieved by adding, respectively, HCl or NaOH were investigated. These values correspond to a negative and a positive surface potential, respectively.

Rheological measurements were made using a Carri-med controlled stress rheometer. This instrument allows stress controlled flow curves, constant stress creep curves and oscillatory stress and frequency sweeps to be determined. Both flow and oscillation data will be reported here. The oscillatory stress results and the flow curves on deflocculated dispersions were measured using a concentric cylinder arrangement, whilst the flow curves on flocculated dispersions were measured with cone and plate geometry.

RESULTS

The range of ionic strength investigated results in double layer extensions of > 10 to < 1 nm enabling values of κa from about 0.5 to > 15 to be explored. Clearly, at the lowest ionic strengths the double layer extension is commensurate with the particle radius itself. The flow data for all the dispersions exhibited shear-thinning character, but the extent of the decrease in viscosity with increasing shear stress varied markedly between the dispersions at ionic strength below 0.01 mol dm^{-3}, which is near to the critical coagulation concentration, and those at higher ionic strength which generally showed viscosities two orders of magnitude higher at all shear stresses.

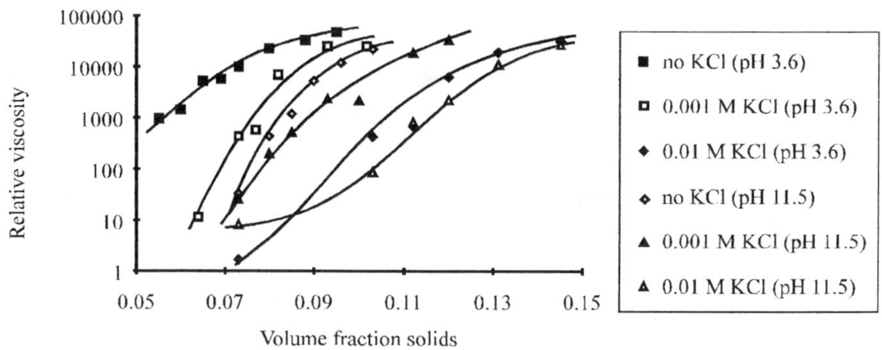

Fig. 1: Relative viscosities of deflocculated Al_2O_3 C slips of various ionic strengths measured at 2 s^{-1} against volume fraction solids.

Figure 1 shows the variation of relative viscosity, measured at 2 s^{-1}, with volume fraction of solids in the low ionic strength (i.e. deflocculated) region at both pH values. The trend for relative viscosities at higher shear rates was similar. The notable feature is that at any volume fraction the relative viscosity decreased with increasing ionic strength in this range. The effect is smaller at high pH, but still evident. This arises from the electroviscous effect for the extended electrical double layers. Essentially the effective volume fraction of particles is considerably larger than the true by an amount that varies with ionic strength. The relative viscosities of dispersions commonly increase exponentially with volume fraction solids as described by the Dougherty-Krieger equation (7)

$$\eta_r = \left(1 - \frac{\phi}{\phi_m}\right)^{-[\eta]\phi_m} \tag{1}$$

where η_r is the relative viscosity and ϕ the volume fraction of solids. $[\eta]$ and ϕ_m are the fitting parameters, intrinsic viscosity and maximum packing fraction, respectively. Fitting the data to this equation gave values for the exponent $[\eta]\phi_m$ of about 10 whereas ideally for non-interacting particles of equiaxed shape $[\eta]$ should approximate to the Einstein value of 2.5 and ϕ_m should be near to 0.6. The discrepancy is attributed to the volume of dispersion liquid excluded by the electrical double layers around the particles.

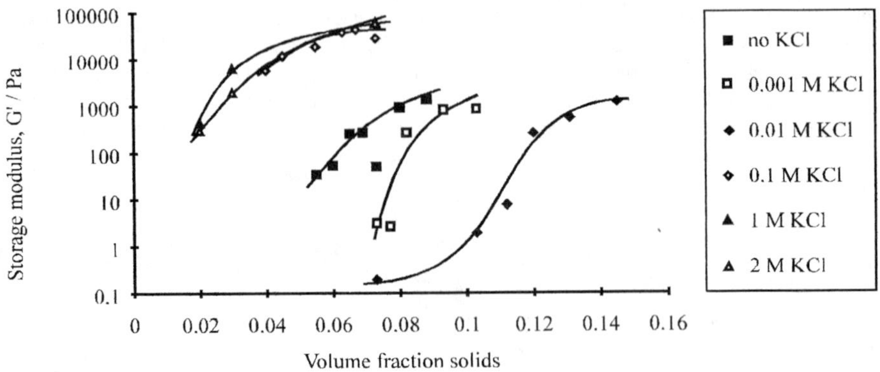

Fig. 2: Storage moduli of Al_2O_3 C slips of various ionic strengths at pH 3.6 against volume fraction solids.

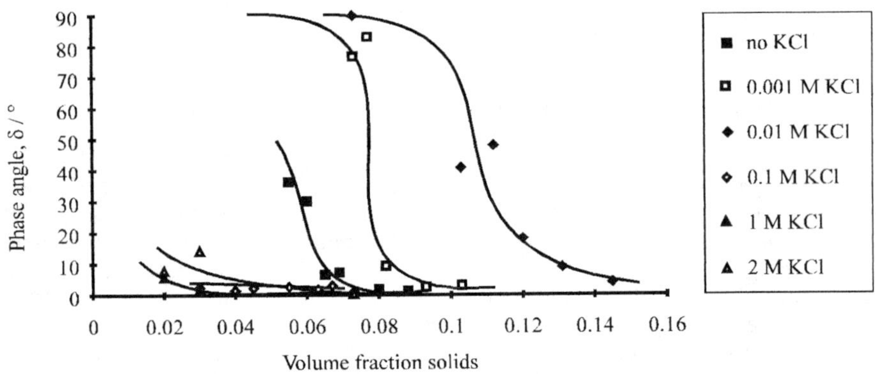

Fig. 3: Phase angle of Al_2O_3 C slips of various ionic strength at pH 3.6 against volume fraction solids.

Storage moduli, G', were followed as a function of frequency and solids concentration over the same range of pH and ionic strength. The flocculated samples at ionic strengths above 0.01 mol dm^{-3} showed little variation with frequency and were generally greater than 1000 Pa over the range studied. The deflocculated samples showed lower storage moduli which decreased with decreasing frequency. Figure 2 shows the high frequency storage modulus as a function of volume fraction of solids for the samples prepared in the acidic pH-range. The data fell into the two groups identified. The flocculated samples showed very similar behavior for all ionic strength levels. G' increased from about 100 Pa by 2 to 3 orders of magnitude over the range from 0.02 to 0.08 ϕ. The G' values for the deflocculated systems started at very low values of approximately 0.1 to 1 Pa but also increased rapidly over a short range of ϕ which shifted progressively to higher ϕ-values with increasing ionic strength up to 0.01 mol dm^{-3}. Thus, the deflocculated samples showed strong elastic effects which are different in character to those shown by the flocculated materials. The stronger frequency dependence indicated that the relaxation times were lower than for the flocculated systems.

The phase angle, δ, is a sensitive measure of the onset of elasticity in the samples. Figure 3 shows that δ was generally lower than 20° for the flocculated systems over the whole solids fractions range studied. The deflocculated dispersions, however, showed a marked drop in δ from 90° at low solids fractions to values below 10° as the solid concentration increased. This transition from viscous to elastic behavior was shifted progressively to higher ϕ-values as the double layer thickness decreased with increasing ionic strength.

DISCUSSION

Elasticity is displayed by all the dispersions studied here. Above the critical coagulation concentration of about 0.01 mol dm^{-3} it arises from the network of attractive links between particles that exists in a flocculated structure. The storage modulus is a function of the number density of links and the interparticle attractive potential. The fact that the dispersions all show similar behavior suggests that there is no great difference between the magnitudes of the interparticle attraction at the different salt concentrations. If there was a significant reduction in the attractive interparticle potential due to hydration interaction as suggested by Velamakanni et al. (8) then both storage moduli and viscosities for the highest salt concentrations should be significantly lower. Since this is not the case, there is no evidence for the existence of such effects with this nano-sized alumina powder.

The most interesting elastic effects are those associated with the deflocculated dispersions which arise from the interpenetration of the extended double layers. At the lowest ionic strengths studied, the value of κa is of the order of 1, effectively doubling the particle radius and thus increasing the effective volume fraction of particles by a factor of 8. The relative viscosity will be increased at all solids concentrations. However, above a critical value determined by the effective particle radius, the particles will be close enough together that the interpenetration of their electrical double layers means that they are effectively sitting in a potential energy well. The steepness of this well will determine the value of the high frequency storage modulus. As the solids concentration increases further the storage modulus will increase and the phase angle should drop sharply from a value close to 90 ° characteristic of a Newtonian fluid.

From the data presented there are two ways of assessing the effective particle radius, a_{eff}, from the relative viscosity data and from the phase angle data. If it is assumed that the Dougherty-Krieger equation describes the relative viscosity data and values assumed for the fitting parameters, then the effective volume fraction ϕ_{eff} can be estimated. The values of $[\eta]$ and ϕ_m were taken as 2.5 and 0.6 respectively. Values around 0.6 are commonly found to give a good fit for hard sphere systems (9). From the values of ϕ_{eff}, a_{eff} can be estimated using the simple expression

$$a_{eff} = a\left(\frac{\phi_{eff}}{\phi}\right)^{\frac{1}{3}} \tag{2}$$

The resulting curves are shown in Figure 4. The data for each ionic strength overlap over most of the range of volume fractions, but except for the pH 3.6 sample with no added electrolyte there is a characteristic volume fraction for each ionic strength below which the effective particle radius becomes a constant. This point in each curve is thought to indicate the onset of double layer overlap and the constant value at low ϕ gives an estimate of the effective particle radius comprising particle radius and the uncompressed double layer thickness. The decrease in a_{eff} at higher volume fractions would consequently be associated with the compression of double layers as neighboring particles are forced closer together.

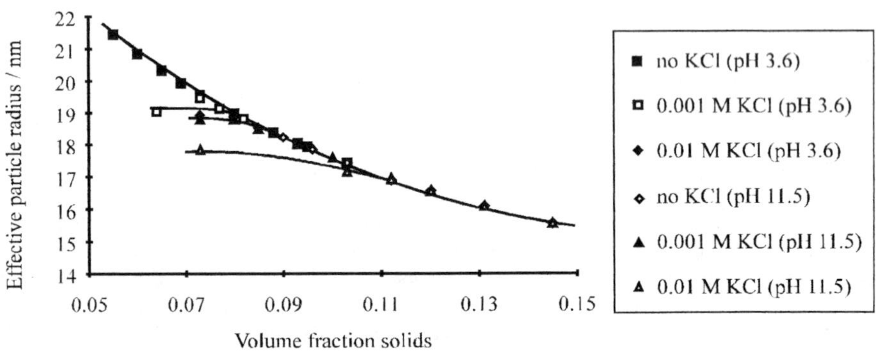

Fig. 4: Effective particle radius of electrostatically stabilized Al_2O_3 C particles calculated from relative viscosity values at 2 s^{-1} against volume fraction solids.

The onset of double layer overlap was also clearly evident in the viscoelastic properties of the suspensions. It is possible to plot the phase angle, δ, against particle surface-to-surface separation, h (Fig. 5) (10):

$$h = 2a\left(\left(\frac{\phi_m}{\phi}\right)^{\frac{1}{3}} - 1\right) \tag{3}$$

Assuming a bcc arrangement ϕ_m assumes a value of approximately 0.68. The surface-to-surface distance at $\delta = 35°$ was taken to calculate an independent value for the effective particle radius:

$$a_{eff} = a + \frac{h}{2} \tag{4}$$

Figure 6 shows the good agreement between the values derived from flow and oscillatory measurements.

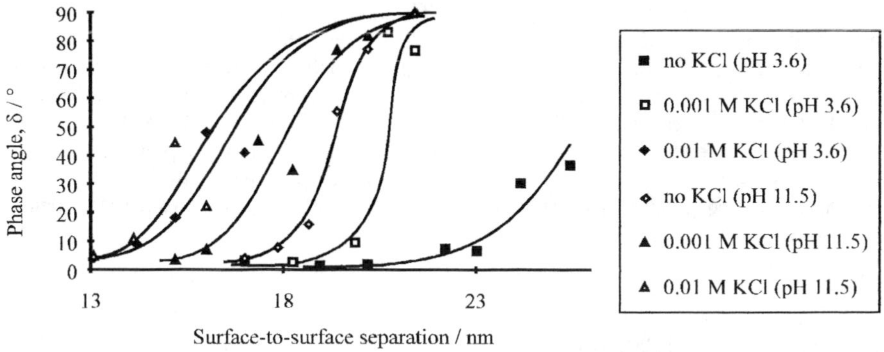

Fig. 5: Phase angle of Al_2O_3 C slips of various ionic strength against surface-to-surface separation.

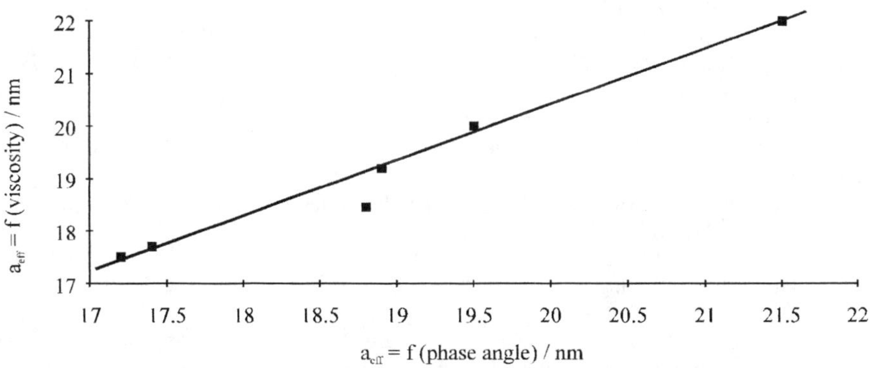

Fig. 6: Correlation between effective radius calculated from viscosity data and from phase angle data for electrostatically stabilized Al_2O_3 C slips.

CONCLUSIONS

It has been shown that rheological measurements in both flow and oscillatory mode are powerful tools in determining the structure in flocculated and deflocculated electrostatically stabilized ceramic nano-suspensions. Dilute deflocculated suspensions approximate to Newtonian liquids whereas both flocculated and concentrated deflocculated systems are characterized by elastic behavior. The onset of double layer overlap in electrostatically stabilized suspensions is accompanied by a liquid-solid phase transition which is evident in a sharp drop in phase angle. This phase transition is progressively shifted to higher solids volume fractions as the ratio of particle radius-to-double layer thickness, κa, increases with increasing ionic strength. The suspensions at rest tend to form crystalline structures where the particles effectively sit in a potential energy well and strongly oppose flow. The high frequency storage modulus gives an estimate of the magnitude of this interparticle repulsion due to interpenetrating double layers in the solid-like state. It was shown that the effective particle radius can be estimated from both the solids volume fraction at the phase transition and the relative viscosity using an approach based on the Dougherty-Krieger equation for hard spheres. The results show that the effective particle radius in 'soft' electrostatically stabilized suspensions, i.e. ones where the double layer thickness is significant with respect to particle size, is a function of both ionic strength and solids fraction. Above the critical coagulation concentration (~ 0.01 mol dm^{-3}) the suspensions flocculate and form a continuous attractive network characterized by high elasticity.

REFERENCES

(1) R. Fries and B. Rand, "Slip-Casting and Filter-Pressing", Chapter 7 in Materials Science and Technology Vol.17: Processing of Ceramics. Edited by R. Cahn. VCH, Weinheim, to be published in 1995.

(2) L. Bergström, "Rheology of Concentrated Suspensions", pp. 193-244 in Surfactant and Colloid Chemistry in Ceramic Processing. Edited by R.J. Pugh and L. Bergström. Marcel Dekker, Inc., New York, 1994.

(3) R. Buscall, J.W. Goodwin, M.W. Hawkins, R.H. Ottewill, Viscoelastic Properties of Concentrated Latices, J. Chem. Soc. Faraday Trans. I, 78, 2873-2899, 1982.

(4) F.F. Lange, Powder Processing Science and Technology for Increased Reliability, J. Am. Ceram. Soc., 72, 3-15, 1989.

(5) R. Fries and B. Rand, Filter-Pressing of Alumina Dispersions, in this Volume.

(6) G. W. Kriechbaum and P. Kleinschmidt, Superfine Oxide Powders - Flame Hydrolysis and Hydrothermal Synthesis, Adv. Mater., 10, 330-337, 1989.

(7) I.M. Krieger and T.J. Dougherty, A Mechanism for Non-Newtonian Flow in Suspensions of Rigid Spheres, Trans. Soc. Rheol., 3, 137-152, 1959.

(8) B.V. Velamakanni, J.C. Chang, F.F. Lange and D.S. Pearson, New Method for Efficient Colloidal Particle Packing via Modulation of Repulsive Lubricating Hydration Forces, Langmuir, 6, 1323-1325, 1990.

(9) C.G. de Kruif, E.M.F. van Iersel, A. Vrij and W.B. Russel, Hard Sphere Colloidal Dispersions: Viscosity as a Function of Shear Rate and Volume Fraction, J. Chem. Phys., 83, 4717-4725, 1986.

(10) Th.F. Tadros and A. Hopkinson, Use of Viscoelastic Measurements for Investigating the Stability/Flocculation of Concnetrated Dispersions, Faraday Discuss. Chem. Soc., 90, 41-55, 1990.

SUSPENSION PARAMETERS OF SiC_w/Al_2O_3 COMPOSITES

Melissa J. Crimp, Mingli Zhang, and Dean A. Oppermann
Department of Materials Science and Mechanics
Michigan State University
East Lansing, MI 48824-1226

ABSTRACT

Research [1,2] in powder ceramic processing has utilized nonaqueous or a mixture of nonaqueous/aqueous suspensions. Bleier [3] theoretically predicted the behavior of SiC powder in water and a variety of nonaqueous suspensions using a dispersibility model. However, a full understanding of the stability of this type of nonaqueous suspension is still not available. In this paper, the stability of SiC whiskers (SiC_w) and Al_2O_3 in ethanol or a mixture of ethanol and water has been evaluated. By varying the ratio of ethanol to water, suspension parameters such as zeta (ζ) potential, viscosity, stability, sedimentation density and microstructure were examined as a function of pH.

EXPERIMENTAL PROCEDURE

Materials

The α-Al_2O_3, type AKP-30, supplied by Sumitomo Chemical Co., has an average particle diameter, reported by the manufacturer, of 0.3μm. The ß-SiC whisker, type TWS-100, supplied by Tokai Carbon Co., has an average aspect ratio of 22.2 with a range of 8.3 to 50, having a whisker diameter of 0.3-0.6μm and a length of 5-15μm, as reported by the manufacturer.

Zeta Potential Measurement

Zeta potential measurements as a function of pH were collected for SiC_w and Al_2O_3 with 0.5vol% solid content and various ethanol/water ratios using a Matec ESA-8000 system. A 0.001N KNO_3 electrolyte concentration was used and the pH adjusted by automatic titration of either KOH or HNO_3.

Stability Calculation

From the zeta potential data versus pH data and the material's Hamaker constants, the stability ratios (W) for homo- and heterocoagulation, $W_{SiC\text{-}SiC}$ (and $W_{Al2O3\text{-}Al2O3}$) and $W_{SiC\text{-}Al2O3}$, respectively, were calculated using Stable Suspension©, a computer program developed by Wilson and Crimp [4,5]. The relative volume fraction of components was 0.5, the electrolyte concentration was 0.001N KNO_3, and the Hamaker constant was calculated from the dielectric constant [6] using a method utilized by Bleier [3].

Viscosity Measurement

Viscosity measurements of 10vol% SiC_w/Al_2O_3 suspensions at 40wt% solid content were made as a function of pH using a Brookfield Digital Viscometer, Model DV-II. The #3 spindle, 100 rpm, and a 0.001N KNO_3 electrolyte were used for all the viscosity measurements. Measurements were then repeated for suspensions of varying ethanol/water ratios.

Sedimentation Experiment

Sedimentation densities for Al_2O_3 and SiC_w, in pure water and 99% pure ethanol, were determined for 5-10vol% solid content suspensions in 0.001N KNO_3 electrolyte. The suspensions were ultrasonically dispersed for 8 minutes and the pH adjusted by addition of either acid (HNO_3) or base (KOH). The test tubes were then covered and the particles in suspension allowed to settle until the sedimentation height reached an equilibrium value. The sedimentation volumes were calculated from the sediment height and used to determine the sedimentation densities.

Slip Casting, Sintering, Microstructure

10vol% SiC_w/Al_2O_3 suspensions (67wt% solid content) for all ethanol/water ratios were ultrasonically dispersed for 10 minutes and slip cast. The green body microstructures were then gold coated and observed using SEM at 15-20 kV.

Green bodies were cold isostatically pressed at 345MPa and partially sintered, to lock in the green structure, at 1650-1700°C, under flowing Ar, for 1-2 hours. The microstructures of the partially sintered samples were then re-examined using SEM.

RESULTS AND DISCUSSION

Zeta Potential

The zeta potentials of SiC_w versus pH, for varying ratios of ethanol/water, are plotted in Figure 1. SiC, which has been shown to have an acidic surface characteristic [7], will absorb hydroxide ions creating a suspension with excess protons. Thus, in 100% water, the iso-electric point (i.e.p.) is in the acidic region of the zeta potential plot (pH=5.2). After increasing the ethanol content, which does not ionize as readily as water, the i.e.p. moved toward higher pH values (pH=7) and disappeared altogether in 99% ethanol suspensions.

The zeta potential for Al_2O_3 versus pH, at varying ratios of ethanol/water, is shown in Figure 2. Al_2O_3 has been shown to have a basic surface characteristic [7]. Al_2O_3 will therefore absorb protons to cause an excess hydroxide ion content. Thus, in 100% water, the i.e.p. is in the basic pH region (pH=8.8). As ethanol, which does not ionize to the same degree as water, is added, the i.e.p. decreases to a more acidic pH (pH=7.3) and, once again, disappears altogether in 99% ethanol suspensions.

The absolute value of the zeta potential for SiC_w and Al_2O_3, as shown in Figures 1 and 2, decreases as the ratio of ethanol to water increases. Since the dielectric constant (D) of ethanol is approximately 1/3 that of water, as the ethanol content increases, the suspension dielectric constant decreases. Therefore, the thickness of the double layer ($1/\kappa$) will decrease with increasing ethanol content. The

Viscosity measurements of 10vol% SiC_w/Al_2O_3 suspensions at 40wt% solid content were made as a function of pH using a Brookfield Digital Viscometer, Model DV-II. The #3 spindle, 100 rpm, and a 0.001N KNO_3 electrolyte were used for all the viscosity measurements. Measurements were then repeated for suspensions of varying ethanol/water ratios.

Sedimentation Experiment

Sedimentation densities for Al_2O_3 and SiC_w, in pure water and 99% pure ethanol, were determined for 5-10vol% solid content suspensions in 0.001N KNO_3 electrolyte. The suspensions were ultrasonically dispersed for 8 minutes and the pH adjusted by addition of either acid (HNO_3) or base (KOH). The test tubes were then covered and the particles in suspension allowed to settle until the sedimentation height reached an equilibrium value. The sedimentation volumes were calculated from the sediment height and used to determine the sedimentation densities.

Slip Casting, Sintering, Microstructure

10vol% SiC_w/Al_2O_3 suspensions (67wt% solid content) for all ethanol/water ratios were ultrasonically dispersed for 10 minutes and slip cast. The green body microstructures were then gold coated and observed using SEM at 15-20 kV.

Green bodies were cold isostatically pressed at 345MPa and partially sintered, to lock in the green structure, at 1650-1700°C, under flowing Ar, for 1-2 hours. The microstructures of the partially sintered samples were then re-examined using SEM.

RESULTS AND DISCUSSION

Zeta Potential

The zeta potentials of SiC_w versus pH, for varying ratios of ethanol/water, are plotted in Figure 1. SiC, which has been shown to have an acidic surface characteristic [7], will absorb hydroxide ions creating a suspension with excess protons. Thus, in 100% water, the iso-electric point (i.e.p.) is in the acidic region of the zeta potential plot (pH=5.2). After increasing the ethanol content, which does not ionize as readily as water, the i.e.p. moved toward higher pH values (pH=7) and disappeared altogether in 99% ethanol suspensions.

The zeta potential for Al_2O_3 versus pH, at varying ratios of ethanol/water, is shown in Figure 2. Al_2O_3 has been shown to have a basic surface characteristic [7]. Al_2O_3 will therefore absorb protons to cause an excess hydroxide ion content. Thus, in 100% water, the i.e.p. is in the basic pH region (pH=8.8). As ethanol, which does not ionize to the same degree as water, is added, the i.e.p. decreases to a more acidic pH (pH=7.3) and, once again, disappears altogether in 99% ethanol suspensions.

The absolute value of the zeta potential for SiC_w and Al_2O_3, as shown in Figures 1 and 2, decreases as the ratio of ethanol to water increases. Since the dielectric constant (D) of ethanol is approximately 1/3 that of water, as the ethanol content increases, the suspension dielectric constant decreases. Therefore, the thickness of the double layer ($1/\kappa$) will decrease with increasing ethanol content. The dependence of $(1/\kappa)$ on dielectric constant can be seen in Equation 1 [8].

$$\frac{1}{\kappa}=\left(\frac{DkT}{8\pi n_o z^2 e^2}\right)^{\frac{1}{2}} \tag{1}$$

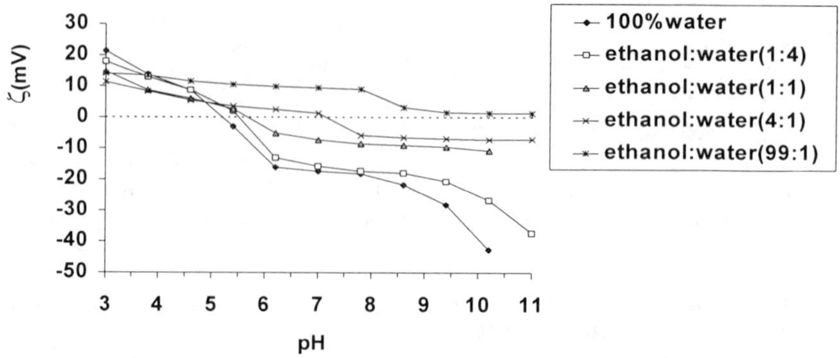

Figure 1. Zeta potential (ζ) vs. pH for SiC_w at a varying ratios of ethanol to water under 0.001N KNO_3 and 0.5vol% solid content.

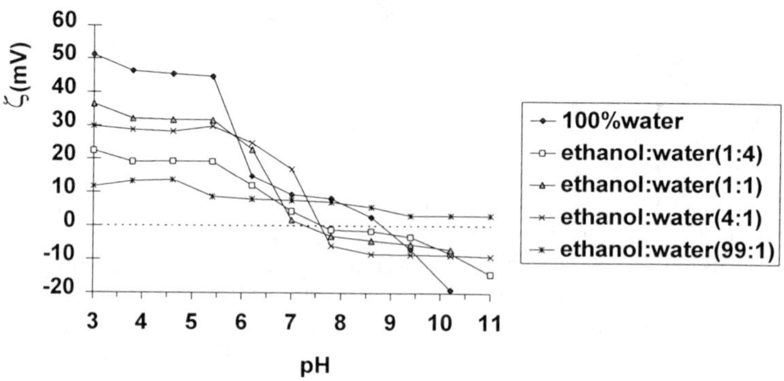

Figure 2. Zeta potential (ζ) vs. pH for Al_2O_3 for varying ratios of ethanol/water with 0.001N KNO_3 and 0.5vol% solid content.

DLVO theory predicts that the particle double layers provide a repulsive force against flocculation [9,10]. Therefore, the zeta potential decreases due to the compression of the double layer. From zeta potential measurements, SiC_w and Al_2O_3 are predicted to become increasingly more unstable as the ratio of ethanol to water increases, as predicted by DLVO theory.

Stability Ratio

Figures 3 to 5 give the stability ratios as a function of pH under constant electrolyte (0.001N KNO_3) concentration and Table I shows the stable pH ranges, as indicated by log(W)>10 [8], for all of the ethanol/water ratios. From Figures 3-5 and Table

I, it is seen that the stability against homocoagulation between SiC_w/SiC_w shifts from pH 3 to 5 and from pH 6 to 11 for pure water and from pH 3 to 8 for 99% ethanol. Stability against homocoagulation between Al_2O_3/Al_2O_3 shifts from pH 3 to 7 and from pH 10 to 11 for pure water and from pH 3 to 5 for 99% ethanol. Stability against heterocoagulation between SiC_w/Al_2O_3 is in regions of pH 3 to 6 and from pH 9 to 11 for pure water suspensions and from pH 3 to 5 for 99% ethanol suspensions.

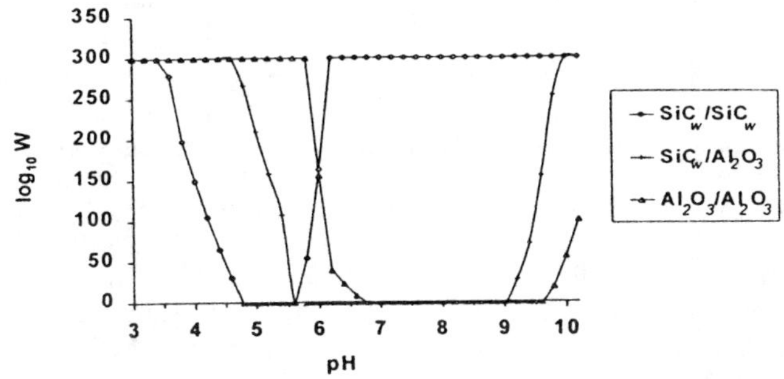

Figure 3: Stability ratio (W) vs. pH for SiC_w/Al_2O_3 at 0.001N KNO_3 of water.

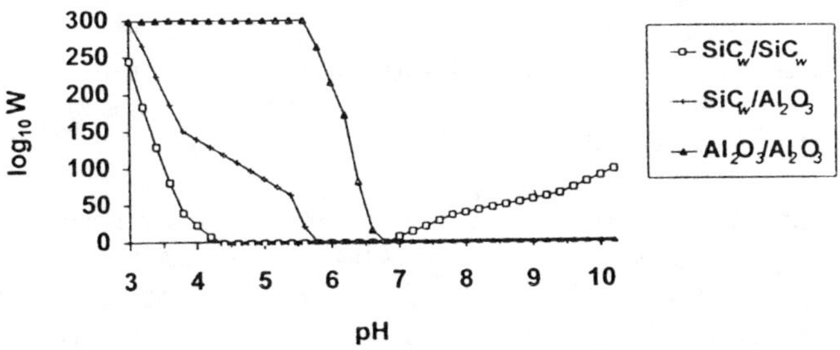

Figure 4: Stability ratio (W) vs. pH for SiC_w/Al_2O_3 at 0.001N KNO_3 of ethanol:water(1:1).

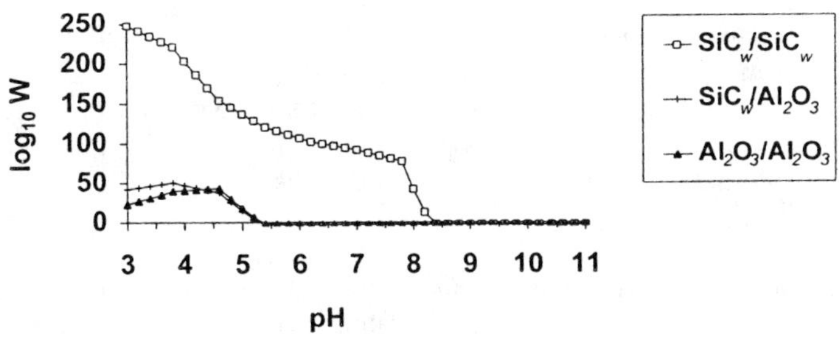

Figure 5: Stability ratio (W) vs. pH for SiC_w/Al_2O_3 at 0.001N KNO_3 of ethanol:water(99:1).

Table I. Stable pH Range

	SiC_w/SiC_w	SiC_w/Al_2O_3	Al_2O_3/Al_2O_3
100%water	3-5, 6-11	3-6,9-11	3-7,10-11
ethanol:water (1:4)	3-5, 6-11	3-5,10-11	3-6, 11
ethanol:water (1:1)	3-4, 7-11	3-6	3-7
ethanol:water (4:1)	3-4, 9-11	3-6	3-7
ethanol:water (99:1)	3-8	3-5	3-5

Analysis shows that the stability generally decreases as the ethanol content increases. At high pH ranges, the stability decreases to zero as the ethanol content increases. However, the stable range between SiC_w and SiC_w increases to pH<8 in the 99% ethanol suspension. The stable range for interactions between SiC_w and Al_2O_3, and Al_2O_3 and Al_2O_3, at 99% ethanol, remains at pH<5. Therefore, at pH<5, interactions are stable for SiC_w to SiC_w, SiC_w to Al_2O_3, and Al_2O_3 to Al_2O_3 in 99% ethanol.

In the case of SiC_w/SiC_w interactions, the stability range increases as the ethanol content is increased. This increased stability can be explained using the DLVO theory which accounts for the total energy of interaction between repulsive interaction energy and van der Waal's attractive energy as:

$$V_T=V_A+V_R\,. \tag{2}$$

Bleier predicted [3] the stability of SiC in ethanol and water respectively by calculation of V_A, and has shown that V_A is higher in ethanol compared to V_A in water. As stated previously, the repulsive energy, V_R, decreases as the ethanol content is increased due to the compression of the double layer. However, the stability results show an increase in the stability ratio at pHs<8 for SiC_w/SiC_w and, thus, an increasing V_T. This increase in V_T is attributed to an increased V_A, predicted by Bleier, that is substantially greater then the decrease in V_R that is evident from zeta potential results.

For SiC_w/Al_2O_3 and Al_2O_3/Al_2O_3 interactions, the stability range decreases as the ethanol content increases. Once again, the zeta potential results show a decrease in V_R with increasing ethanol content. Since V_T must also decrease, decreased stability, the effect of V_A must be less pronounced for the SiC_w/Al_2O_3 and Al_2O_3/Al_2O_3 interactions than in the SiC_w/SiC_w case.

Viscosity

Figure 6 shows the viscosity of 10vol% SiC_w/Al_2O_3 suspensions with 40 wt% solid loading at varying ratios of ethanol/water. For 100% water, the suspension viscosity is low at pHs<8 and pHs>11 and higher from pH 9 to 10. When the ethanol content increases, the viscosity increases dramatically at high pHs, but at pHs<4, the 10vol% SiC_w/Al_2O_3 suspensions have low viscosities even in 99%

ethanol. These measurements support the stability predictions, where high stability coincides with lower viscosities and low stability coincides with higher viscosities.

Sedimentation Bulk Density

Figure 7 shows a plot of sedimentation bulk densities as a function of pH for Al_2O_3 and SiC_w in water and ethanol suspensions. Comparing Al_2O_3 sedimentation densities in water and in ethanol, shows that the sedimentation density in water is higher than that in ethanol. This is consistent with the stability predictions where Al_2O_3 is more stable in water throughout the entire pH range. Comparing SiC_w sedimentation densities in water and in ethanol, shows that in ethanol there is a higher sedimentation density at lower pHs and a lower sedimentation density at higher pHs. This, again, is consistent with stability results where stable suspensions at low pHs have high sedimentation densities and unstable suspensions at high pHs have low sedimentation densities.

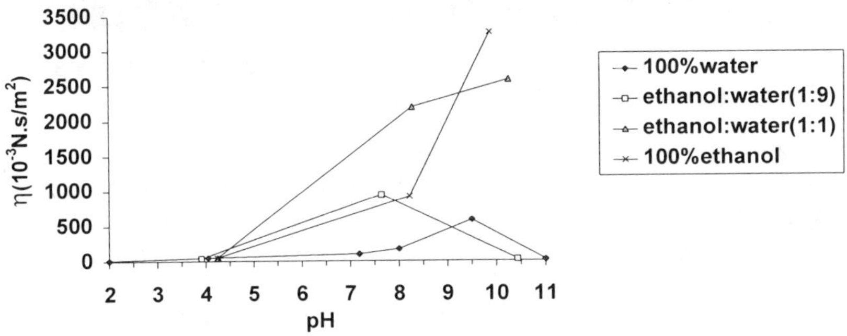

Figure 6. Viscosity (η) of 10vol% SiC_w/Al_2O_3 with 40wt% solid content at varying ratios of ethanol to water.

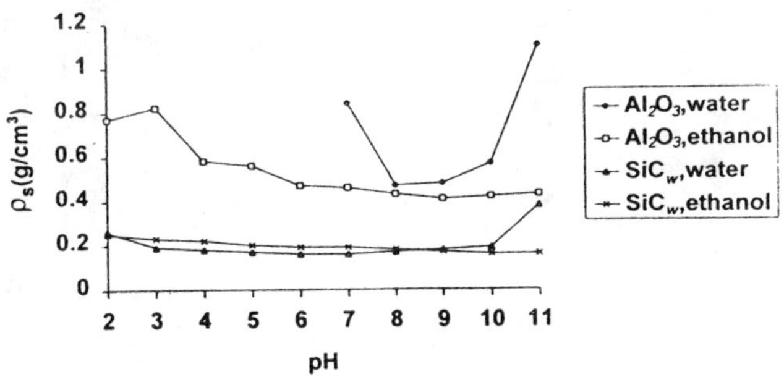

Figure 7. Sedimentation bulk density (ρ_S) vs. pH for Al_2O_3 and SiC_w in water and ethanol respectively (0.001N KNO_3).

Microstructures

The green body microstructure (shown in Figure 8) shows the SiC_w to be evenly distributed and unagglomerated in water, for mixtures of ethanol and water (1:1), and in ethanol. There is no obvious heterocoagulation between SiC_w and Al_2O_3 and only slight homocoagulation between Al_2O_3 particles at 100% ethanol is observed. This result is consistent with the stability predictions.

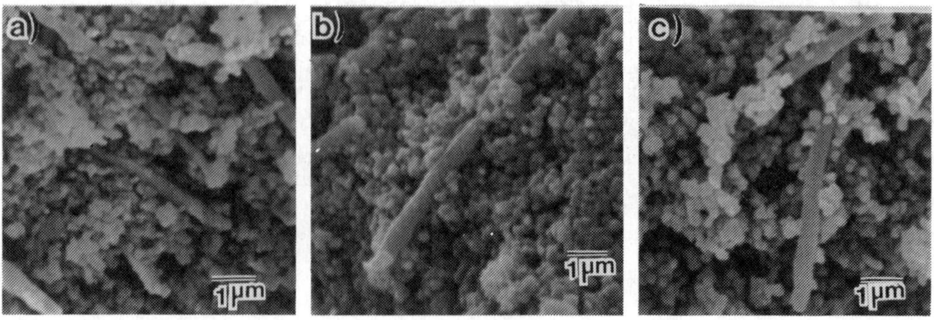

Figure 8. SEM micrographs of 10vol% SiC_w/Al_2O_3 green bodies at pH 4, slip cast in a) water, b) ethanol:water(1:1), and c) ethanol.

In Figure 9, the microstructures of partially sintered 10vol% SiC_w/Al_2O_3 samples show that SiC_w is well distributed within the partially sintered Al_2O_3 matrix. The SiC_w and Al_2O_3 are uniformly distributed throughout for suspensions in 100% water and ethanol:water(1:1).

Figure 9. SEM micrographs of 10vol% SiC_w/Al_2O_3 partially sintered samples processed in a) water, b) ethanol:water (1:1).

CONCLUSIONS

As the ethanol content increases, absolute values of the zeta potential for SiC_w and Al_2O_3 decrease. The i.e.p. shifts to more basic pHs for SiC_w and to more acidic pHs for Al_2O_3. Ethanol, or a mixture of ethanol to water, reduces the dielectric constant and compresses the double layer resulting in decreased zeta potentials. This led to a decrease in the stability for both materials, as verified by viscosity and sedimentation experiments. Green body microstructures revealed that the SiC_w and

Al_2O_3 were well dispersed in both water and ethanol:water(1:1) which supported the stability calculations. However, there was slight homo-coagulation of Al_2O_3 particles apparent in the 99% ethanol samples. After partial sintering, microstructural analysis again demonstrated good dispersion of SiC_w.

REFERENCES

1. R.E. Mistler, D.J. Shanefield and R.B. Runk, "Tape Casting of Ceramics", p.411-48, Ceramic Processing Before Firing Eds. G. Y. Onoda, Jr. and L. L. Hench, Wiley & Sons, New York, 1978.
2. F. Takao, W.R. Cannon, and S.C. Danforth, "Nonaqueous Colloidal Processing Reaction-Bonded Silicon Nitride-SiC Whisker Composites", Advances in Ceramics, Vol. 21, p. 699, (1987).
3. A. Bleier, "Fundamentals of Preparing Suspensions of Silicon and Related Ceramic Powders", J. Am. Ceram. Soc., 66, C79(1983).
4. B.A. Wilson and M.J. Crimp, "Prediction of Composite Colloidal Suspension Stability Based Upon the Hogg, Healy, Fuerstennau Interpretation", Langmuir, 9, p. 2836 (1993).
5. B.A. Wilson, M. S. Thesis, "Prediction of Colloidal Suspension Stability for SiC/Si_3N_4 and $FeAl/Al_2O_3$ Fiber Systems Using Material and System Parameters", Michigan State University, Results, p.65, (1992).
6. CRC Handbook of Chemistry and Physics; 70th ed., Eds. R.C. Weast, CRC Press INC., Boca Raton, Fl., 1989.
7. B.I. Lee and U. Paik, "Dispersion of Alumina and Silica Powders in Non-aqueous Media: Mixed-Solvent Effects", Ceramic International, 19, 214(1993).
8. R.J. Hunter, Foundations of Colloid Science, Vol 1, Clarendon Press, Oxford, 1987.
9. B.V. Deraguin and L.D. Landau "Theory of the Stability of Strongly Charged Lyophobic Sols and the Adhesion of Strongly Charged Particles in Solutions Of Electrolyte", Acta Physichim., URSS, 14, p.633 (1941).
10. E.J.W. Verwey and J.Th.G. Overbeek, Theory of the Stability of the Lyophobic Colloids, Elsevier, Amsterdam (1948).

RHEOLOGICAL CHARACTERIZATION OF TEMPERATURE EFFECTS IN PARTICLE-FILLED POLYMER FLUIDS

J. A. Horn and B. R. Patterson
Department of Materials Science and Engineering
J. Zhou
Department of Chemistry
The University of Alabama at Birmingham
Birmingham, AL 35294

ABSTRACT

Rheological properties of ceramic powder - polymer suspensions have been measured. Steady shear and oscillatory properties were found to depend strongly upon temperature as well as the state of oxidation of the polymers. It was observed that the suspensions in oxidized polymers underwent a reversible transition from essentially Newtonian to shear-thinning behavior when heated. When most of these suspensions were heated, it was found that the low-shear rate viscosities increased and phase angles at low oscillatory strain decreased.

INTRODUCTION

The rheological behavior of certain suspensions of ceramic powders in polymer melts is found to have a strong temperature dependence. Suspensions of oxide powders in oxidized polybutene, at solids loadings of 30 to 40 vol%, can be changed reversibly between states of viscous, nearly Newtonian behavior and pseudoplastic behavior. This paper presents some of the data and observations regarding the phenomenon and possible theoretical explanations.

MATERIALS

The materials used in this study are listed in Table I. With the exception of the polyisoprene liquid, the polymers and ceramic powders used are commercially available products. The polyisoprene liquid was made at the University of Alabama at Birmingham (1). The properties listed in Table I are taken from published literature for the commercial products.

The term "oxidized" is used in this paper to describe the condition of polymer liquids that have been exposed to air, either for long periods of time (years) at room temperature, or for short periods of time (days) at 100°C. Oxidation was recognized by a color change in the polymer from transparent (when reduced) to yellow (2(a)). In addition, FTIR spectroscopy (Fig. 1) revealed an increase in adsorption at wavelengths near 1735 cm^{-1} which likely indicates the existence of C=O bonds. The extent of oxidation has not been quantified yet, and the presence of other species introduced by "aging" the polymer has not been ruled out.

Oxidation of the polybutene did not cause any significant change in its molecular weight distribution. A comparison of GPC results for reduced versus oxidized H35 polybutenes (in terms of both M_n and M_w) showed the two samples differed by only 1%. The rheology of the polymers (unfilled) was also unaffected by oxidation.

Table I. Materials used in this study

Ceramic Powder or Polymer Liquid	Powder Size or Polymer MW	Supplier and Tradename
TiO_2	~3μm	Alfa-Aesar
Al_2O_3	~1.5μm	Alcoa, XA1520
Al_2O_3	~3.5μm	Alcoa, XA4000
ZrO_2	~1μm	Magnesium Elektron, SC105
Oxidized Polybutene	M_n~420	Amoco, L50
Reduced Polybutene	M_n~660	Amoco, H35
Oxidized Polybutene	M_n~660	Amoco, H35
Oxidized Polybutene	M_n~965	Amoco, H100
Reduced Polyisoprene	M_w~2500	ref. (1)
Oxidized Polyisoprene	M_w~2500	ref. (1)

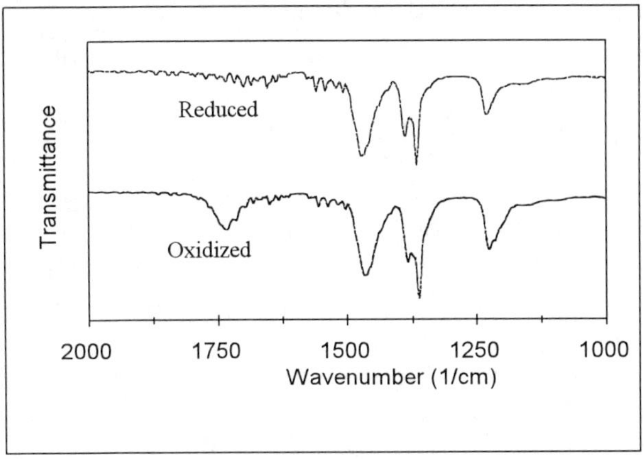

Figure 1. FTIR spectra of reduced and oxidized polybutene (Amoco H35).

EXPERIMENTAL PROCEDURES

All suspensions were prepared by hand-mixing ceramic powders into polymer liquids. After mixing, samples were allowed to sit overnight before they were re-stirred and tested.

Rheology measurements were made on a Rheometrics RFSII controlled-strain rheometer, fitted with parallel plates. On each sample, tests were run in the following order: frequency sweep at 1% strain; strain sweep at 10rad/s; steady shear rate sweep, beginning at the highest rate and sweeping down. Temperature was increased after each set of tests and the tests run again.

RESULTS

Amoco polybutenes were used as liquid media for powder suspension studies because of their near-Newtonian behavior (Fig. 2(a&b)) and low chemical reactivity (2(b)). The linear relationship between shear stress and shear strain rate (invariant viscosity) indicates Newtonian behavior; and phase angles near $\delta = 90°$ for a range of frequencies and 1%strain confirm that there is no association among the flow units (3). These polymers had been used previously in systems that served as models of metal injection molding feedstock (4). Those suspensions of coarse metal powders (11μm to 55μm diameter spheres) were observed to be nearly Newtonian at room temperature, which was the only temperature used in those studies. The same polybutenes were used again in the present study of ceramic powder suspensions (1-3μm diameter).

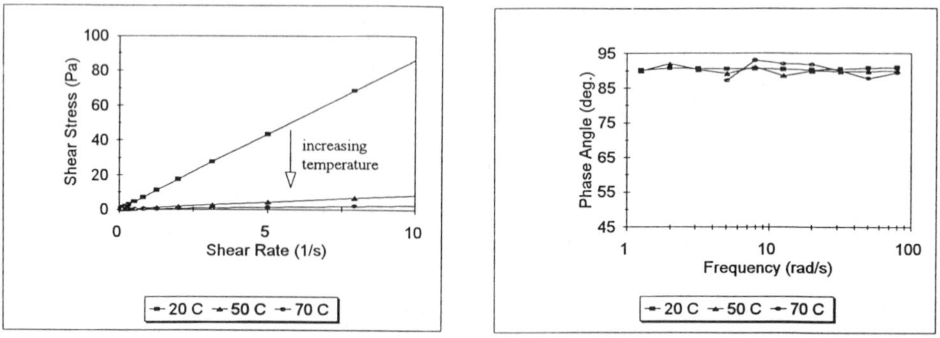

(a) (b)

Figure 2. Rheological data for unfilled polybutene liquid (Amoco H35) at 20°C, 50°C and 70°C.

A suspension of TiO_2 (30vol%) was prepared in a polybutene liquid (Amoco H35) left from previous studies (in storage for approximately 7-10 years). The suspension was Newtonian at 20°C but shear thinning with a yield point at 50°C. Figure 3(a) shows the change in rheological behavior under steady shear conditions. Figure 3(b) shows the increased elasticity (decreasing phase angles, δ) with increasing temperature for oscillatory shear at 1%strain. Phase angles of $45° < \delta < 90°$ indicate weak to very weak association among the flow units, and angles $0° < \delta < 45°$ indicate strong association (3). Returning to 20°C restored the original Newtonian behavior.

It should be noted that the initial heating of the TiO_2 suspension (either in a closed container over boiling water, or in an open container in a recirculating air furnace) caused a structural change that was easily observed macroscopically. The suspension at room temperature was fluid and pourable, while the heated suspension became stiff, resembling an aqueous system coagulated with a high concentration of salt. Upon subsequent cooling the suspension maintained a rigid appearance (i.e. it maintained a shape independent of the shape of the container) until it was stirred, under which action it became fluid again.

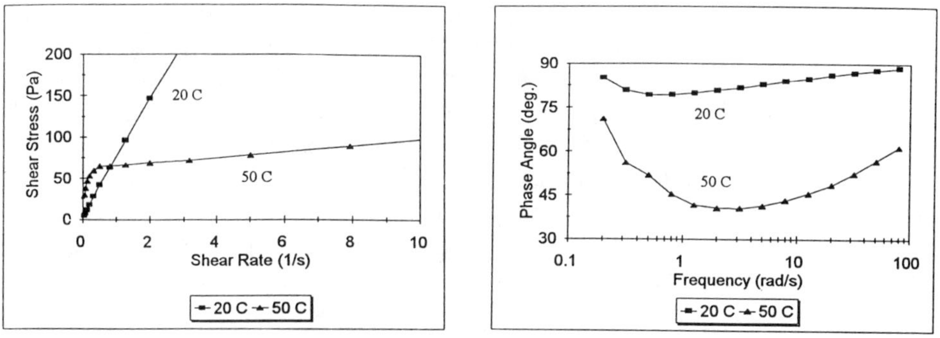

(a) (b)

Figure 3. Rheological data for 30vol% TiO_2 suspension in polybutene (Amoco H35) at 20°C and 50°C..

The rheological data above demonstrate a transformation in the suspension from a state of stable dispersion to a state of association among the flow units. At low shear rates ($\dot{\gamma} < 1s^{-1}$) the viscosity of this suspension was higher at the higher temperature. The increase in viscosity was in contrast with the behavior of the polymer liquid alone which decreased with the same temperature increase (Fig. 2).

Repeated heating and cooling, or extended time at high temperatures, caused the high temperature viscosities to decrease.The degree of shear thinning (the slope of the viscosity-shear rate curve) decreased with time, especially at high rates of shear. The suspensions, however, remained non-Newtonian at low shear rates when heated, and the phase angles at 1% oscillatory strain remained low.

The same general behavior that occurred with the TiO_2 suspensions also occurred with Al_2O_3 and ZrO_2 powders suspended in oxidized polybutene (Amoco H35). The absolute values of the viscosities and dynamic parameters differed, but the change from Newtonian to non-Newtonian flow upon heating and the accompanying decrease in oscillatory phase angle were qualitatively similar.

Table II contains a few of the rheological parameters for suspensions of several types of ceramic powders in oxidized H35 polybutene. The polybutene used in these suspensions was prepared by placing new, reduced polymer in a recirculating air oven at 100°C for one week. Each phase angle reported here was measured at approximately 1 rad/s during a sweep of oscillatory frequencies (0.1 - 100 rad/s at 1%strain) at the given temperature. The last column gives the temperature range within which the low-shear-rate viscosity of each suspension began to increase with increasing temperature. The torque generated by the unfilled polymer was below the range of accurate detection at 70°C. The ZrO_2 suspension changed from nearly Newtonian to shear thinning with increasing temperature, but the viscosity at low shear rates consistently decreased.

Table II. Rheological Properties of 30vol% Powder Suspensions in Oxidized H35 Polybutene

Powder	Phase angle @ 1rad/s, 1%strain (for three temperatures)			Temperature range of positive $\Delta\eta(T)$
	20°C	50°C	70°C	
Unfilled	90°	90°	-	-
3µm TiO_2	85°	73°	42°	20-50°C
3.5µm Al_2O_3	90°	88°	85°	60-80°C
1.5µm Al_2O_3	88°	59°	31°	5-10°C
1µm ZrO_2	84°	63°	44°	-

Suspensions of 30vol% TiO_2 (3µm) were prepared in each of the six polymer liquids. While all six of the liquids are similar in nature, they differ slightly according to at least one of the following: chain structure (polyisoprene or polybutene), molecular weight, and extent of oxidation (reduced or oxidized). Table III contains phase angles for these suspensions. Note that the behavior of the suspension in reduced polybutene did not change in a consistent manner with changes in temperature. The suspension in reduced polyisoprene produced torque beyond the range of the rheometer transducer.

Table III. Rheological Properties of 30vol% TiO_2 Suspensions in Various Polymer Liquids

Polymer	Phase angle @ 1rad/s, 1%strain (for three temperatures)		
	20°C	50°C	70°C
Reduced Polybutene, H35	36°	50°	45°
Oxidized Polybutene, H35	85°	73°	42°
Oxidized Polybutene, L50	53°	31°	30°
Oxidized Polybutene, H100	88°	82°	63°
Reduced Polyisoprene	-	-	-
Oxidized Polyisoprene	53°	36°	20°

DISCUSSION

It is likely that the near-Newtonian behavior of the room temperature suspensions is the result of dispersion stabilization imparted by polybutene molecules adsorbed onto the particle surfaces. Such stabilization might be described as "elastic steric" stabilization of the type reported by Smitham and Napper (5,6). In elastic steric stabilization, flocculation is prevented by the physical interaction of adsorbed molecules in the "interpenetration plus compression" range of interparticle separation. The bulk polymer liquid may be thought of as a solvent for the adsorbed chains, in which case the bulk polymer should behave approximately as a theta-solvent (5,6). Israelachvili and Kott have reported the apparent "pinning" of polybutadiene (chemically similar to the polymers in the present study) onto mica surfaces in a polymer melt, creating a short range repulsion between approaching surfaces (7).

Since the suspensions in the reduced polymer liquids exhibited decidedly non-Newtonian behavior at all temperatures used, it is apparent that chemically-altered groups in the oxidized polymer enhanced adsorption onto the powders. The increased adsorption may have resulted from polymer-surface interactions of an acid-base type (8). The ceramic powders used in this study have not yet been ranked according to acidity, but if the oxidized polymers indeed contained carbonyl (C= O) bonds of a basic nature, it is likely that the solid surfaces were behaving as acids and bonding to the polymers.

Smitham and Napper (6) reported a maximum particle size that can be stabilized in a polymer melt of a certain molecular weight. In an analogous manner, the present data indicate that, for one particle size (3μm TiO_2) there is a minimum molecular weight polymer required for stabilization. The suspension prepared with polybutenes H35 and H100 (M_n ~ 660 and 965) behaved more nearly Newtonian and less elastic than the suspension prepared with the L50 polybutene (M_n ~ 420). The minimum chain length needed for the stabilization of these particles corresponds to a molecular weight somewhere that of the L50 and H35 liquids.

The reason for the "thickening" at higher temperatures is not obvious. When suspensions are stabilized by polymers in a solvent, flocculation is known to occur if the thermodynamic conditions become such that the solvent is no longer a good one for the polymer (9). In the present case, however, since the "solvent" molecules are identical to the stabilizing molecules, no changes in solubility is expected.

Since the neat polymer liquids did not become non-Newtonian with temperature change, it is apparent that this thickening behavior stems from interactions involving the suspended particles (either the particles themselves or the polymer molecules adsorbed to them). It appears that the suspension of the finer Al_2O_3 powder was less stable than that of the coarser powder. The phase angles for the finer Al_2O_3 suspension were consistently lower, and the viscosity increase with heating occurred at lower temperatures. If van der Waals forces, causing interparticle association, were responsible for the thickening effect, it might be expected that larger particles would be less stable against flocculation. As mentioned above, Smitham and Napper (6) found that there was an upper limit to the size of particles that could be stabilized in a polymer melt of a certain molecular weight. It is therefore likely that the "thickening" seen here is due to the behavior of the adsorbed polymer molecules.

ACKNOWLEDGEMENTS

This research was supported by the Alabama DOE/EPSCoR Program.

For the donation of experimental materials, the authors express their appreciation to Alcoa (alumina powders) and Amoco Chemicals (polybutenes).

REFERENCES

1. J. Mays and J. Zhou, Department of Chemistry, The University of Alabama at Birmingham.
2. (a) Amoco Chemical Company, Research and Development Department (private communication).
   (b) Amoco Chemical Company, Amoco Polybutene, Bulletin 12-N, 1994.
3. C. L. Rohn, "Rheological Characterization of Coatings for Fabrics and Fibers", J. Coated Fabrics, **19**(1), 181-92 (1990).
4. M. K. Agarwala, B. R. Patterson, and P. E. Clark, "Rheological Behavior of Powder Injection Molding Model Slurries", J. Rheology, **36**(2), 319-34 (1992).
5. D. H. Napper, Chapter 14, Polymeric Stabilization of Colloidal Dispersions, Academic Press, Inc., San Diego, 1983.
6. J. B. Smitham and D. H. Napper, "Elastic Steric Stabilization in Polymer Melts", JCIS, **54**(3), 467-70 (1976).
7. J. N. Israelachvili and S. J. Kott, "Liquid Structuring at Solid Interfaces as Probed by Direct Force Measurements: The Transition From Simple to Complex Liquids and Polymer Fluids", J. Chem. Phys., **88**(11), 7162-66 (1988).
8. F. M. Fowkes and M. A. Mostafa, "Acid-Base Interactions in Polymer Adsorption", Ind. Eng. Chem. Prod. Res. Dev., **17**(1), 3-7 (1978).
9. D. H. Napper, Chapter 5, Polymeric Stabilization of Colloidal Dispersions, Academic Press, Inc., San Diego, 1983.

USING POLYMER BLENDS TO FINE-TUNE BINDER PERFORMANCE

David W. Whitman, Xun Tang, Donald I. Cumbers and Scott A. Ibbitson
Rohm and Haas Company, Spring House, PA 19477

ABSTRACT

The glass transition temperature (T_g) of polymeric binders strongly affects ease of forming and mechanical properties of green ceramic parts. The T_g of commercially available binders spans a wide range of temperatures, but the ideal T_g for an application may not be readily available. In this study we demonstrate how blending of polymers with different T_gs can be used to fine tune binder performance in dry pressing and tape casting.

INTRODUCTION

Polymeric binders are used to enhance green strength in ceramic formulations. Different ceramic forming methods have very different binder requirements, which are compounded by differences in powders and equipment available at each manufacturing location. Although a wide range of binders is commercially available, it may not be possible to find a binder with the exact balance of properties needed for a particular application.

Although many polymer properties affect binder performance, glass transition temperature (T_g) is the key polymer variable that must be optimized for each application. If the T_g is too low the binder is very soft and strength is reduced. If the T_g is too high, the binder becomes brittle and difficult to deform.

Plasticizers have been added to formulations for tape casting to increase tape flexibility and enhance lamination[1,2,3] and to dry pressing formulations to facilitate compaction.[4,5] Polyethylene glycol has been added to polyvinyl alcohol-based formulations for dry pressing, to reduce granule hardness.[6] Ambient humidity has been found to plasticize polyvinyl alcohol and reduce its effective T_g.[7,8,9,10,11]

In this study, we report results of blending acrylic latex polymers of similar compositions but different T_gs to adjust binder performance. Examples are presented in the use of acrylic latex blends in dry pressing and tape casting.

EXPERIMENTAL PROCEDURE

Materials

A proprietary alumina powder used for all dry pressing experiments. Particle size was 2.9 μm (Coulter LS-130).

The powder used for tape casting was Alcoa A-16SG alumina. Surface area was 9.7 m^2/g (nitrogen BET) and mean particle size was 0.5 μm (Coulter LS-130).

The binders used for studies of neat polymer and for dry pressing were Duramax™ B-1001 (T_g = -6 °C) and Duramax™ B-1007 (T_g = 43 °C) ceramic binders. The binders used for tape casting were Duramax™ B-1000 (T_g = -26 °C) and Duramax™ B-1035 (T_g = -40 °C) ceramic binders. All the binders were emulsion polymers from Rohm and Haas Company.

The dispersant used was Duramax™ D-3005 dispersant, a low molecular weight ammonium polycarboxylate solution polymer from Rohm and Haas Company.

Preparation of Polymer Blend Films

Films of polymer blends were prepared by mixing varying amounts of the two binders, then drying at 60 °C in 9-cm diameter plastic Petri dishes. The dishes were coated with a mold release agent (Crown #6075N Dry Film Lubricant) prior to use.

Evaluation of Polymer Blend Films

Differential scanning calorimetry (TA model 2920) was used to observe glass transition(s) in the dried binder samples. Samples were annealed at 150 °C prior to measurement. Experiments were run from -80 °C to 150 °C at 20 °C per minute.

Tensile properties of the films were measured by recording stress-strain curves on a universal testing machine (Instron Model 1122) with an automated data collection system (Instron Series IX version 1.01). Standard tensile bar samples were cut with a die, giving a 0.64 cm tensile width and a 1.91 cm gauge length. Sample thickness was measured to ± 3 μm using a dial micrometer (Ames model 56212.5). Samples were tested using a crosshead speed of 5 cm per minute. Reported values are averages of at least 5 determinations.

Preparation of Dry Pressed Alumina Samples

Slurries were prepared by charging an 000 jar mill that was half-filled with grinding cylinders with alumina (63.0 g), dispersant (0.3% on alumina) and enough deionized water to bring slurry solids to 65 wt. %. The mixture was milled at 120 rpm for 40 minutes, then decanted. Binder blends were added at a total binder level of 2.7% on alumina. The mixtures were stirred mechanically for 10 minutes, then spray dried (Büchi model 190). Granule size was about 10 μm.

Samples were pressed in a 12.7-mm diameter cylindrical uniaxial die, mounted in a laboratory press (Carver model C). The die was cleaned, polished with jeweler's rouge, and lubricated with a solution of 2% stearic acid in acetone prior to pressing each sample. Samples were pressed at 100 MPa for 15 seconds. Sample weight was 1.0 g, giving pellets with final diameter to height ratio of approximately 4.

Evaluation of Dry Pressed Alumina Samples

Green density was calculated from the weight and dimensions of a minimum of 4 pellets. Pellet dimensions were measured with a digital caliper accurate to ± 0.01 mm. Pellet weight measurements were accurate to ± 0.0001 g and were corrected for organic content of the sample.

Green strength was measured by diametral compression in an automatic loading device (Soiltest G-900 Versaloader). Samples were compressed at a loading rate of 0.13 mm per minute. Strength was calculated using Eq. (1), where σ is the maximum tensile stress or strength, p is the applied load at fracture, D is the sample diameter and L is the sample length.[12]

$$\sigma = \frac{2p}{\pi DL} \tag{1}$$

Preparation of Tape Cast Alumina Samples

Slurries were prepared by charging an 0 jar mill that was half-filled with grinding cylinders with alumina (1200 g), dispersant (0.3% on alumina) and enough deionized water to bring the slurry solids to 82 wt. %. The mixture was milled at 80-100 rpm for 3 hours, then decanted. Binder blends were added at a total binder level of 12 wt. % on alumina. The mixtures were stirred mechanically for 30 minutes, then filtered through a 100 mesh sieve.

Green tapes were cast with a doctor blade onto Mylar sheets pre-coated with a thin layer of ammonium stearate. Casting speed was 12 cm per minute and wet thickness was 250 µm. The wet tapes were dried at 150 °C in a forced air oven for 20 minutes. Dry thickness was 130-150 µm depending on formulation.

Evaluation of Alumina Green Tape Samples

The tensile properties of the green tapes were measured by the method of Cannon, et al.[1] Stress-strain curves were recorded on a universal testing machine (Instron Model 1122) with an automated data collection system (Instron Series IX version 1.01). Standard tensile bar samples were cut with a die, giving a 0.64 cm tensile width and a 1.91 cm gauge length. Sample thickness was measured to ± 3 µm using a dial micrometer (Ames model 56212.5). Samples were tested using a crosshead speed of 0.25 cm per minute. Air-actuated grips with constant pressure were used to eliminate tearing at the grip section. Data was discarded from runs where tensile bars fractured on either ends of the narrow tensile section or fractured with apparent tearing. Reported values are averages of at least 10 determinations.

RESULTS AND DISCUSSION

Polymer Films

Four different polymers were used as binders in this study. Dried films of these polymers were analyzed using differential scanning calorimetry to determine their respective T_gs. Results are summarized in table 1.

Table I: T_g of polymers studied

Binder	T_g (°C)
Duramax™ B-1035	-40
Duramax™ B-1000	-26
Duramax™ B-1001	-6
Duramax™ B-1007	43

Figure 1 shows DSC traces for binders with T_g = -6 and 43 °C, and a 1:1 blend of the two binders. Two distinct glass transitions are seen in the blended sample, corresponding to transitions seen in the pure samples. The polymers in the blend maintained their microscopic chemical identity even when intimately mixed.

Bulk mechanical properties of films prepared from blends of the T_g = -6 and 43 °C binders were examined by tensile testing. Figures 2 and 3 show tensile moduli and breaking strengths of the films.

Tensile modulus measures the stiffness of a polymer; binders with lower moduli deform more readily. As shown in Figure 2, tensile modulus varied smoothly as the proportions of the blend were changed. Increasing the amount of high T_g binder increased the modulus of the blend.

Strength also varied smoothly as blend proportions changed, increasing as high T_g binder was added. Bulk properties like modulus and strength reflected weighted averages of the properties of the blend components.

The observed smooth variation in bulk properties will not necessarily hold for all polymer blends. The polymers tested are mutually compatible, and their molecular composition is very similar. Insufficiently similar polymers may show non-linear behavior, and incompatible blends may show discontinuous properties due to phase separation.

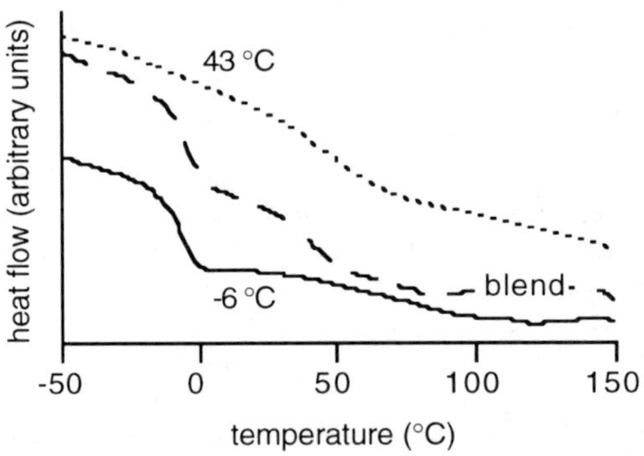

Figure 1: DSC results for binders with T_g = -6 and 43 °C, and a 1:1 blend.

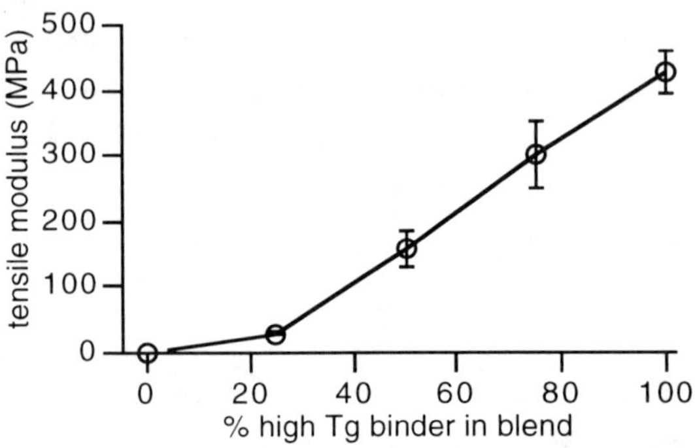

Figure 2: Tensile moduli of films of binder blends with T_g = -6 and 43 °C.

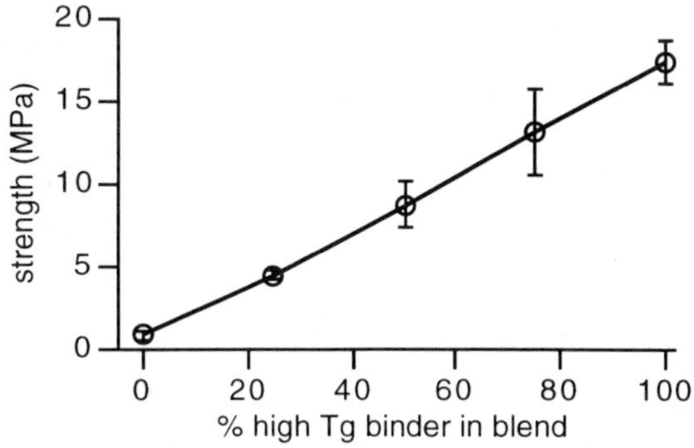

Figure 3: Breaking strengths of films of binder blends with T_g = -6 and 43 °C.

Dry Pressing

Alumina samples spray dried with blends of T_g = -6 and 43 °C binders were pressed at 100 MPa. All samples were prepared with 2.7% binder. Green density and green strength of the samples are reported in Figures 4 and 5.

Alumina samples dry pressed with the high T_g binder gave lower green density and higher green strength than those with the low T_g binder. These observations are consistent with previous studies that varied T_g by plasticizer addition[4, 5, 6] and humidity.[7, 8, 9, 10, 11]

Blends of the two binders gave intermediate dry pressing performance. The observed smooth variation of green strength and density in the blends can be understood in terms of mechanical properties of the binder blends. Blends with high levels of low T_g binder have low tensile moduli, which result in easily deformable granules. However, blends rich in low T_g binder have low tensile strengths, which is reflected in the reduced green strength observed in dry pressed samples.

In using binder blends to fine-tune performance, formulators need to strike a balance between maximizing green strength and green density. One good strategy is to set a target green density and use a blend as rich in high T_g binder as will allow achieving this density on the available pressing equipment. The smooth variation in properties observed in these blends means that only a few experiments are needed to determine a suitable mixture.

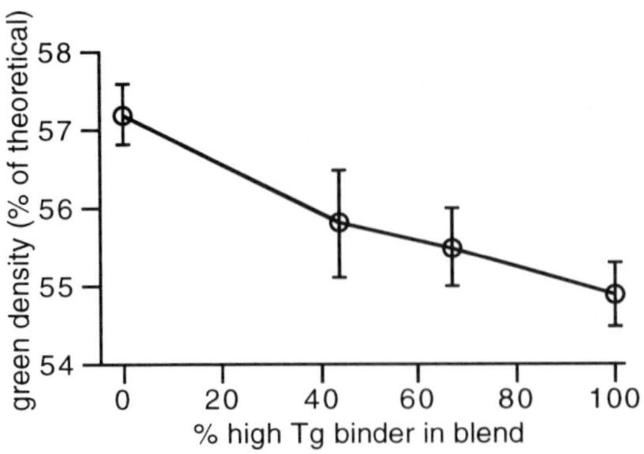

Figure 4: Green densities of alumina samples dry pressed with blends of binders with T_g = -6 and 43 °C.

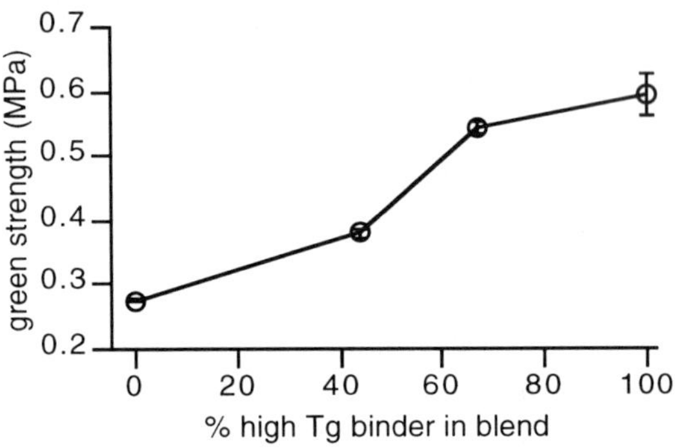

Figure 5: Green strengths of alumina samples dry pressed with blends of binders with T_g = -6 and 43 °C.

Tape Casting

Alumina samples were tape cast with blends of binders with T_g = -40 and -26 °C. All samples had a binder level of 12%, and no plasticizer was used. Figures 6 and 7 show the results of tensile testing of the cast tapes.

As in the dry pressed samples, green strength of cast tapes increased with added high T_g binder.

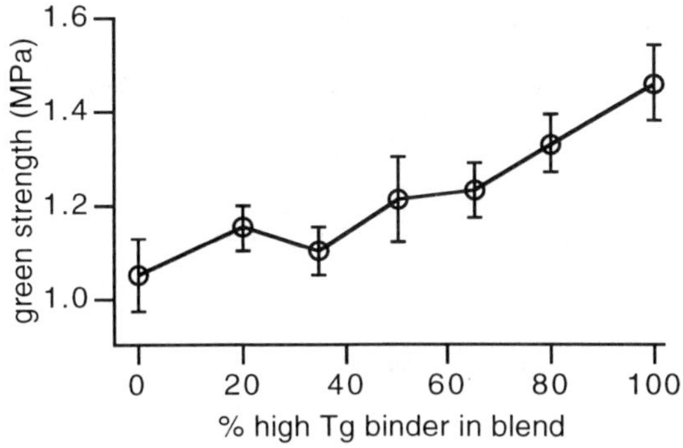

Figure 6: Tensile strength of unfired alumina tapes cast with blends of binders with T_g = -40 °C and -26 °C.

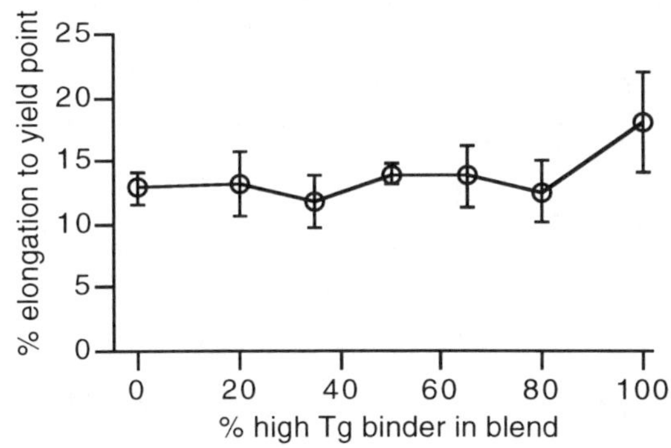

Figure 7: Yield point of unfired alumina tapes cast with blends of binders with T_g = -40 °C and -26 °C.

As shown in Figure 7, proportions of the two binders had little effect on the strain tapes could withstand before yielding. This observation means that strength could be increased without rendering the tape prone to irreversible deformation due to stresses in processing. This result is probably not general for all polymer blends. In particular, for blends of polymers having T_gs that are respectively above and below room temperature, decreases in yield point are likely with added high T_g polymer due to the abrupt property changes associated with the glass transition.

To fine-tune binder performance for tape casting, formulators need to trade off green strength with flexibility and (if required) ease of lamination. Increasing the proportion of high T_g binder will increase strength, but may increase tape stiffness and may require higher lamination pressures.

CONCLUSIONS

Blends of polymers retain distinct glass transitions of the individual blend components. However, for compatible polymers, many macroscopic properties such as tensile strength and modulus vary smoothly with the relative proportions of high and low T_g polymers.

This observed smooth variation in properties can be used to fine-tune polymer performance when polymer blends are used as binders for ceramic processing. This capability has been demonstrated for dry pressing and tape casting, and is expected to be general for other forming techniques.

Green strength can be increased by adding high T_g polymer to a formulation.

Deformability can be increased by adding low T_g polymer to a formulation. Increased deformability leads to higher density in dry pressing, and to enhanced tape flexibility.

REFERENCES

1 A. Karas, T. Kumagai and R. Cannon, "Casting Behavior and Tensile Strength of cast $BaTiO_3$ Tape," Adv. *Ceram. Mat.* **3** [4], 347-77 (1988).

2 E.P. Hyatt, "Making Thin, Flat Ceramics - A Review," *Am. Ceram. Soc. Bull.*, **65** [4] 637-8 (1986).

3 J.B. Blum and W.R. Cannon, "Tape Casting of $BaTiO_3$," *Mat. Res. Soc. Symp Proc.* **40**, 77-81 (1985).

4 D.C.C. Lam, "Effects of Binder and Compaction Pressure on Strength and Fracture Origins in Bodies Pressed from Granules," *J. Ceram. Soc. Japan*, **102** [11], 1010-1015 (1994).

5 B. Nyberg, E. Carlström, M. Persson and R. Carlsson, "Uniform Distribution of a Pressing Aid," Ceramic Processing Science, Proc. of 2nd International Conference, Berchtesgoden, Germany (1988).

6 C.W. Nies and G.L. Messing, "Effect of Glass Transition Temperature of Polyethylene Glycol-Plasticized Polyvinyl Alcohol on Granule Compaction," *J. Am. Ceram. Soc.*, **67** [4], 301-304 (1984).

7 R.A DiMilia and J.S. Reed, "Dependence of Compaction on the Glass Transition Temperature of the Binder Phase," *Am. Ceram. Soc. Bull.*, **62** [4] 484-488 (1983).

8 R.A. DiMilia and J.S. Reed, "Effect of Humidity on the Pressing Characteristics of Spray-Dried Alumina," in Advances in Ceramics, Vol. 9, 38-46 (1984).

9 R.G. Frey, J.W. Halloran, "Compaction Behavior of Spray-Dried Alumina," *J. Am. Ceram. Soc.*, **67** [3], 199-203 (1984).

10 J.A. Brewer, R.H. Moore and J.S. Reed, "Effect of Relative Humidity on the Compaction of Barium Titanate and Manganese Zinc Ferrite Agglomerates Containing Polyvinyl Alcohol," *Am. Ceram. Soc. Bull.*, **60** [2] 212-220 (1981).

11 D.W. Whitman, D.I. Cumbers and X.K. Wu, "Humidity Sensitivity of Dry Press Binders," *Am. Ceram. Soc. Bull.*, (in press).

12 A. Rudnick, A.R. Hunter, and F.C. Holden, "An Analysis of the Diametrical Compression Test," *Mater. Res. Stand.*, **3** [4], 283-89 (1963).

Duramax is a trademark of the Rohm and Haas Company.

Powder Processing—Forming and Consolidation

A COMPARISON OF PRESSURE COMPACTION AND DIAMETRAL COMPRESSION TESTS FOR DETERMINING GRANULE STRENGTHS

S. Jill Glass and Clay Newton, Sandia National Labs, Albuquerque, NM, 87185

ABSTRACT

Pressure compaction (P-C) tests were used to characterize the strength of ceramic granules. When compaction data are plotted as relative density of the compact versus the compaction pressure two linear regions are generally observed. The intersection of these regions, which is known as the "breakpoint," has been used as a semi-quantitative measure of granule strength. Comparisons were made between the P-C breakpoint and strengths of 150-200 μm diameter ZnO, TiO_2 (rutile), and lead magnesium niobate - lead titanate (PMN-PT) granules, where the strengths were determined by diametral compression (D-C) tests. At high compaction pressures the compliance of the die itself is significant and was accounted for in the analyses. Tests were conducted at different compaction rates, and with different aspect ratio compacts. High aspect ratios and loading rates decrease the slope of the second linear portion of the compaction curve and produce higher apparent P-C breakpoints. Comparison of the P-C breakpoint to the average D-C strength indicates that the D-C strength is approximately fifty percent higher for PMN-PT granules. To eliminate the uncertainty in the results due to irregular granules sizes and shapes, comparisons were also made for uniform size (210 μm) glass spheres. In this case the average D-C strength coincided with a second breakpoint in the P-C data, which occurred after compaction by a mechanism of bridge formation and collapse had ceased.

INTRODUCTION

Lightning strikes can cause structural damage, ignite flammable materials, and produce circuit malfunctions in missiles, aircraft, and ground systems. Lightning arrestor connectors (LACs) are used to divert harmful lightning energy away from these systems by providing less destructive breakdown paths. Ceramic granules in the size range of 150-200 μm are used in LACs to provide physical and electrical separation of contacts (pins) from the surrounding metal web, and to control the voltage breakdown level. ZnO varistor materials, TiO_2, and lead magnesium niobate - lead titanate (PMN-PT) have been used as the granule material. LACs can be subjected to vibrational stresses that may exceed the granules' fracture strength and lead to a degradation in LAC performance; therefore granule strength is a critical material parameter. The two tests selected to determine granule fracture strength were a pressure compaction (P-C) test,[1-3] and a diametral compression (D-C) test for individual granules.[4-6] The objectives of the study were to determine how to obtain reproducible results from pressure compaction tests, and to determine the relationship between the P-C breakpoint and the D-C strength. Tests were conducted on ZnO, rutile, and PMN-PT granules. For the PMN-PT granules, tests were conducted on materials fired at temperatures between 800 and 1200°C to correlate granule strength with processing history.

LITERATURE REVIEW

Compaction Testing

Pressure compaction tests have been widely used to characterize the strength of powders, agglomerates, granules, and aggregates.[2,3] The compaction test can be used as a quality control tool to determine if powders and granules display acceptable compaction behavior for subsequent forming by dry pressing or isostatic pressing. Bruch was one of the first to use compaction curves to interpret ceramic consolidation behavior.[7] The compaction process depends on the deformation properties of the powders or granules, compaction parameters such as loading rate, die design, and the dimensions and shape of the compacted body. Much effort has been devoted to defining general compaction equations, studying density and pressure distributions in dies,[8] and characterizing the compaction behavior of particulate materials.[9] Mahoney and Readey review compaction models in another paper in this volume.[10]

Compaction Mechanisms

Compaction of ceramic granules occurs in three general stages: 1) granule rearrangement and sliding without fracture; 2) local deformation and fracture of granules; and 3) sliding and rearrangement of particles or fracture fragments within granules.[11] Although the compaction mechanisms have been described as distinct stages, there are components of each during the entire compaction process.[12] When load-deflection data from a compaction test are plotted as relative density vs. log compaction pressure, the data often exhibit distinct linear regimes as shown schematically in Fig. 1. Walker (1923) first observed this behavior for both metallic and non-metallic powders.[13] In many compaction models, including that of Reed and Runk,[1] regions 1) and 2) are distinguished from each other by a parameter known as the breakpoint. Before the breakpoint there is very little density increase with pressure as the granules undergo rearrangement without fracture. The second linear region represents granule rearrangement in conjunction with granule fracture. The breakpoint is a semi-quantitative indicator of the fracture stress of the granule, and the different linear regions of the compaction diagram are thought to represent the pressure regimes where different compaction mechanisms predominate.

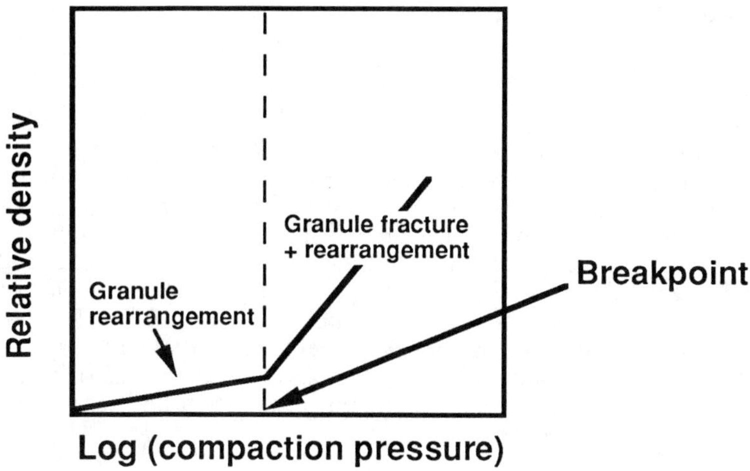

Fig. 1. Semilogarithmic compaction curve exhibiting characteristic linear regions separated by the breakpoint.

Experiments

To obtain meaningful data from compaction tests, they must be conducted under standardized conditions and experimental errors due to contributions from the deformation of the testing equipment must be minimized. Data from tests where these experimental errors have not been accounted for or minimized often exhibit nonsensical behavior on compaction plots, making it impossible to identify material characteristics and compaction mechanisms. Matsumoto pointed out that errors can occur due to contributions from the load-frame, die, compact, and load cell, but in his tests the only significant contribution was from the load cell.[14] Baconegra showed that die compliance can also be significant.[15] Experimental approaches for minimizing such errors have been described by Mort et al.[16]

Diametral Compression Strength Measurements

Granule and agglomerate strengths have also been measured in diametral compression tests, where single granules are loaded in compression between platens until fracture occurs. The granule tensile strength, σ_t, is calculated from the fracture load, P_{max}, and the granule diameter, d, using Hiramatsu's and Oka's equation:[6]

$$\sigma_t = \frac{2.9\ P_{max}}{\pi d^2} \qquad (1)$$

Equation 1 has been used in numerous studies.[4,17,18]

EXPERIMENTAL PROCEDURE

Pressure Compaction (P-C) Tests

A cylindrical die with a 3.2 mm diameter opening was used for pressure compaction tests. Tests were conducted at ambient temperature and humidity with no die lubrication. Unlike typical spray-dried ceramic granules, all of the granules in this study had been fired to sufficiently high temperatures to remove all organic materials. All had also undergone some degree of densification during firing. Three or more samples were tested from each batch of granules. Nominally 0.06 g of PMN-PT granules were compacted to final aspect ratios, L_f/D, of 0.36 to 0.42 (where L_f=final compact height and D=compact diameter). For the rutile tests three masses were used that corresponded to final aspect ratios of $L_f/D=0.69$ for 0.06 g, $L_f/D=1.18$ for 0.10 g, and $L_f/D=2.58$ for 0.20 g. The die was filled using a funnel or creased weighing paper, and no attempt was made to increase the tap density by shaking or vibrating the die.

P-C tests were conducted by loading the filled die in a mechanical testing load-frame. The crosshead was moved incrementally to make contact with the top plunger of the die until a positive load of approximately one to two pounds was indicated. The displacement gauge was then zeroed and the crosshead was activated at a rate of 0.127 mm/min. The test was continued until a compressive load of approximately 422 MPa (60 ksi) was reached. Tests on rutile granules were also conducted at a rate of 0.635 mm/min. Load cell and displacement gauge indications were collected one to two times per second for the P-C tests.

Errors due to the deformation of the loadframe and load cell were eliminated by measuring the displacement of the top plunger relative to the die body. To account for the error due to the deformation of the top plunger of the fixture, the displacement of the empty fixture as a function of load was subtracted from the displacement of the granule-filled die. Displacement data were converted to relative densities using granule mass, initial compact height, measured displacement (corrected for empty die displacement), and theoretical density. Relative density and compaction pressure were plotted as relative density vs. log of the compaction pressure. When displayed in this manner, the data exhibit the expected linear regions. Best-fit lines were drawn through the data in the two linear regions and the pressure at the intersection of the best-fit lines was recorded as the breakpoint.

Diametral Compression Tests

To measure granule strength individual granules were compressed between platens on a mechanical testing load-frame. The granule diameter, d (assumed to be the height of the granule as it was positioned between the platens), was measured using a microscope, camera, and video monitor. The granule was fractured by lowering the upper platen at 0.0508 mm/min or 0.127 mm/min. Twenty or more granules of each material were fractured. Granule tensile strength was calculated using Eq. 1.

RESULTS AND DISCUSSION

Pressure Compaction Results

Figure 2 shows a typical plot of how compact relative density changes with increasing compaction pressure for ZnO granules. As expected, two linear regions, corresponding to Stages 1 and 2 in powder compaction are observed. The results also show that die plunger displacement can contribute to additional deformation at high compaction pressures, which can lead to erroneous interpretation of the data. For example, the uncorrected data indicate an erroneously high relative density. At the highest compaction loads the die plunger deformation accounts for approximately 9% of the total measured displacement of the plunger for a filled die. The uncorrected data also incorrectly indicate a lower apparent breakpoint, and more granule fracture and rearrangement after the breakpoint. Thus the correction for die plunger deformation is necessary in order to make valid measurements of how changes in the granule properties themselves affect the compaction behavior and parameters such as the breakpoint.

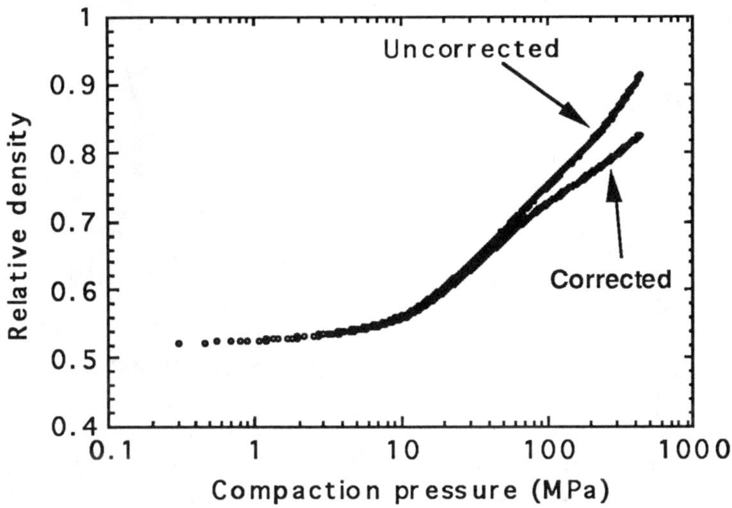

Fig. 2. A typical compaction curve for ZnO granules. Comparison of uncorrected data to data that accounts for deformation of the fixture shows that die plunger deformation at high pressures represents ~9% of the total displacement. This error produces erroneously high densities and a lower apparent breakpoint.

The effect of aspect ratio on compaction behavior for rutile granules is shown in Fig. 3. Tests conducted at $L_f/D = 0.7$ and 1.2 gave similar results; however, $L_f/D = 2.6$ gave a lower final density, a smaller slope in the second linear region of the compaction curve, and a higher apparent breakpoint. Die wall friction effects, which lead to stress gradients in powder and granular compacts, and which are known to be more extreme in higher aspect ratio compacts,[19] are the source of this effect. The decreased slope is a clear indication that a significant number of the granules are being subjected to stresses less than the applied stress. In the presence of significant stress gradients, higher applied pressures must be exerted before many of the granules in the compact are subjected to their failure stress. Because both of the lower L_f/D values produced similar curves, it appears that these values are below the limit necessary to ensure uniform stress transmission in the compact for these types of material.

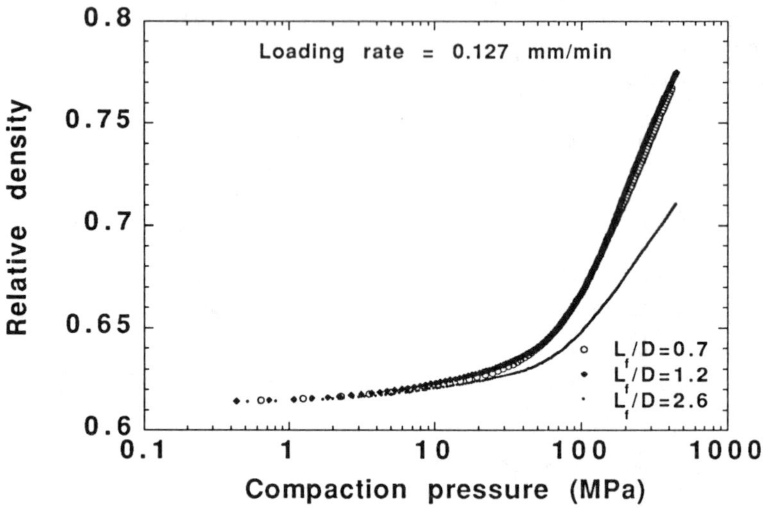

Fig. 3. Effect of aspect ratio, L_f/D on the compaction behavior of rutile granules. For $L_f/D \leq 1.2$ similar compaction curves are obtained. At $L_f/D = 2.6$ the final density is lower, and the slope of the second linear region is smaller, producing a higher apparent breakpoint.

The effect of loading rate on the compaction behavior of rutile granules is shown in Fig. 4. Although the loading rates used in this study are very slow relative to those used in commercial pressing operations, there is a noticeable effect of the higher loading rate. The increased loading rate decreases the slope of the second linear region of the compaction curve, increases the apparent breakpoint, and produces a lower final density. The effect is likely due to the fact that the higher loading rate provides less time for sliding and rearrangement of granule fragments following fracture. Two previous studies showed that higher loading rates (0.508 to 5.08 mm/min,[9] and 0.25 and 5.0 mm/min,[2]) retard the onset of compaction by granule fracture and rearrangement, but produced a higher slope once this compaction mechanism began. The higher slope is opposite to the effect that was observed in the present study; however, in both of these studies the loading rate effect was associated with viscous deformation of powder-binder systems. The granules in the present study contain no binder.

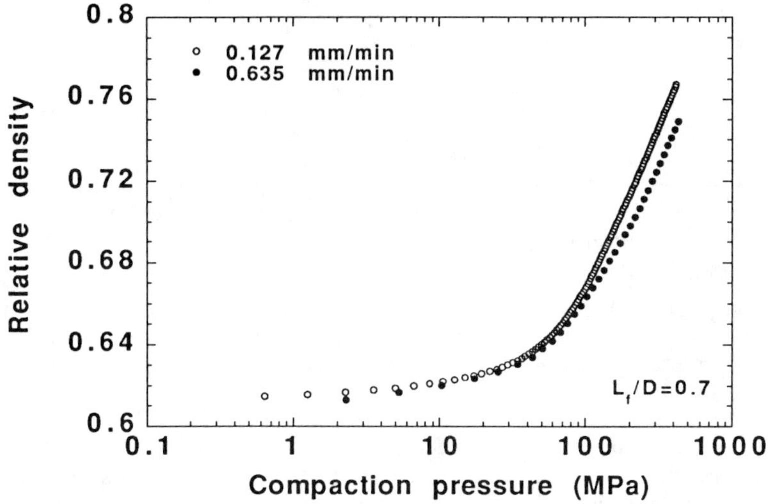

Fig. 4. Effect of loading rate on the compaction behavior of fired rutile granules. The higher loading rate results in a lower final density and the slope of the second linear region is smaller, resulting in a higher apparent breakpoint.

Compaction tests on PMN-PT granules fired at temperatures between 800 and 1200°C were conducted at the slower loading rate of 0.127 mm/min, with masses that produced aspect ratios less than $L_f/D = 0.5$. The objective of this part of the study was to identify the firing temperature that produced granules with the highest strength for use in the lightning arrestor connector application. The compaction behavior as a function of firing temperature is shown in Fig. 5. As the firing temperature increases the initial density increases, the breakpoint initially increases and then decreases, and the final density generally increases. The breakpoint increase between 800 and 1000°C, which is interpreted as an increase in granule strength, is likely related to increased granule density. At temperatures greater than 1000°C further density increases are limited; however, grain growth within the granules becomes significant leading to a decline in the breakpoints and strength.

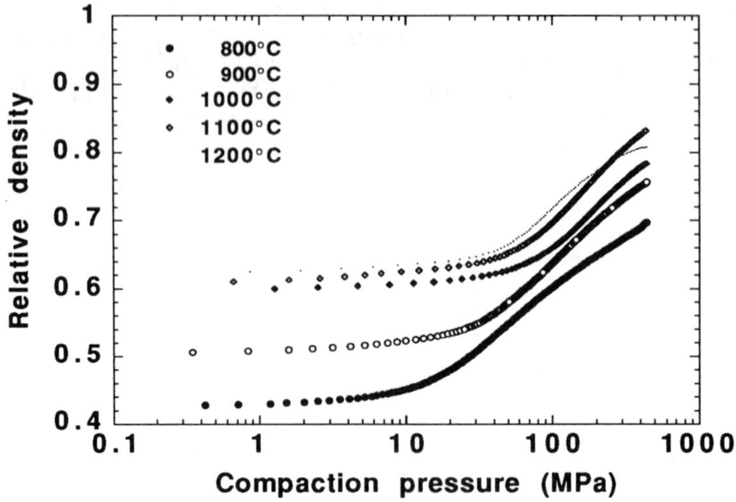

Fig. 5. Compaction behavior of lead magnesium niobate - lead titanate (PMN-PT) granules as a function of firing temperature. Initial density increases as firing temperature increases and breakpoint increases to a maximum at 1000°C.

Compaction results are shown in Fig. 6 for the uniform size 210 µm glass spheres. The final aspect ratio, L_f/D, was 0.68. This compaction data exhibits three linear regions with two breakpoints, the first at 86 MPa (12.4 ksi) and the second at 171 MPa (24.9 ksi). Examination of the data between the two breakpoints reveals that jumps in density are accompanied by significant load drops, consistent with major rearrangements. One compaction test was terminated at approximately 138 MPa (20 ksi). The presence of broken glass spheres indicated that the density increases are due to a combination of sphere fracture and rearrangement, and not just due to major rearrangements.

Diametral Compression Results

The average D-C strength data for some of the PMN-PT granules are shown in Fig. 7 along with the P-C breakpoints for the same materials. The D-C strengths are on average ~50% higher than the breakpoints. Similar results were obtained for rutile granules and PMN-PT granules of different compositions. These results are in contrast to those of Song and Evans who found that the D-C strength of 20-40 µm spray-dried ZrO_2 agglomerates was ~52% of the breakpoint.[18] Although we have not used the approach of Song and Evans, whereby the area fraction of agglomerates that transmits the force through the assembly is accounted for, this correction would lower our breakpoint values and move them farther from our D-C strengths. The reason the results of this study are in contrast to those of Song and Evans is presently unknown; however, it seems logical that the breakpoint should be lower than the average D-C strength because the breakpoint is the stress near the onset of granule fracture. Hence a better comparison would be made between the P-C breakpoint and the lowest D-C strengths, rather than the average D-C strength.

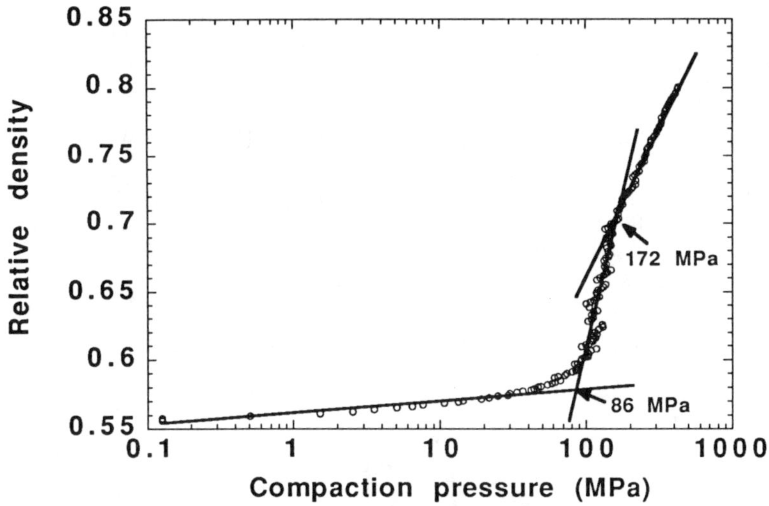

Fig. 6. Compaction behavior of uniform 210 μm diameter glass spheres. Three linear regions and two breakpoints are observed.

The D-C strength data for PMN-PT granules are shown in Table I along with the breakpoints from Fig. 6. Due to the inherent brittle nature of ceramics, the range of granule sizes, and the irregular shape of the PMN-PT granules (with the consequent difficulty in determining a granule diameter), there is considerable scatter in the D-C data; however, the lowest values in the D-C strength distribution correspond more closely to the breakpoints than the average D-C strengths do.

To eliminate the uncertainty associated with the irregularity of the PMN-PT granules, D-C strengths were also measured for uniform size (210 μm) glass spheres and compared to their P-C breakpoint. The glass spheres also have a uniform spherical shape, which minimizes stress concentrations that can lead to lower apparent strengths for irregularly shaped granules. The glass sphere D-C strength data are also shown in Table I.

Table I. Diametral compression strength data for PMN-PT granules and uniform size glass spheres.

Material	Avg. Strength (MPa)	Range (MPa) Min-Max	Breakpoint (MPa)	
PMN-PT, 900°C	47 ± 16	23 - 80	24	
PMN-PT, 1000°C	94 ± 35	32 - 181	56	
PMN-PT, 1100°C	60 ± 19	16 - 102	50	
PMN-PT, 1200°C	46 ± 16	13 - 79	34	
			1st	2nd
Glass (210 μm)	379 ± 166	127 - 584	86	172

The glass sphere average D-C strength of 379 MPa (55 ksi) greatly exceeds both breakpoint strengths; however, it appears that the lowest D-C strengths in the distribution correspond most closely to the second P-C breakpoint of 172 MPa (24.9 ksi). Because spheres are breaking at compaction stresses between the first and second breakpoints, at stresses less than the lowest strength spheres, some appear to sustain stresses much higher than the applied stress.

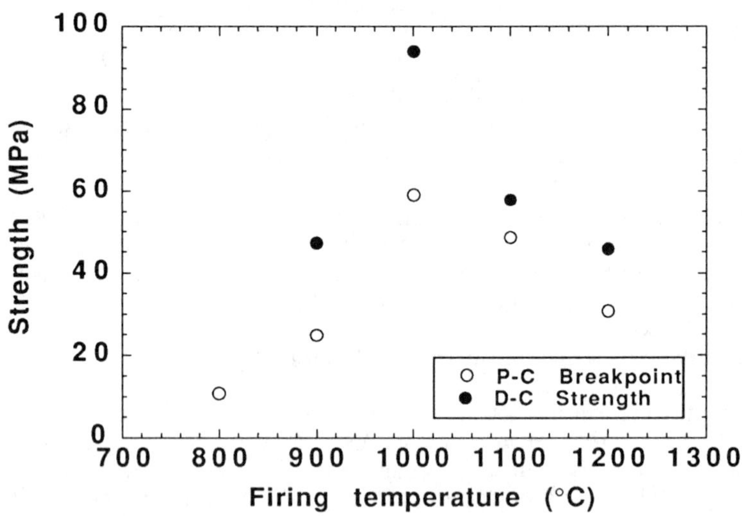

Fig. 7. Diametral compression (D-C) strength and pressure compaction (P-C) breakpoint as a function of firing temperature for lead magnesium niobate - lead titanate (PMN-PT) granules.

A possible mechanism for this is the formation of load-bearing bridges or chains. This mechanism has also been observed by discrete element modeling (DEM).[20] The bridge mechanism also explains the P-C results in Fig. 6. When a sphere in a load-bearing bridge fractures, significant rearrangement of the sphere assemblage occurs. The load is also no longer supported by a bridge and it decreases until another bridge forms. This is consistent with the density increases and accompanying load drops observed in the second linear region of the compaction curve in Fig. 6.

At stresses of ~ 172 MPa (25 ksi), the mechanism of bridge formation and collapse appears to end, and compaction of spheres under a more uniform applied stress begins. The stresses at the onset of this compaction mechanism are close to the low range of the D-C strength distribution. Thus in compaction tests where bridge formation and collapse occurs, the compaction stress that corresponds most closely to the granule strength is the second breakpoint.

The compaction mechanism of bridge formation and collapse was not observed in tests on ZnO, rutile, or PMN-PT granules. This may be due to the fact that these granules exhibit a range of sizes, and fill space in such a way that bridges do not form. The irregular shape of these granules and the possibility that small pieces of the granule break off under pressure, rather than the entire granule fracturing in half or shattering may also play a role in their tendency not to form bridges. Fracture of small regions of the granule would not allow the major rearrangements that are typical following the fracture of a glass sphere. Absence of bridge formation and collapse may also be related to the fact that significant rearrangement cannot occur following granule fracture, due to the roughness of the surface of the polycrystalline granules and their irregular shape. Granules appear to become locked in place at much lower pressures.

CONCLUSIONS

To identify compaction mechanisms, to make valid comparisons between one material and another, and to obtain reproducible results, experimental errors must be accounted for, and compaction test conditions must be standardized. In this study the error due to die deformation was 9% at the highest loads (420 MPa) and results in a higher density and a lower apparent breakpoint. Final aspect ratios, $L_f/D \leq 1.2$ produce consistent compaction results for rutile granules. $L_f/D=2.6$ values decreased the final compaction density, and decreased the slope of the second linear region of the curve. This effect is due to stress gradients in compacts with high aspect ratios. Increased loading rates affect the compaction curve in the same manner as high aspect ratios. This effect is due to the fact that granules have less time to rearrange at higher loading rates. The average D-C strengths of PMN-PT granules are approximately 1.5 times higher than the P-C breakpoints; however, the low range of the D-C strength distribution is close to the breakpoint stress. This is consistent with the breakpoint representing the stress near onset of granule fracture, and not the average granule strength. Glass sphere compaction data exhibited two breakpoints; the first represents the stress at the onset of compaction due to bridge formation and collapse, and the second represents the onset of compaction due to granule fracture and rearrangement under a more uniform applied stress. Bridge formation and collapse does not appear to play a significant role in compaction of ZnO, rutile, or PMN-PT granules.

ACKNOWLEDGMENTS
This work was supported by the U.S. Dept. of Energy under contract No. DE-AC04-94AL85000 at Sandia National Laboratories.

REFERENCES

1. J. S. Reed and R. B. Runk, "Dry Pressing," pp. 71 in Treatise on Materials Science and Technology, Vol. 9, Ceramic Fabrication Processes, Edited by F. F. Y. Wang, Academic Press, NY, 1976.
2. G. L. Messing, C. J. Markhoff, and L. G. McCoy, "Characterization of Ceramic Powder Compaction," J. Am. Ceram. Soc., 61 [8] 857 (1982).
3. B. J. McEntire, "Dry Pressing," pp. 141-146 in Vol. 4, Engineered Materials Handbook, Ceramics and Glasses, ASM International, 1991.
4. J. Y. Wong, S. E. Laurich-McIntyre, A. K. Khaund, and R. C. Bradt, "Strengths of Green and Fired Spherical Aluminosilicate Aggregates," J. Am. Ceram. Soc., 70 [10] 785-91 (1993).
5. P. T. Bertrand, S. E. Laurich-McIntyre, and R. C. Bradt, "Strengths of Fused and Tabular Alumina Refractory Grains," Am. Ceram. Soc. Bull., 67 [7] 1217-22 (1988).
6. Y. Hiramatsu and Y. Oka, "Determination of the Tensile Strength of Rock by a Compression Test of an Irregular Test Piece," Int. J. Rock Mech. Min. Sci., Vol. 3, pp. 89-99 (1966).
7. C. A. Bruch, "Problems in Die-Pressing Submicron Size Alumina Powder," Ceram. Age, 83 [10] 44-53 (1967).
8. R. A. Thompson, "Mechanics of Powder Pressing I-III," Am. Ceramic. Soc., Bull., 60 [2] pp. 237-245 (1981).
9. R. A. Youshaw and J. W. Halloran, "Compaction of Spray-Dried Powders," J. Am. Ceram. Soc., 61 [2] 227-230 (1982).
10. F. M. Mahoney and M. J. Readey, "Ceramic Compaction Models: Useful Design Tools or Simple Trend Indicators?," Proceedings of the Symposium on "Science, Technology and Commercialization of Powder Synthesis and Shape Forming Processes," Apr. 30-May 4, 1995, ACerS Annual Meeting, Cincinnati, Ohio.
11. R. A. DiMilia and J. S. Reed, "Dependence of Compaction on the Glass Transition Temperature of the Binder Phase," Am. Ceram. Soc. Bull., 62 [4] 484-488 (1983).
12. A. R. Cooper, Jr. and L. E. Eaton, "Compaction Behavior of Several Ceramic Powders," J. Am. Ceram. Soc, 45 [3] pp. 97-101 (1962).
13. C. L. Huffine and C. F. Bonilla, "Particle-Size Effects in the Compression of Powders," AIChE J., 8 [4] 490-93 (1962).
14. R. L. K. Matsumoto, “Generation of Powder Compaction Response Diagrams,” J. Am. Ceram. Soc., 69 [10] C246-247 (1986).

15. M. H. Bocanegra, "The Break Point in the Pressure-Density Curve of Magnesia Prepared by the Sea Water Magnesia Process," J. Mater. Sci., 28 (1993), 3467-3472.
16. P. R. Mort, R. Sabia, D. E. Niesz, and R. E. Riman, "Automated Generation and Analysis of Powder Compaction Diagrams," Powder Tech., 79, pp. 111-119, (1994).
17. H. Kamiya, K. Isomura, and G. Jimbo, and T. Jun-ichiro, "Powder Processing for the Fabrication of Si_3N_4 Ceramics: I, Influence of Spray-Dried Granule Strength on Pore Size Distribution in Green Compacts," J. Am. Ceram. Soc., 78 [1] 49-57 (1995).
18. J-H. Song and J. R. G. Evans, "A Die Pressing Test for the Estimation of Agglomerate Strength," J. Am. Ceram. Soc., 77 [3] 806-14 (1994).
19. A. W. Janssen, "Versuche uber Getreiderdruck in Silozellen (Experiments on Grain Pressure in Silos)," Z. Ver. Dtsch. Ing., 39, 1045-49 (1895).
20. G. G. W. Mustoe, G. L. DePoorter, and D. R. Greening, "Methodology for Discrete Numerical Modeling of Powder Compaction Processes," Proc. of the First Inter. Conf. on Adv. Synthesis of Engineered Materials, San Francisco, CA, 31 Aug. - 2 Sept., 1992, pp. 73-79, ASM International.

SPRINGBACK IN DRY PRESSING: TIME, ADDITIVE, AND PROCESSING EFFECTS

Kevin J. Moeggenborg, Nalco Chemical Company, One Nalco Center, Naperville, IL 60563

ABSTRACT

A study of the effects of time and various processing parameters on the springback of an alumina green body were investigated. Data indicate that most springback occurs within 10 minutes of ejection of a green body from the die. Radial springback of cylindrical green bodies was seen to increase with increasing aspect ratio, binder loading, binder hardness, and powder surface area.

INTRODUCTION

Springback is the expansion of a ceramic green body upon release of pressing pressure and is observed, to some extent, in all dry pressed ceramic materials. Pressing of a ceramic powder causes both plastic and elastic deformation of a ceramic powder. Release of elastic strain upon reduction of pressing pressure is the cause of springback. Although some springback may be necessary to affect separation of the green body from the punch surfaces, excessive springback contributes to cracking and lamination of the body on ejection from the die.[1] Since the pellet expands radially during ejection from the die, one other potential problem related to springback is the loss of dimensional tolerance that would be associated with springback that occurred over extended periods of time. Many ceramic green bodies are machined or shaped after pressing. Springback which occurred over a long period of time would result in ceramic bodies with differing dimensions depending upon the time between pressing and machining. The present work was initiated to determine the effects of various processing factors on green body springback. In addition, the variation of springback with time was monitored for a series of processing conditions.

EXPERIMENTAL PROCEDURES

All powder samples were prepared from slurries dispersed by Nalco 1389 dispersant at a 0.25 wt% (actives/powder) dosage. After addition of binder polymers to the slurries, viscosity was adjusted to <1000 cP by water addition as needed. All slips were spray dried on a Yamato DL-41 laboratory spray dryer. Powders were dried to <0.30% moisture in dessicators prior to use. Dried powders were classified through 100 and 325 mesh screens prior to pressing. The -100, +325 fraction was used for pressing studies. Powders were uniaxially pressed using a Carver hydraulic lab press with electronic speed and hold-down controls. Two dies were used for the pressing. The first was a tungsten carbide lined die with carbide punches. The second was a hardened tool steel die with steel punches.

Separation of the press platens was measured with a dial caliper mounted to the bottom platen. The caliper recorded the minimum separation of the platens during pressing. The height of the empty die at pressure was measured prior to each experiment. The difference between these two values gave the height of the pellet at pressure. The height of the die after pressing (zero pressure) was measured using a dial micrometer on a test stand for each pellet. The difference between the empty die height and its height after pressing a disk gave the height of the disk in the die cavity. The pellet height and diameter were measured immediately after ejection from the die and periodically thereafter. Pellets were stored in either a desiccator or a 50 % relative humidity chamber between measurements.

RESULTS AND DISCUSSION

Initial Work

The effects of a number of factors on the springback of green bodies were determined. Alcoa A152SG alumina powders containing 4 wt% of several different binder systems were evaluated. Springback was measured in two dimensions for the pressed pellets; axial (height) and radial (diameter). Early work was performed in the tungsten carbide dieset which had a polished surface. It was observed that the axial springback in the die was quite large immediately upon release of press pressure as shown in Figure 1. Upon ejection from the die (t=0), pellets shrank in the axial dimension while expanding radially. Reduction in height continued over time while the pellets expanded radially for t>0. It is also apparent, however, that most of the springback of a pellet occurs within the initial seconds after pressing and that little or no change occurs beyond the first hour after ejection from the die. Measurements beyond the first 24 hours showed no variation beyond experimental error at up to 130 hours.

The effect of atmospheric humidity on the springback of pressed pellets was measured as a function of time. If springback were to occur over a period of time, it would be expected that humidity differences would affect the rate of springback by plasticizing the binder. This weakening (and, possibly, swelling) of the binder would allow elastic springback to occur at a faster rate. Pairs of identical pellets were pressed and stored under either 0% or 50% relative humidity conditions. No difference in springback was observed for the two conditions. This result is consistent with the data of Figure 1 which show most or all springback to occur very early after ejection of the pellet from the die.

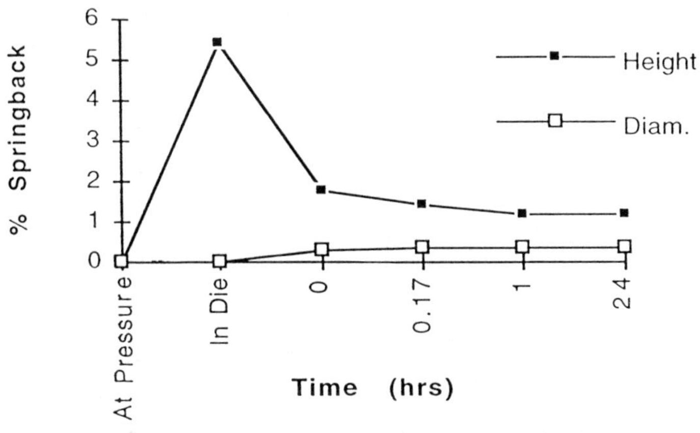

Figure 1: Height and Diametral springback effects before and after ejection from the die press.

The effects of pellet aspect ratio and pressing pressure on springback were also evaluated. Results are shown in Figures 2 and 3, respectively. The data represented for the aspect ratio are for pellets pressed at both 5 and 25 kpsi while data for the effect of pressure represent pellets at both 0.1 and 0.6 aspect ratios. The two figures are quite similar and indicate the pellets pressed at a high aspect ratio or low pressure undergo much less springback than pellets pressed at high pressure and low aspect ratios. These results relate closely to the density differences between pellets pressed under the different conditions. Pellets pressed at the higher press pressure attained an average corrected density of 2.47 g/ml while those at the lower pressure only attained densities of 2.23 g/ml. Similarly, high aspect ratios yielded pellets of average density 2.30 as compared to 2.40 g/ml for the lower aspect ratio. Density differences between these conditions are attributable to frictional losses in the powder during pressing.[2]

Additive and Powder Effects

Initial studies investigated the effects of several potential factors on the springback of a ceramic green body. Most manufacturers, however, are constrained to a given green density range in which their pressing process must operate. To simulate this more closely, studies were conducted in which all powders were pressed to a given corrected green density of 2.25 +/- 0.02 g/ml. Within this constraint the effects of powder surface area, binder dosage, and binder type were evaluated. Two Alcoa alumina powders, A152SG and A16SG, with specific surface areas of 3 and 9 m^2/g, respectively were used for the experiment. The binder systems tested were PEG 8000/PVA in a 50:50 wt ratio and Nalco 93QC215/94QC231 in an 80:20 ratio. The PEG/PVA system is a relatively hard binder while the Nalco system at the 80:20 plasticizer/binder ratio is much softer. Each system was evaluated at 2 and 4 wt% dosages.

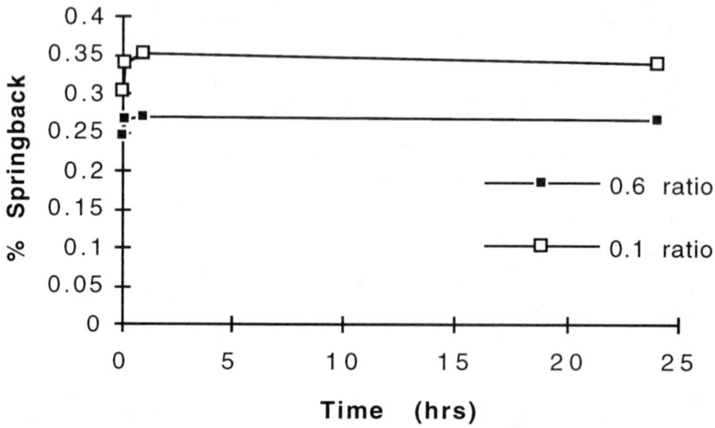

Figure 2: The effect of green body aspect ratio on springback for bodies pressed to equal pressure.

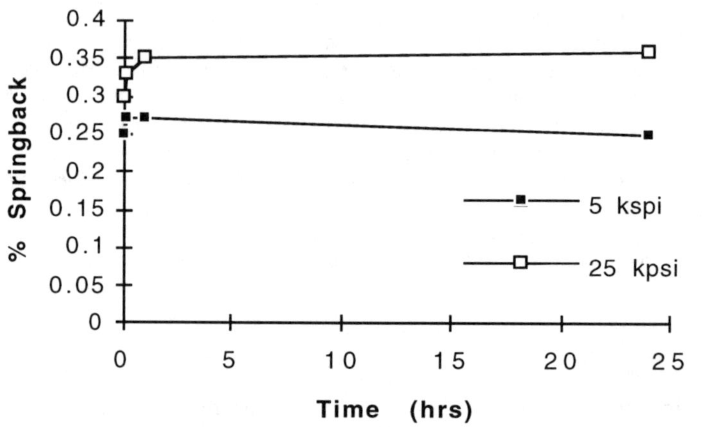

Figure 3: The effect of press pressure on springback of green bodies.

The pressure required to reach the target green density varied substantially with each of the four variables. Data on pressure requirements are given in Table I. As expected, the softer binder system (Nalco 93QC215/94QC231) attained the density requirement at much lower pressures than did the PEG/PVA system. Up to 230% higher pressing pressure was required to reach green density with the PEG/PVA system. Pressing pressure was also much higher for samples containing the higher surface area powders due to increased frictional losses in these systems. Additionally, pressure increased with aspect ratio and binder load for both binder systems.

TABLE I: Pressures required to achieve green density for experimental powders.

Powder	Dosage (%)	Aspect Ratio	Pressure Nalco	Pressure PEG/PVA
A16SG	2	0.1	10.0 ksi	23.0 ksi
		0.6	23.0	35.0
A16SG	4	0.1	16.0	25.0
		0.6	25.0	48.0
A152SG	2	0.1	3.5	4.0
		0.6	5.0	6.0
A152SG	4	0.1	4.0	8.5
		0.6	7.0	13.0

A plot of diametral springback versus time for pellets of different aspect ratios but the same density is shown in Figure 4. When pressed to equal density, the higher aspect ratio yielded higher springback values than did the low aspect ratio pellets This is opposite of the effect observed for pellets pressed to equal pressure. We believe this to be related to die wall and powder friction effects. Pellets pressed at a higher aspect ratio experience more friction with the die wall, and poorer pressure transmission through the height of the pellet. Upon release of press pressure, die wall friction prevents the pellet from fully releasing internal stresses by axial springback. As a result, more stress must be relieved by radial springback upon ejection of the green body. In pellets with larger aspect ratios, this effect would be intensified. Unfortunately, inaccuracies in our measurement method for determining axial springback did not allow this hypothesis to be confirmed experimentally.

The effect of powder size (surface area) on springback was quite pronounced as shown in Figure 5. Pellets made using the high surface area A16SG powder underwent almost twice as much springback as did A152SG powders. Again, the A16SG powders possess three time the surface area (9 vs. 3 m^2/g) as the A152SG. Since the binder dosage is the same, on a weight basis, for both powder systems, the A16SG has less binder coverage to mitigate the frictional forces. As a result, the higher surface area powder cannot undergo rearrangement to higher packing efficiencies as readily as the low surface area powder. A larger portion of the applied pressure, therefore, is used to elastically deform the powder particles. This strain is released by springback once the green body is ejected from the die.

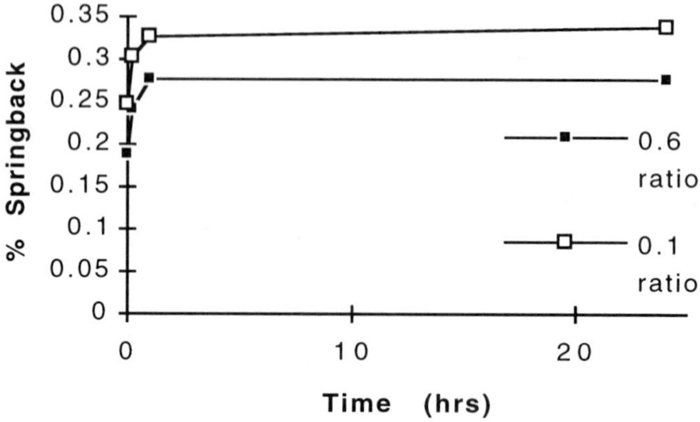

Figure 4: The effect of green body aspect ratio on springback for bodies pressed to equal density

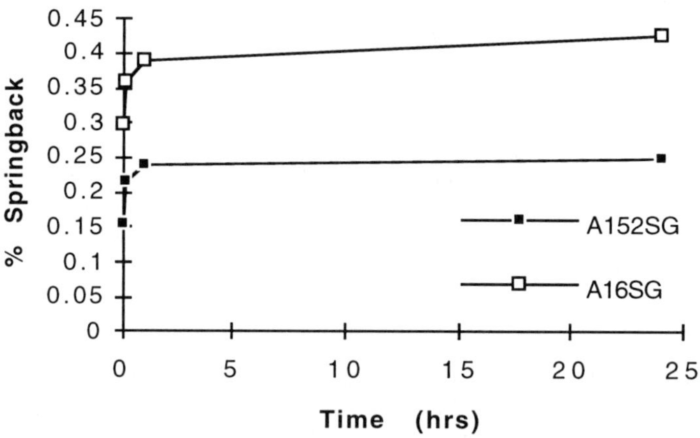

Figure 5: Springback as a function of time for green bodies from high (A152SG)and low (A16SG) surface area powders.

A comparison of springback of powders containing the two different binder systems is given in Figure 6. As can be seen, the harder, less lubricating,

PEG/PVA binder yielded higher springback values than did the 93QC215/94QC231 system. The effect of binder dosage showed that springback values were higher at the 4% dosage for both binder systems. This indicates that a portion of the springback exhibited by these powders is due to elastic deformation of the binder systems themselves. Data for the 24 hour springback values for the different binder, dosage, and powder values are given in Table II. Comparing the values, it can be seen that changes in binder dosage and powder surface area have less effect on springback for powders containing the 93QC215/94QC231 system than for powders containing the PEG/PVA binder. We believe this is because the Nalco binder is both more lubricating and softer than the PEG/PVA system. As a result, the Nalco binder allows more densification by particle packing and rearrangement than the PEG/PVA system. Additionally, the binder itself deforms more plastically than the PEG/PVA system leading to softer more easily deformed granules.

TABLE II: 24 hour springback values for various experimental samples pressed to equal corrected density.

Powder	Dosage (%)	Aspect Ratio	Spring Nalco	Spring PEG/PVA
A16SG	2	0.1	0.245%	0.245 %
		0.6	0.245	0.280
A16SG	4	0.1	0.175	0.175
		0.6	0.210	0.175
A152SG	2	0.1	0.350	0.455
		0.6	0.420	0.560
A152SG	4	0.1	0.245	0.315
		0.6	0.350	0.455

CONCLUSIONS

This study has shown that the majority of springback in alumina ceramics takes place within the first ten minutes after ejection of a green body from the die. Beyond the first hour, little or no springback occurs. Because springback is so rapid, atmospheric humidity below 50 % does not affect either the rate or magnitude of springback. Pressing pressure has a strong influence on the springback of a green body due to its effect on green body density.

Ceramic powder surface area, and the green body aspect ratio both increase the radial springback of green bodies pressed to equal density. This is thought to be due to increasing frictional forces within the compacting powder and between the powder and the die wall. Springback also increases with binder loading, presumably due to elastic deformation of the binder itself. This effect is smaller for a softer binder system than for a hard binder. Additionally, the softer binder reduces the effects of powder surface area and aspect ratio on springback.

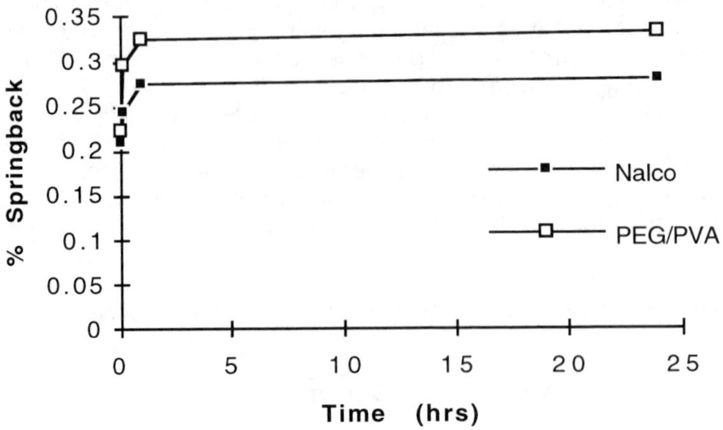

Figure 6: Effect of binder on springback of green bodies pressed to equal density.

REFERENCES

1. J. S. Reed, *Introduction to the Principles of Ceramic Processing,* Wiley-Interscience, New York, 1988.

2. M. I. Blazquez and B. J. Briscoe, "Friction Effects on the Compaction of Ceramic Powders", *Proceedings of the Third Intl. Conference on Powder Processing Science*, 1990.

CONTROL OF DRY PRESSING PERFORMANCE VIA BINDER OPTIMIZATION

David W. Whitman, Xianliang K. Wu and William C. Finch
Rohm and Haas Company, Spring House, PA 19477

ABSTRACT

The glass transition temperature (T_g) of polymeric binders is one of the most important parameters controlling binder performance during dry pressing. Binder T_g strongly affects the ease of forming and quality of dry pressed ceramic parts. Binder T_g can be easily varied during acrylic binder synthesis to allow fine tuning binder performance. In this paper we compare granule flow, green strength, green density, compaction behavior, springback and microstructure of granule prepared with 3 acrylic binders synthesized to have similar structures, but differing T_gs. Results are found to be in agreement with previous studies which varied T_g via plasticization.

INTRODUCTION

Polymeric binders are added to formulations for dry pressing ceramics to facilitate pressing, and to provide sufficient green strength for handling of pressed parts. Considerable work has been done in screening off-the-shelf materials for improved performance in dry pressing.[1,2,3,4,5,6,7,8]

The glass transition temperature of a polymer is a fundamental property which strongly influences the suitability of the polymer for different end uses. At sufficiently low temperatures, all polymers behave as glassy, brittle solids. Upon warming above its T_g, a polymer will change from this brittle state to a soft, rubbery state. This transition is typically accompanied by an abrupt change in modulus, tack and many other properties.

The T_g of polymeric binders is one of the key parameters controlling binder performance during dry pressing and strongly affects the quality of dry pressed ceramic parts. Binder T_g has been shown to affect the green density and green strength of dry pressed ceramic parts.[9,10]

Polymer T_g can be shifted to lower temperatures by incorporation of suitable plasticizers. For some polymers, water acts as a plasticizer, allowing T_g adjustment at the cost of sensitivity to humidity variation.[11,12,13,14] The inherent (unplasticized) T_g of most polymer systems is fixed, so that optimization is difficult. However, by using different pendant side chain groups, the T_g of acrylic polymers can be smoothly varied from -80 to +170 °C. The richness of acrylic chemistry allows fine tuning the T_g of acrylic binders for different forming processes.

In this paper, we examine the effect of varying the T_g of a series of otherwise similar acrylic polymers by changing the relative proportions of hard and soft monomers used. We report results on the effect of binder T_g on granule flow, green strength, green density, compaction behavior, die sticking, spring back and microstructure. A companion paper examines the effect of blending polymers of different T_g.[15]

EXPERIMENTAL PROCEDURE

Materials

The powder used in this study was Alcoa A-16 SG alumina. Surface area was 9.7 m^2/g (nitrogen BET) and mean particle size was 0.5 µm (Coulter LS-130). The dispersant used was Duramax™ D-3005 dispersant from Rohm and Haas Company.

The binders used were Duramax™ B-1020 binder from Rohm and Haas Company, and two experimental analogs prepared using an identical process. All three binders were prepared with an identical combination of monomers, but proportions of the monomers were varied to achieve different polymer T_gs. All the binders were aqueous emulsions and were used as received.

Polymer Characterization

Polymer T_g was measured by differential scanning calorimetry (TA model 2920). Samples were annealed at 150 °C prior to measurement. Experiments were run from -80 °C to 150 °C at 20 °C/minute.

Granule Preparation

Slurries were prepared by charging a #3 ball mill that was half-filled with grinding cylinders with alumina (7920 g), dispersant solution (0.3% based on alumina, solids basis), and enough water to bring slurry solids to 60%. The mixture was milled at 50 rpm for 20 minutes, then decanted. Binder emulsion (237 g solids basis, 3% based on alumina) was added with stirring.

Slurries were spray dried on an Anhydro spray drier with an inlet temperature of 240 °C and an outlet temperature of 130 °C. Prior to testing, granules were stored a minimum of 72 hours over a saturated solution of calcium nitrate, which maintained a constant relative humidity of 50%. Duramax B-1020 binder has been shown to be insensitive to ambient humidity[14]; this step was included only for consistency with other published work.

Granule Characterization

All granule tests were carried out at room temperature (18-20°C).

Granule flow was evaluated using ASTM method B-213, using a funnel with a 3-mm opening. Reported flow times are averages of at least 6 determinations.

Particle size distribution of dried granules was measured using a Coulter LS-1000 laser light scattering instrument with a dry powder module.

Samples were pressed in a 12.7-mm diameter cylindrical uniaxial die, mounted in a laboratory press (Carver model C). The die was cleaned, polished with jeweler's rouge, and lubricated with a solution of 2% stearic acid in acetone prior to pressing each sample. Samples were pressed at 70 MPa for 15 seconds. Sample weight was 1.0 g, giving pellets with a final diameter to height ratio of approximately 4.

Pellet dimensions were measured with a digital caliper accurate to ± 0.01 mm. Samples were allowed to relax a minimium of 1 hour after pressing before measurements were taken. Pellet weight was accurate to ± 0.0001 g and was corrected for organic content of the sample. Springback was evaluated by comparing the diameter of the die cavity to the diameter of pressed pellets. A minimum of 4 pellets was measured, and results are expressed as a percentage. Green density was calculated from the weight and dimensions of a minimum of 4 pellets.

Green strength was measured by diametral compression in an automatic loading device (Soiltest G-900 Versaloader). Samples were compressed at a loading rate of 0.13 mm per minute. Strength was calculated using Eq. (1), where σ is the maximum tensile stress or strength, p is the applied load at fracture, D is the sample diameter and L is the sample length.[16]

$$\sigma = \frac{2p}{\pi DL} \tag{1}$$

Continuous compaction diagrams were generated using the method of Niesz, et al.[17] Samples were pressed in a 6.4-mm diameter die mounted in a universal testing machine (Instron model 1122) at a crosshead speed of 0.5 mm per minute. Sample weight was 0.25 g, giving pellets with diameter to height ratio of approximately 4. The die was cleaned, polished with jeweler's rouge, and lubricated with a solution of 2% stearic acid in acetone prior to each sample. Reported curves were corrected for organic content of the samples and for elastic deformation of the test apparatus. Granule yield points were calculated as the intersection of least squares fits of the two linear regions of the compaction diagrams.[18, 19] Standard deviation of yield point determination was estimated as ± 0.04 MPa.

Microstructure of fracture surfaces of green parts were examined by SEM (Amray 1200).

RESULTS AND DISCUSSION

Polymer characterization

Duramax™ B-1020 binder is a commercially available material that shows good dry pressing performance.[20] Analogs were synthesized using ingredients and processes identical to this binder, but with different ratios of co-monomers to vary the T_g of the resulting polymers. The T_g of each polymer was measured by differential scanning calorimetry, and results are reported in table I.

Table I, Binder T_g via DSC

Polymer	T_g (°C)
low T_g analog	-16
Duramax™ B-1020	29
high T_g analog	75

Granule Characterization

No significant effect of binder Tg was noted on the granule size distribution. The median granule size was 82 µm. This result was as expected, since emulsion binders have very little effect on slurry rheology.

Fast granule flow is needed for rapid die fill in high speed presses. Flow behavior for alumina spray dried with the 3 binders is shown in Figure 1.

A statistically significant decrease in flow time was noted with increase of binder T_g. The largest change was noted between the T_g = -16 and 29 °C binders. The -16 °C binder is above its T_g at room temperature, while the other binders are still in the glassy state. This low T_g binder is therefore tackier than the other polymers under the experimental conditions. The observed change is caused by the large increase in polymer tack associated with the glassy to rubbery phase transition.

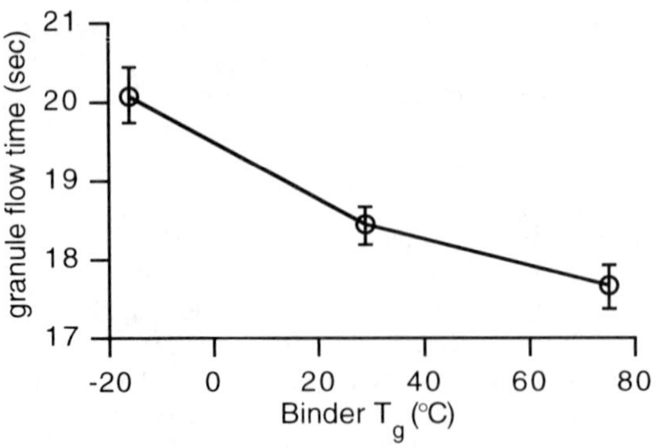

Figure 1: Effect of binder T_g on granule flow time.

The green densities of alumina samples pressed at 70 MPa are shown in Figure 2. Green density of samples decreased with increasing polymer T_g, consistent with previous studies.[9, 10] Lower binder T_g gave higher density, presumably because the softer binder deforms more easily.

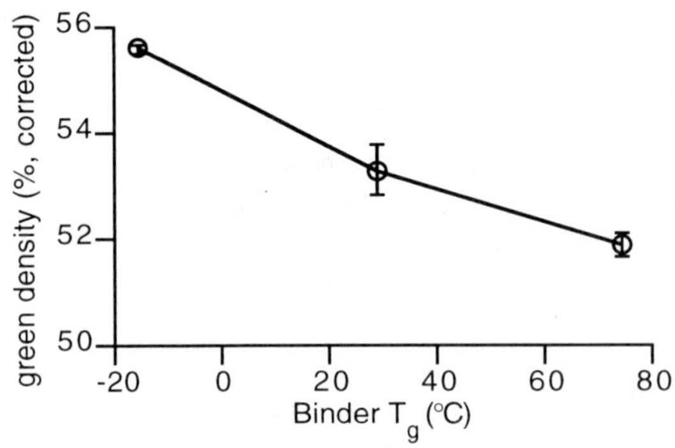

Figure 2: Effect of binder T_g on green density of samples pressed at 70 MPa.

Excessive springback can lead to lamination and end-capping problems.[21, 22] Depending on die shape, springback may either hinder or facilitate part ejection. Springback results for samples pressed at 70 MPa are reported in Figure 3.

Lowest springback was obtained with the T_g = -16 °C binder. Since this binder was above its T_g at the testing temperature, plastic deformation allowed stress relief during pressing. The other binders were below their T_gs at testing temperature, and gave very similar springback results.

Continuous compaction curves for granules prepared with the three binders are given in Figure 4. Binder T_g had a dramatic effect on compaction, shifting the curves towards higher pressures as T_g increased. Figure 5 plots yield point pressure derived from the compaction curves, and shows that yield point increased with binder T_g. Yield point pressure is a measure of granule hardness, and the observed results are consistent with previous studies of the effect of T_g on compaction.[9, 10]

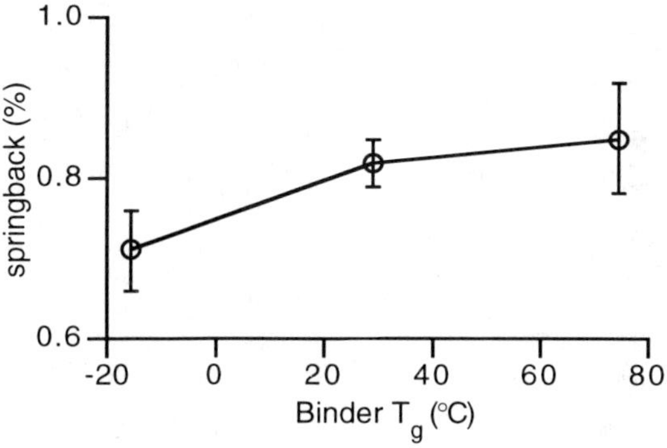

Figure 3: Effect of binder T_g on springback of samples pressed at 70 MPa.

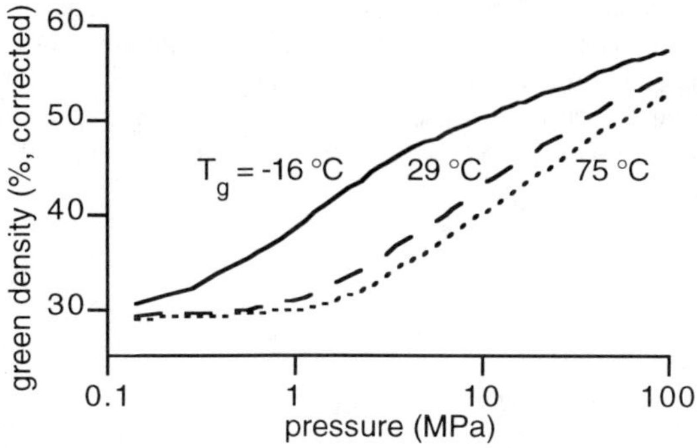

Figure 4: Effect of binder T_g on compaction response.

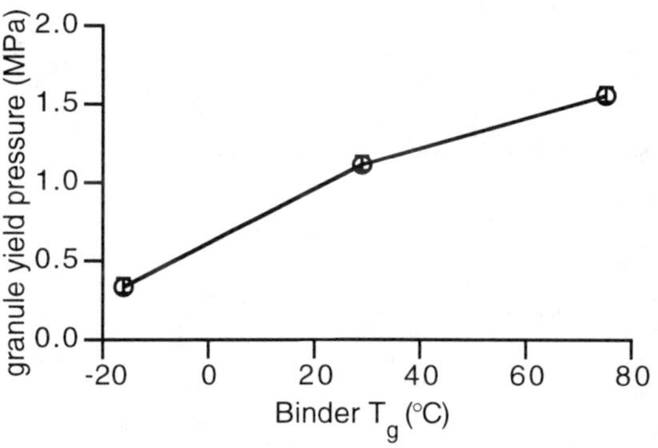

Figure 5: Effect of binder T_g on granule yield point.

Using the compaction curves for guidance, pellets were pressed at appropriate pressures to achieve various green densities. Figure 6 plots the green strengths of alumina pellets pressed with each binder. Increased binder T_g gave higher strength at all densities studied.

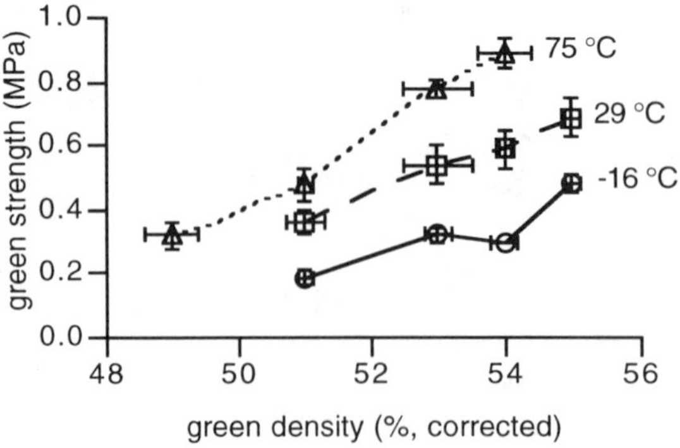

Figure 6: Effect of binder T_g on green strength at various green densities.

Fracture surfaces of samples pressed at 100 MPa were examined using SEM to probe the elimination of granule boundaries during pressing. Granule boundaries were totally eliminated in the T_g = -16 °C sample, presumably because this soft binder deformed readily during pressing and allowed the granules to collapse. Pronounced granule boundaries are visible in the T_g = 75 °C sample, consistent with the brittle state of the binder under the experimental conditions. Remnants of granule boundaries are visible in the T_g = 29 °C sample, but much less pronounced than in the high T_g analog.

Figure 7: Microstructure of compact pressed at 100 MPa with T_g = -16 °C binder.

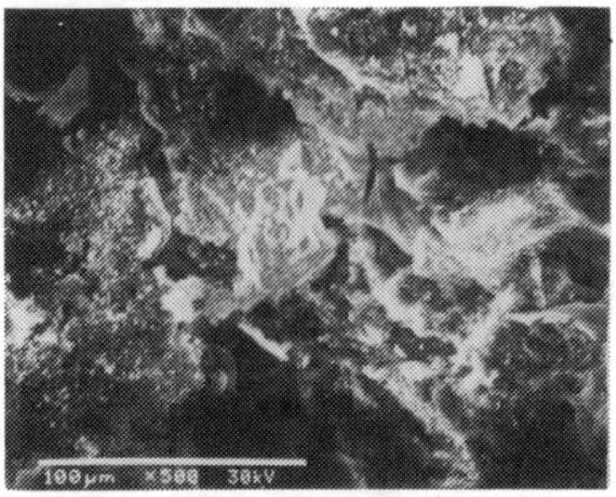

Figure 8: Microstructure of compact pressed at 100 MPa with T_g = 29 °C binder.

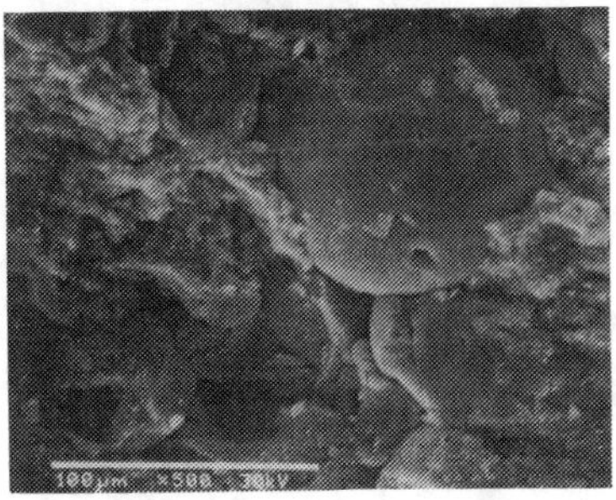

Figure 9: Microstructure of compact pressed at 100 MPa with T_g = 75 °C binder.

CONCLUSIONS

Dry pressing performance has been reported for a series of acrylic emulsion binders of similar composition, but varying T_g. No single T_g was found to optimize all binder properties. High binder T_g gave enhanced green strength and faster granule flow. Low binder T_g gave higher green density, easier granule deformation and reduced springback.

ACKNOWLEDGMENTS

D.I. Cumbers, A.T. Donato and Kirti S. Deshmukh for technical assistance.
S.Y. Kwong and R.R.L. Loh of Advanced Cerametrics Inc. for spray drying.

REFERENCES:

1 R. Bast, "Organic Additives for Dry Pressing," *Interceram* **39** [6], 13-14 (1990).

2 F. Aldinger and H.J. Kalz, "Importance of Chemistry in the Development of High-Performance Ceramics," *Angew. Chem. Int. Ed. Engl.* **26** [5], 371-381 (1987).

3 G.J.J. Beckers and K. Zeilstra, "Influence of Additives on the Pressing and Sintering Behavior of Spray-Dried Submicron Alumina," Euro-Ceramics, 1989, 99-103.

4 R.G. Frey, J.W. Halloran, "Compaction Behavior of Spray-Dried Alumina," *J. Am. Ceram. Soc*., **67** [3], 199-203 (1984).

5 J. Zheng and J. Reed, "Particle and Granule Parameters Affecting Compaction Efficiency in Dry Pressing," J. *Am. Ceram. Soc.* **71** [11] C-456-C-458 (1988).

6 D.C.C. Lam, "Effects of Binder and Compaction Pressure on Strength and Fracture Origins in Bodies Pressed from Granules," *J. Ceram. Soc. Japan*, **102** [11], 1010-1015 (1994).

7 J.W. Harvey and D.W. Johnson Jr., "Binder Systems in Ferrites," *Am. Ceram. Soc. Bull*., 59[6], 637-645 (1980).

8 T. Quadir, "Use of Ethylene/Vinyl Acetate Polymer Binders as Drying Pressing Aids for Ceramic Powders," US Patent 5,368,795 (1994).

9 C.W. Nies and G.L. Messing, "Effect of Glass Transition Temperature of Polyethylene Glycol-Plasticized Polyvinyl Alcohol on Granule Compaction," *J. Am. Ceram. Soc*., **67** [4], 301-304 (1984).

10 R.A. DiMilia and J.S. Reed, "Dependence of Compaction on the Glass Transition Temperature of the Binder Phase," *Am. Ceram. Soc. Bull*., **62** [4] 484-488 (1983).

11 R.G. Frey, J.W. Halloran, "Compaction Behavior of Spray-Dried Alumina," *J. Am. Ceram. Soc*., **67** [3], 199-203 (1984).

12 R.A. DiMilia and J.S. Reed, "Effect of Humidity on the Pressing Characteristics of Spray-Dried Alumina," in Advances in Ceramics, Vol. 9, 38-46 (1984).

13 J.A. Brewer, R.H. Moore and J.S. Reed, "Effect of Relative Humidity on the Compaction of Barium Titanate and Manganese Zinc Ferrite Agglomerates Containing Polyvinyl Alcohol," *Am. Ceram. Soc. Bull*., **60** [2] 212-220 (1981).

14 D.W. Whitman, D.I. Cumbers and X.K. Wu, "Humidity Sensitivity of Dry Press Binders," *Am. Ceram. Soc. Bull.*, (in press).

15 D.W. Whitman, X. Tang, D.I. Cumbers and S.A. Ibbitson, "Using Polymer Blends to Fine-Tune Binder Performance," paper SXIX-37, Annual Meeting of the American Ceramic Society, Cincinnati, OH (1995).

16 A. Rudnick, A.R. Hunter, and F.C. Holden, "An Analysis of the Diametrical Compression Test," *Mater. Res. Stand.*, **3** [4], 283-89 (1963).

17 P.R. Mort, R. Sabia, D.E. Niesz and R.E. Riman, "Automated Generation and Analysis of Powder Compaction Diagrams," American Ceramic Society Annual Meeting , Cincinnatti, OH (1993).

18 D.E. Niesz, R.B. Bennett and M.J. Snyder, "Strength Characterization of Powder Aggregates," *Am. Ceram. Soc. Bull.*, **51** [9] 677-680 (1972).

19 S.J. Lukasiewicz and J.S. Reed, "Character and Compaction Response of Spray-Dried Agglomerates," *Am. Ceram. Soc. Bull*., **57** [9] 798-805 (1978).

20 X.K. Wu, D.I. Arnold, W.L. Kaufell and D.W. Whitman, "Comparison Study of Polyvinyl Alcohol, Polyethylene Glycol and Polyacrylic Binders for Dry Pressing Ceramics," Annual Meeting of the American Ceramics Society (1994).

21 R.A. Thompson, "Mechanics of Powder Pressing: II, Finite-Element Analysis of End-capping in Pressed Green Powders," *Am. Ceram. Soc. Bull*., 60 [2], 244-247 (1981).

22 A.B. Van Groenou, "Compaction of Ceramic Powders," Powder Technol., **28**, 221-228 (1981).

Duramax is a trademark of the Rohm and Haas Company.

GREEN-STATE DEFORMATION OF BOEHMITE-COATED SILICON NITRIDE COMPACTS

W.-H. Shih, D. J. Farrell, and W. Y. Shih, Department of Materials Engineering, Drexel University, Philadelphia, PA 19104.

ABSTRACT

A water-based processing method for silicon nitride without polymeric additives was investigated. Silicon nitride particles were coated with boehmite via a sol-gel process to modify the long-range as well as short-range interactions between the silicon nitride particles. The coated suspensions gelled with time and the green-state deformation of coated and uncoated Si_3N_4 bodies were studied. It is shown that a 1" diameter coated green-state Si_3N_4 cylinder can be deformed with pressure to a 3" diameter disk without visible cracks after drying. Furthermore, it is shown that coated Si_3N_4 green bodies can be extruded without much resistance. None of these properties were observed for the uncoated Si_3N_4 green bodies. Boehmite coating is shown to be an excellent forming aid in the aqueous colloidal processing of silicon nitride.

INTRODUCTION

Silicon nitride-based ceramics have a wide variety of applications such as cutting tools, bearing balls, turbocharger rotor, gas turbine rotors etc., depending on the application temperature.[1] The superior properties of silicon nitride result from the strong covalent bonding between Si and N atoms. However, the strong covalent bonding also makes the sintering and forming of silicon nitride difficult. Diffusion of atoms is limited and the deformation under external influence is restricted. In addition, due to the complex surface chemistry of silicon nitride particles, the solids loading of silicon nitride in water is relatively low compared to that of oxide ceramics such as alumina. To overcome these processing difficulties, several research groups have used a coating approach to introduce processing aids more uniformly and at the same time modify the surface chemistry of silicon nitride. Mah et al.,[2] Shaw and Pethica,[3] and Garg and De Jonghe,[4] coated sintering aids on silicon nitride particles using sol-gel precursors. Liden et al.[5] coated sintering aids on silicon nitride particles using colloidal particles.

Although the distribution of sintering aids has been studied as indicated above, the improvement on the consolidation and forming has not been addressed. In a recent study,[6] we coated boehmite (AlOOH) on silicon nitride using a sol-gel coating approach and showed that the consolidation and forming of silicon nitride were significantly improved. In this paper, results on the green-state deformation are presented. Green-state deformation has been shown to improve the density and sintering behavior of green bodies.[7] In addition, the complex-shape forming capability of green bodies is investigated. The green-state deformation behavior of compacts of coated and uncoated silicon nitride were compared. The deformation study includes green-state pressing and extrusion. It was found that the coated silicon nitride green bodies can deform under lower pressure and they exhibited a greater amount of deformation than uncoated silicon nitride green bodies. The coated slurries can also be used to fill a spherical mold and maintained their shape after gelation. The gelation of boehmite suspensions is similar to that of clay suspensions.[8] The gelation of clay is due to the surface-edge attraction of platelet particles.[8] The gelation of boehmite is thought to be the same as that of clay except that the signs of charges on the boehmite surfaces and edges are opposite to those of clay.[9]

EXPERIMENTAL PROCEDURE

Silicon nitride powders (UBE, SN-E03, surface area 3.6 m^2/g) were coated with boehmite using a sol-gel approach developed by Yoldas.[10] The coating of silicon nitride was done by preparing a 76.65 ml of 6 vol% aqueous silicon nitride suspension first. After the suspension was sonicated for 15 minutes and heated to 90°C, aluminum secondary butoxide of 10.2 ml, which corresponds to forming 0.04 mole of boehmite, was then added. Hydrochloric acid was added for peptization and the pH of the solution was adjusted to 3. The suspensions of coated powders were then heated to evaporate the solvent. The coated powders were collected, and washed with deionized water several times, and subsequently ground.

Aqueous suspensions of both coated and uncoated silicon nitride were made with 15 minutes of sonication. A very small amount of HCl (corresponding to 0.012M) was added to suspensions to adjust the suspension pH values to 3. The volume fraction of uncoated silicon nitride in suspension was calculated using the density of silicon nitride, 3.24 g/cm^3. The volume fraction of coated silicon nitride in suspensions was calculated using the density of the coated powder, 3.2 g/cm^3, which was obtained by assuming all the boehmite is uniformly distributed on spherical silicon nitride particles of radius 0.26 μm estimated from the surface area data. The layer thickness of boehmite on silicon nitride was calculated to be 14.5 nm. More detailed information about the coating and the preparation of suspensions can be found in our previous study.[6] The gelation of boehmite-coated Si_3N_4 is easily obtained at pH=3 and therefore, in this paper, all suspensions were kept at pH=3.

Once the slurries were given enough time to gel (3 hours), they were tested in a press with paraffin film on top and bottom of the green bodies to prevent sticking to the press. The height and width of the compact were measured as a function of the applied load. Grease was placed in between the slurries and the paraffin films to act as a lubricant to reduce horizontal shearing during pressing. The extrusion tests were conducted on the Instron to obtain stress versus strain data for a constant crosshead speed of 20 mm/min. Similar extrusion tests of boehmite-alumina mixtures had previously been studied.[11] The green body was first placed in the extrusion apparatus, as shown in Fig.1, and was pushed by the crosshead at a constant speed through the die. The crosshead pressure and displacement were recorded.

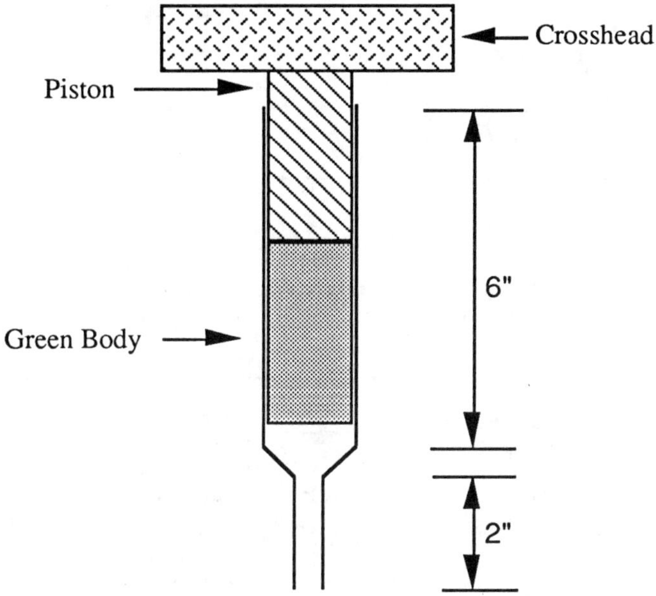

Fig.1. Experimental setup for extrusion study.

RESULTS

Fig.2 is a plot of the height of green body, h, vs. pressure in the pressing test. It can be seen that the height obtained for the coated compacts was smaller than that of uncoated silicon nitride compacts. This result indicates that the coated silicon nitride green bodies are easier to deform. Fig.3 is a plot of the width of the compact as a function of the applied pressure. It is shown that the coated Si_3N_4 green bodies exhibit a great deal of extension in the radial direction of the green body under load whereas the deformation of the 45 vol% uncoated Si_3N_4 green body was very limited. The uncoated green body also exhibited cracking while being deformed under the load.

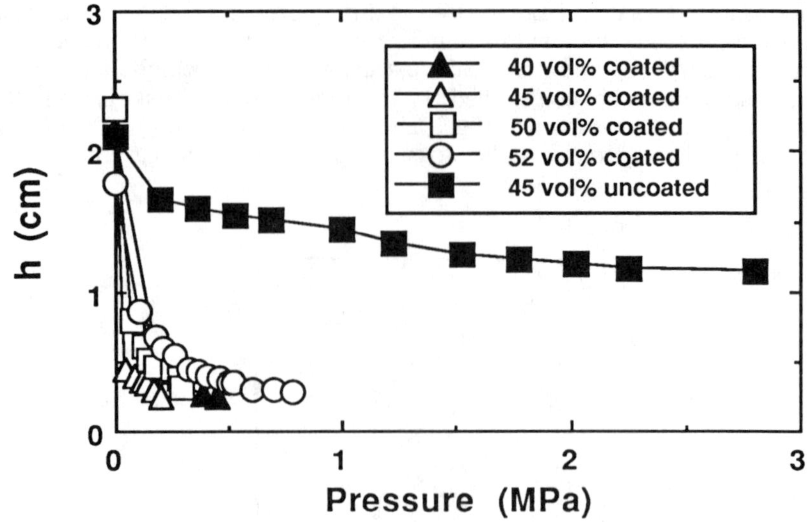

Fig.2. The height of green bodies as a function of the applied pressure for several coated and one uncoated Si_3N_4 cylinders.

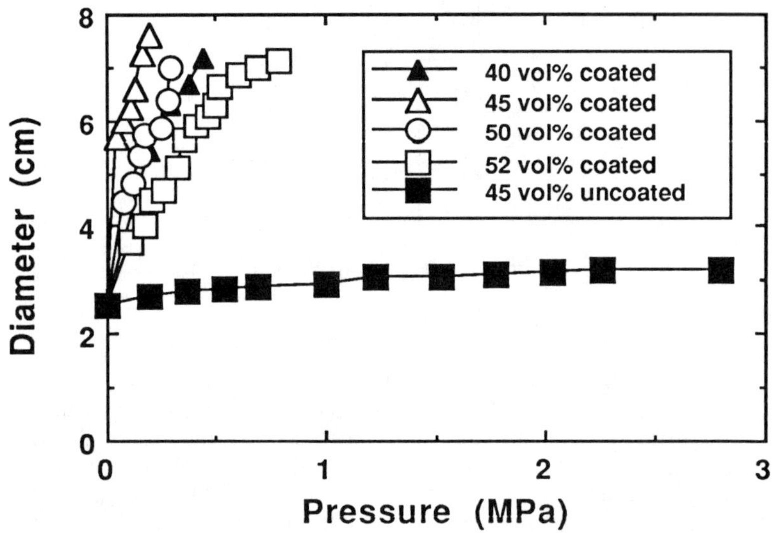

Fig.3. The width of green bodies as a function of the applied pressure for several coated and one uncoated Si_3N_4 cylinders.

It is clear that the coated silicon nitride green bodies deform much easier than the uncoated green bodies. To see if this deformation involves a change in volume, the volume was plotted against pressure in Fig.4. It is shown in Fig.4 that the volume is relatively constant for most coated silicon nitride green bodies except for 40 vol% and uncoated bodies. There was a 20% volume decrease in the 40 vol% coated and 45 vol% uncoated green bodies after pressing.

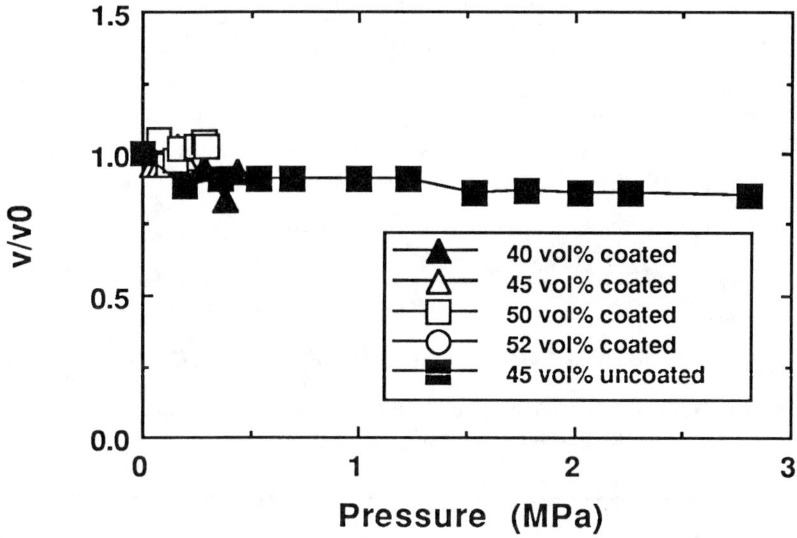

Fig.4. The volume of green bodies as a function of the applied pressure for several coated and one uncoated Si_3N_4 cylinders.

In the extrusion study, the crosshead speed is maintained at 2 cm/min and the pressure on the crosshead was recorded. The pressure on the crosshead as a function of crosshead displacement for coated silicon nitride green bodies of various concentrations is shown in Fig.5. The pressures plotted for the 40 and 45 vol% coated green bodies are shown to have a gradual increase as the crosshead displacement increases. Unlike the lower volume percentages, the 50 and 52 vol% coated green bodies exhibit an initial rapid increase then followed by a gradual increase for the remainder of the curve. The initial increase in the pressure can be related to the modulus on a typical stress/strain curve, whereas the flattened region of the curve is the region representative of yielding or flow. All of the coated green bodies tested showed yielding behavior as denoted by the slow change of pressure vs. displacement.

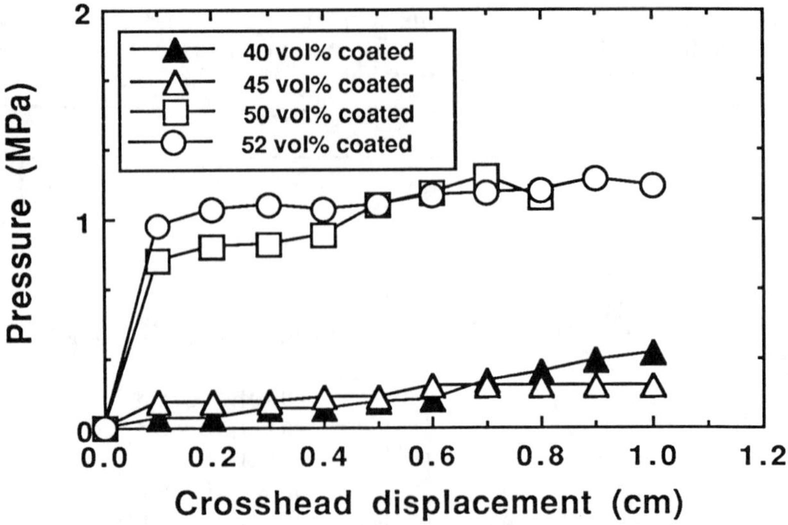

Fig.5. The pressure on the crosshead as a function of the crosshead displacement in extrusion of coated Si_3N_4 green bodies.

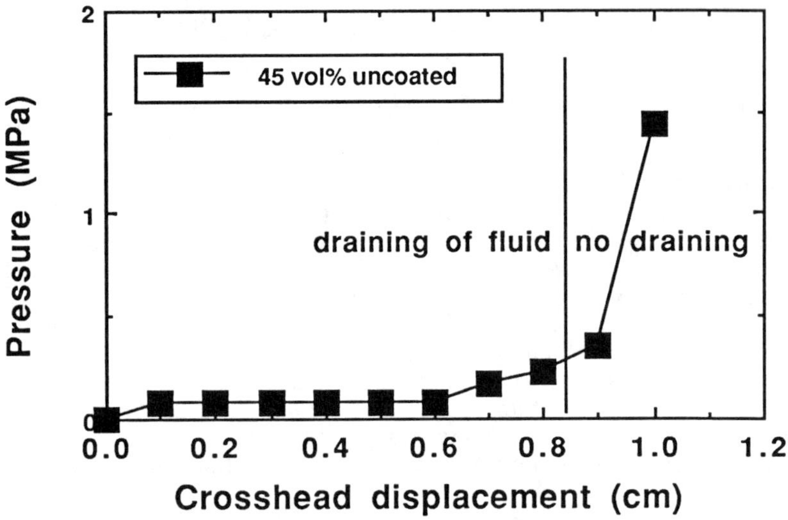

Fig.6. The pressure on the crosshead as a function of the crosshead displacement in extrusion of uncoated Si_3N_4 green bodies.

The pressure-crosshead displacement behavior for the uncoated silicon nitride green body is shown in Fig.6. Unlike the coated green bodies, the 45 vol% uncoated green body did not exhibit plastic flow behavior. For the values of displacement lower than 8 mm, it is shown that there is a gradual increase in stress. These stress values were very small and were in the range of the values obtained for the 40 and 45 vol% coated Si_3N_4 green bodies. However, observations at these stresses showed that there was a separation of phases between the ceramic and the water. The excretions initially coming from the extrusion device were very watery and had a milky color. But as the displacement increased above 8 mm, the stress began to increase rapidly as the extruder attempted to push out the higher concentrated slurry. This is shown by the sharp increase in stress from 9 to 10 mm in Fig.6.

DISCUSSION

The pressure applied to a green body is calculated as $\frac{\text{load}}{\text{area}}$. Since the area increases during loading, the amount of pressure that can be applied to the coated Si_3N_4 green bodies is limited. Fig.7 shows the applied pressure as a function of the applied load to the green bodies. Because the area of the uncoated Si_3N_4 green bodies does not change very much, the pressure-load relationship is close to be a straight line. On the other hand, the pressure-load relationship for coated Si_3N_4 green bodies is not a straight line since the area under the load is changing.

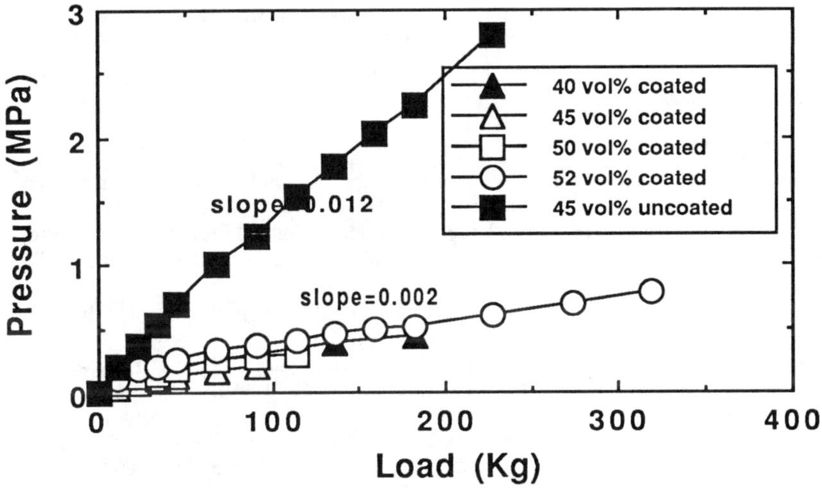

Fig.7. The pressure on deformed green bodies as a function of the applied loads.

The extrusion experiment is a constant-strain-rate uniaxial rheological test. When the crosshead pressure approaches a constant and independent of the crosshead displacement, the test samples reach a fluid state. The data in Fig.5 indicate that the coated Si_3N_4 green bodies behave like fluids in that the pressure increases very slowly with the displacement. On the other hand, the data in Fig.6 shows different behavior for small and large displacements. At small displacements, the uncoated green body showed drainage behavior with watery fluid coming out of the extruder. The milky color of the excretions indicates that some boehmite particles were in the water. A phase separation between a dilute suspension and the concentrated slurry occurred. At larger displacements, the drainage of fluid stopped and the pressure increases drastically with displacement. The rapid increase of pressure results from the fact that the uncoated Si_3N_4 green body can not deform and pass through the die in the extruder.

The deformed green bodies of coated Si_3N_4 did not show any visible cracks after drying. The uncoated green bodies was difficult to deform and showed visible cracks while being deformed under the load. The better drying behavior is probably related to the gelation of boehmite. Since water is trapped within the gel network, it is speculated that the evaporation of water would be slower and the drying stress is reduced. A spherical mold was made and the coated Si_3N_4 slurries were filled and held in the mold for 3 hours. Spheres with smooth surface finish can be obtained.

CONCLUSIONS

It is shown in this paper that boehmite coating on Si_3N_4 significantly improved the green-state deformation behavior of Si_3N_4. A 1" diameter coated green-state Si_3N_4 cylinder can be deformed under pressure to a 3" diameter disk without visible cracks after drying. Furthermore, it is shown that coated Si_3N_4 green bodies can be extruded with the applied pressure of the crosshead remaining almost constant. None of these properties were observed for the uncoated Si_3N_4 green bodies. Boehmite coating is shown to be an excellent forming aid in the aqueous colloidal processing of silicon nitride.

REFERENCES

1. T. Yamada, "Preparation and Evaluation of Sinterable Silicon Nitride Powder by Imide Decomposition Method," Am. Ceram. Soc. Bull. **72**[5], 99-106 (1993).
2. T.-I. Mah, K. S. Mazdiyasni, and R. Run, "Characterization and Properties of Hot-Pressed Si3N4 with Alkoxy-Derived CeO_2 or Y_2O_3 as Sintering Aids," Am. Ceram. Soc. Bull., **58** [9], 840 (1979).
3. T. M. Shaw and R. A. Pethica, "Preparation and Sintering of Homogeneous Silicon Nitride Green Compacts," J. Am. Ceram. Soc., **69** [2], 88 (1986).
4. A. K. Garg and L. C. De Jonghe, "Microencapsulation of Silicon Nitride Particles with Yttria and Yttria-Alumina Precursors," J. Mater. Res., **5**[1], 136-42 (1990).
5. E. Liden, M. Persson, E. Carlstrom, and R. Carlsson, "Electrostatic Adsorption of a Colloidal Sintering Agent on Silicon Nitride Particles," J. Am. Ceram. Soc., **74**[6], 1335 (1991).
6. W.-H. Shih, L.-L. Pwu, and A. A. Tseng, "Boehmite Coating as a Consolidation and Forming Aid in Aqueous Silicon Nitride Processing," J. Am. Ceram. Soc., **78**[5], 1252-60 (1995).
7. C. P. Cameron and R. Raj, "Better Sintering through Green-State Deformation Processing," J. Am. Ceram. Soc., **73**, 2032-37 (1990).
8. H. van Olphen, *Introduction to Clay Colloidal Chemistry* (Wiley, New York 1980).
9. A. C. Pirre and D. R. Uhlmann, "Comparison of Clay-Water Systems with Aluminum Hydroxide Gels Prepared in Acidic Media," Mat. Res. Soc. Symp. Proc. **Vol.73**, 481 (1986).
10. B. E. Yoldas, "Alumina Sol Preparation from Alkoxides," Am. Ceram. Soc. Bull. **54**, 289 (1975).
11. A. Y. Chen and J. D. Cawley, "Extrusion of α-Al_2O_3-Boehmite Mixtures," J. Am. Ceram. Soc., **75**, 575-79 (1992).

PLASTIC SHAPING OF AQUEOUS ALUMINA SUSPENSIONS WITH SACCHARIDES AND DICARBOXYLIC ACIDS

H. Goel, C. H. Schilling, S. B. Biner, J. A. Moore, and B.K. Lograsso

Ames Laboratory* and Department of Materials Science and Engineering, Iowa State University, Ames, IA 50011

ABSTRACT

Traditional methods for the shape-forming of engineering ceramics entail plastic deformation of powder slurries containing hazardous organic liquids as suspending media. Replacing these organics with aqueous media requires the development of environmentally-benign, water-soluble additives which serve as plasticizers and binders. Fundamental studies were performed with aqueous suspensions of colloidal α-Al_2O_3 to evaluate the role of sucrose, maltodextrin, and oxalic acid on viscosity, sedimentation and filtration characteristics, plastic flow behavior of filter cakes, and sinterability. Maltodextrin and oxalic acid systems exhibited superior results, including filtration to high packing-densities and clay-like plasticity with minimal cracking.

INTRODUCTION

Two major problems faced by the ceramic industry today are ductile-brittle transitions during plastic shaping operations and

---

* Ames Laboratory is operated by Iowa State University under the contract number W-7405-eng-82 with the U. S. Department of Energy.

environmental hazards associated with the debinding of organics. Ductile-brittle transitions occur when batch materials undergo a transition from plastic behavior to a stiff and brittle consistency as the packing density exceeds a critical maximum value (which is typically on the order of 40-55 volume percent in mono-sized spherical particle systems). In order to avoid this transition, the packing density must be maintained below this critical maximum value. This forces the use of relatively large concentrations of inter-particle liquid media, which in turn leads to numerous problems such as large drying and sintering shrinkages, cracks and warpage, long debinding times, and porosity defects and carbonaceous residues resulting from the gaseous decomposition of organics.[1-4]

The key to achieving high packing densities and clay-like plasticity in nonclay ceramic suspensions is to develop chemical additives producing short-range interparticle repulsion forces that aid particle rearrangement. This phenomenon was recently demonstrated by adsorbing small, organic molecules on oxide powder surfaces: particles in flocculated suspensions become weakly attractive but non-touching.[5-9] A similar phenomenon is recently suggested to be a major source of the plasticity in clays, although instead of arising from the steric hindrance of adsorbed macromolecules, clays are believed to exhibit a short-range, *hydration repulsion* due to a layer(s) of adsorbed water.[10-14] Interparticle forces in suspensions of submicron ceramic powders without adsorbed molecules can be described using DLVO theory, which sums the attractive van der Waals potential with the repulsive, electrostatic potential produced by charged surface sites partially shielded by counterions in suspension.[15] Without surface-adsorbed molecules, particle contacts are strongly adhesive in suspensions that are flocculated near the isoelectric point, because of the presence of only the van der Waals potential. In turn, this reduces the maximum packing density attainable during filtration.

We have recently shown that maltodextrin enhances the consolidation densities and plasticity of aqueous alumina suspensions.[16] In this paper, we report similar data and include studies on oxalic acid as an additive which exhibits superior results. We now understand that maltodextrin and oxalic acid enhance consolidation and plasticity as a result of the adsorption of these molecules on alumina surfaces.[17] In addition, we also show that only small amounts of these additives are required which burn out cleanly.

EXPERIMENTAL APPROACH

Aqueous suspensions were prepared for two different powder systems: α-alumina (A16 SG, average particle size 0.4 μm, Alcoa Corporation, Bauxite, AR) and kaolin (Pioneer Airfloated Clay, average particle size 1.0-1.2 μm, Dry Branch Kaolin Company, Dry Branch, GA). For the alumina systems, two types of suspensions were prepared: weakly flocculated and strongly flocculated.[18] For weakly flocculated suspensions, optimum concentrations of various additives (maltodextrin, glucose, oxalic acid etc.) were dissolved in deionized water, and alumina powders were gradually added to each stirring solution. Strongly flocculated suspensions of the same alumina powder were also prepared near the isoelectric point without adding any of the above additives; the pH of these suspensions was raised to 8.6 using reagent grade NH_4OH. Optimum concentrations of these additives were obtained by subjecting alumina suspensions, with additives in varying concentration, to gravity sedimentation experiments. In these experiments, the optimum additive concentrations corresponded to those producing the maximum sediment density.[17] Kaolin suspensions were prepared by the gradual addition of powders to mechanically stirring deionized water. No additives were used in kaolin suspensions.

Viscosity measurements were performed on slurries with a concentration of 20 volume percent powder using a coaxial cylinder viscometer (Haake Type VT500, Paramus, NJ). Ceramic suspensions were consolidated using *expression* in a filter press at a constant pressure.[19,20] Liquid-saturated cakes were removed from the filter press (Figure 1) and the green densities were immediately measured using an oil-immersion technique based on the Archimedes principle.

The liquid saturated filter cakes that were obtained by pressure filtration were too viscous to be analyzed by a conventional cone and plate or couette-type viscometer. Therefore, a squeeze flow device was used that produces equibiaxial extensional loading.[21,22] This state of loading was chosen because it easily produces tensile cracks of controlled orientation in samples. Liquid saturated ceramic cakes were axially compressed between smooth platens of an Instron testing machine (Instron Corporation, Type 4204, Canton, MA) to a pre-determined strain at a constant speed of 0.5 millimeters per minute. Load and displacement were monitored using a computerized data acquisition system. The top and bottom surfaces of the samples were lubricated with mineral oil in order to reduce shear stresses.[21,22] Finally, each deformed specimen was photographed to record the formation of cracks during the test.

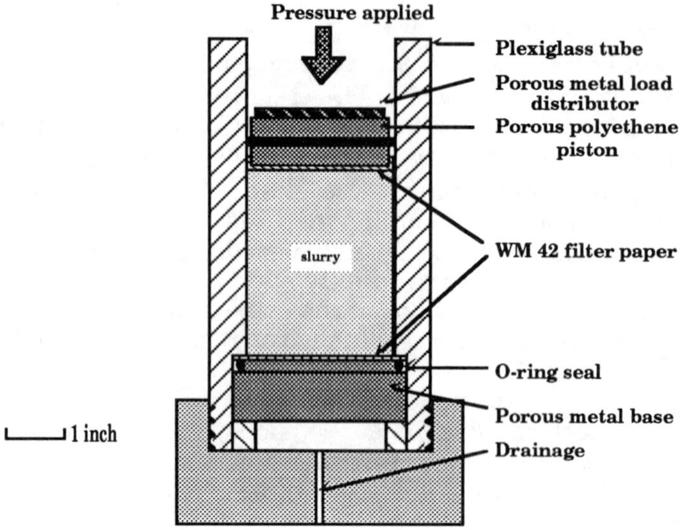

Figure 1: Schematic Illustration of the Pressure-Filtration Apparatus. Suspensions were filtered in a plexiglass acrylic tube measuring 5.08 cm inside diameter. Drainage occured through filter paper membranes contacting the piston and the porous metal base.

Sintering studies were also performed on alumina samples consolidated by slip casting. These filter cakes were subsequently dried at 100°C and sintered for four hours at various peak temperatures. The average density of each sintered sample was evaluated by the Archimedes principle.[23]

Simultaneous differential thermal analysis and thermogravimetric analysis (DTA/TGA) were performed in order to evaluate the pyrolysis behavior of different additive systems. Two types of slip cast alumina samples were analyzed: (*i*) strongly-flocculated samples prepared using the above procedure by raising the pH to 8.6 with NH_4OH, and (*ii*) samples containing maltodextrin. The latter sample was prepared using the optimum concentration of maltodextrin that was previously determined from the sedimentation studies. Results were compared with baseline samples containing maltodextrin alone. Each specimen weighed approximately 10 mg and was heated in flowing air at 50 mL per minute. The heating rate was maintained at 40°C per minute from room temperature to 1000°C. Two samples of each composition were evaluated in order to verify reproducibility.

RESULTS AND DISCUSSION

Adding alumina to deionized water without any additive resulted in a strongly flocculated condition, with sediment densities, ø, of 0.13 volume fraction solids. Sediments containing glucose or oxalic acid did not exhibit high densities (ø= 0.14 and 0.25 respectively).[17] In contrast, the maltodextrin system displayed a very high density of 0.46 volume fraction solids. Smith observed that these maximum sediment densities occurred when the surface adsorption of maltodextrin and oxalic acid reaches a maximum.[17] Adsorption isotherms measurements were not performed on the glucose system.

Viscosity measurements indicated psuedoplastic behavior for all systems (Figure 2). Suspensions that were prepared with maltodextrin exhibited the lowest viscosities, while the kaolin suspensions and the alumina suspensions containing oxalic acid were most viscous. The strongly flocculated system and the glucose systems exhibited intermediate viscosities.

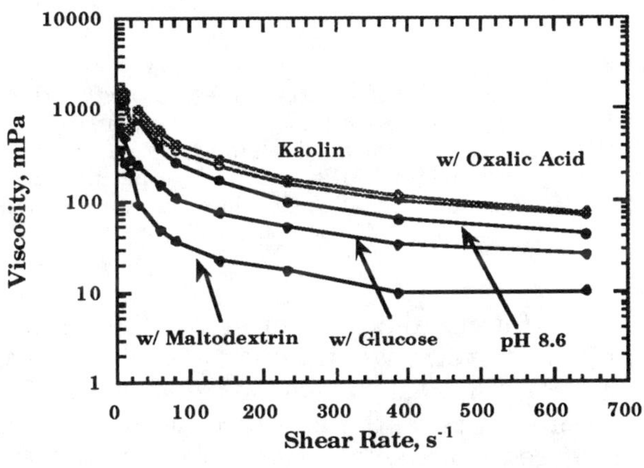

Figure 2: Viscosity Measurements on Suspensions Containing 20 Volume Percent Solids Loading. The maltodextrin-alumina suspensions exhibited the lowest viscosities, while kaolin and oxalic acid-alumina systems were the most viscous.

Consolidation by expression at a pressure of 0.54 MPa showed that the highest packing density of 60 vol% solids was achieved by the kaolin system, while the alumina systems containing maltodextrin and oxalic acid exhibited a slightly lower density of 57 vol% solids (Figure 3). This was an unexpected result, because of the opposite behavior exhibited by the oxalic acid system in sedimentation and viscosity studies. Future studies will address this anomalous behavior.

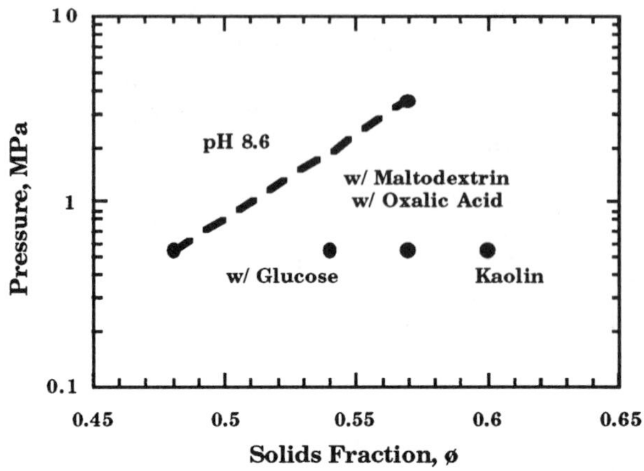

Figure 3: The Role of Suspension Chemistry and Filtration Pressure on Packing Densities in Aqueous, Liquid-Saturated Filter-Cakes. 2.5 wt% additions of glucose, maltodextrin, and oxalic acid successively increased the densities of alumina cakes attainable at 0.5 MPa filtration pressure. The alumina sample at pH 8.6 required a much higher filtration pressure of 3.5 MPa to achieve a packing density as great as that of the maltodextrin-alumina or oxalic acid-alumina samples consolidated at the lower pressure.

The strongly flocculated system at pH 8.6 showed the lowest packing density of 49 vol% when consolidated by expression at a pressure of 0.54 MPa. In order to achieve densities of 57 vol% solids in this strongly flocculated case, the pressure had to be increased to 3.5 MPa. This difference in the behavior of the alumina suspensions with and without additives is believed to be a result of additive adsorption on alumina surfaces. Adsorption isotherms measurements by Smith[17] suggest that these additives coat the alumina powders, forming short-range, steric repulsion forces that in turn reduce interparticle friction and adhesion.

As seen in Figure 4, kaolin exhibited highly plastic behavior in the squeeze-flow tests. A smooth and continuous decrease in slope with increasing strain corresponded with the fact that no cracking was observed. However, in the strongly flocculated case, the behavior was very dependent on the solids loading. For example, suspensions with 49 vol% solids loading showed a high degree of plasticity with minimal cracking, but as the density increased to 57 vol%, much higher loads were required to deform the cake to the same strain. Also, abrupt changes in the slope of the loading curve were noticed for the strongly flocculated system that was 57 vol% dense. These changes in slope are likely attributed to the cracks that were observed in the photographed samples (Figure 4). Another interesting feature of these photographs was that the cracks were oriented in radial and tangential directions, indicating that the samples were subjected to equibiaxial extensional loading.

Equibiaxial extension tests indicated that the maltodextrin-alumina system, with a density of 57 vol%, showed a higher degree of plasticity than the strongly flocculated system at the same density (Figure 4 and 5). The degree of cracking shown by this maltodextrin system was also much less intense than the strongly flocculated system having the same density. However, the load necessary to produce a given strain in both systems was comparable and much greater than corresponding loads exhibited by the kaolin system. In contrast, the glucose-alumina system, with a packing density of 54%, exhibited a greater amount of cracking, although these cracks were smaller and fewer than those observed in the strongly flocculated case with 57% solids loading.

Highly concentrated alumina suspensions (57 vol% solids loading) with oxalic acid exhibited the most clay-like stress-strain behavior as compared to the other systems mentioned above (Figure 6). A striking result was that not only do the clay system and the oxalic acid-alumina system have comparable solids loadings, they also required very similar magnitudes of loads to produce a given strain. However unlike clay, the oxalic acid-alumina system did show a small amount of cracking (Figure 6). It should be noted, however, that the degree of cracking in these samples was much less than the maltodextrin or strongly flocculated systems with comparable solids loadings.

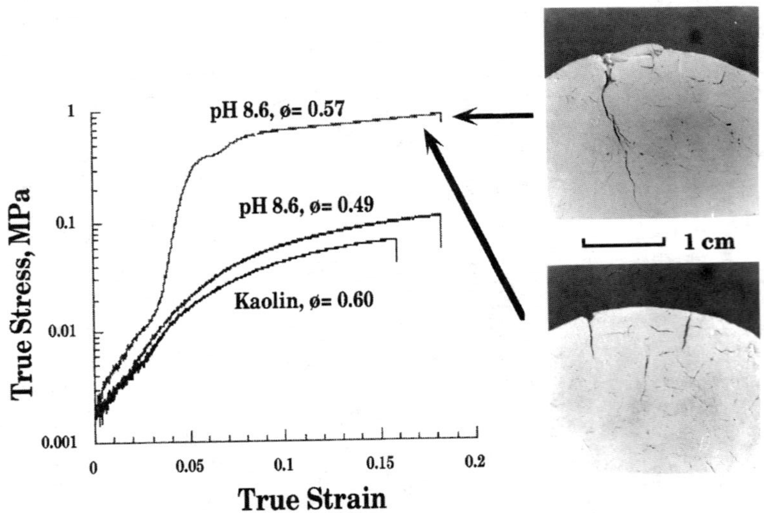

Figure 4: Stress-Strain Curves During Equibiaxial Extension of Kaolin and pH 8.6 Alumina Samples. Kaolin exhibited ideal plastic behavior, with a smooth curve and no cracks visible. The pH 8.6 sample exhibited similar behavior only if its packing density (49 vol%) was much lower than that of kaolin (60 vol%). The pH 8.6 sample exhibited brittle, cracking behavior if its packing density (57 vol%) approached that of the kaolin.

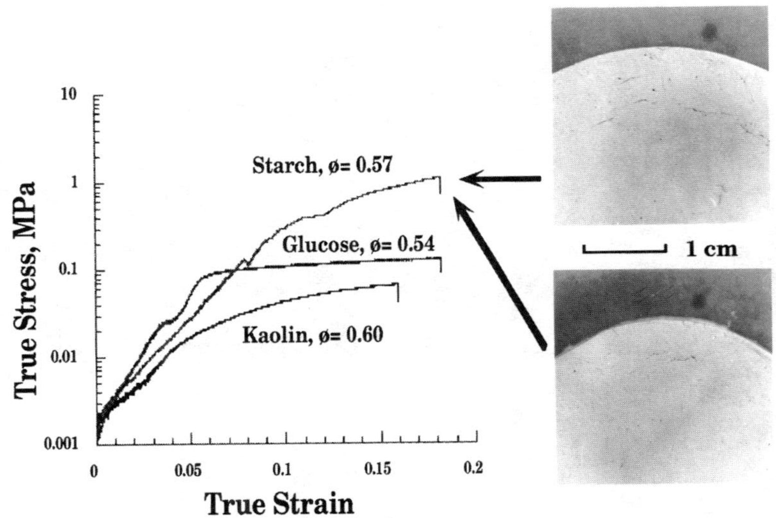

Figure 5: Stress-Strain Curves during Equibiaxial Extension of Kaolin, Glucose-Alumina, and Maltodextrin-Alumina Samples. The starch sample (maltodextrin) exhibited clay-like plasticity, although small cracks resulted. The glucose sample exhibited brittle behavior with a larger degree of cracking.

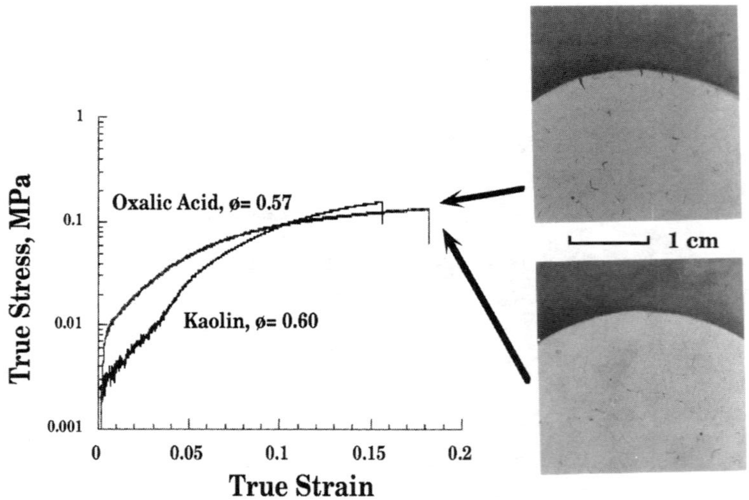

Figure 6: Stress-Strain Curves in Equibiaxial Extension of Kaolin and Oxalic Acid-Alumina Systems. The oxalic acid system exhibited the most clay-like plasticity relative to all other alumina specimens tested.

Slip casting with maltodextrin or oxalic acid produced high packing densities in alumina systems, which in turn enhanced the sinterability. Figure 7 illustrates that the maltodextrin-alumina and the oxalic acid-alumina systems resulted in significantly greater densities at a given sintering temperature than the alumina sample prepared without any additive.

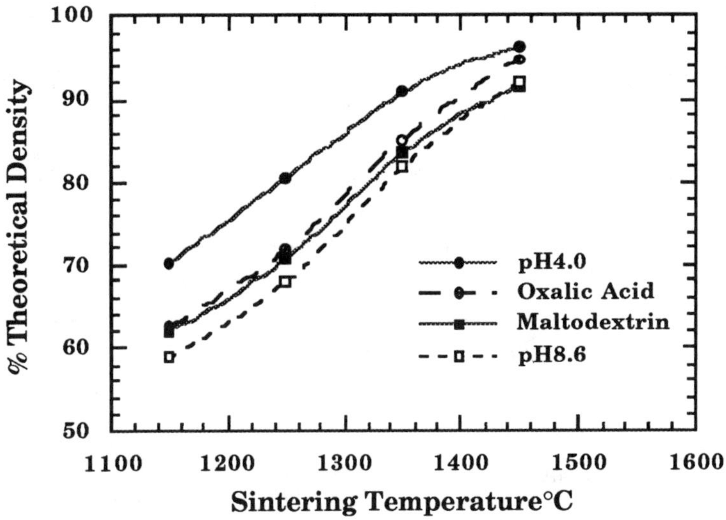

Figure 7: The Role of Maltodextrin and Oxalic Acid on the Sinterability of Slip-Cast Alumina. Alumina samples prepared with maltodextrin or oxalic acid exhibited significantly higher densities at all sintering temperatures relative to strongly flocculated samples at pH 8.6 that did not contain any additive.

Maltodextrin samples began decomposing at approximately 290°C and exhibited complete burnout at 550°C, based on the DTA/TGA results shown in Figure 8. Maltodextrin samples that were prepared with alumina present exhibited similar behavior, and exhibited a much smaller weight loss of 1.7% upon heating to 1000°C. The alumina samples that were slip cast at pH 8.6 exhibited a weight loss of only 0.1% over the same temperature range. These results indicate that very little carbonaceous residue remained after heat treatment of the alumina/maltodextrin samples. Similar behavior is expected in the oxalic acid-alumina system.

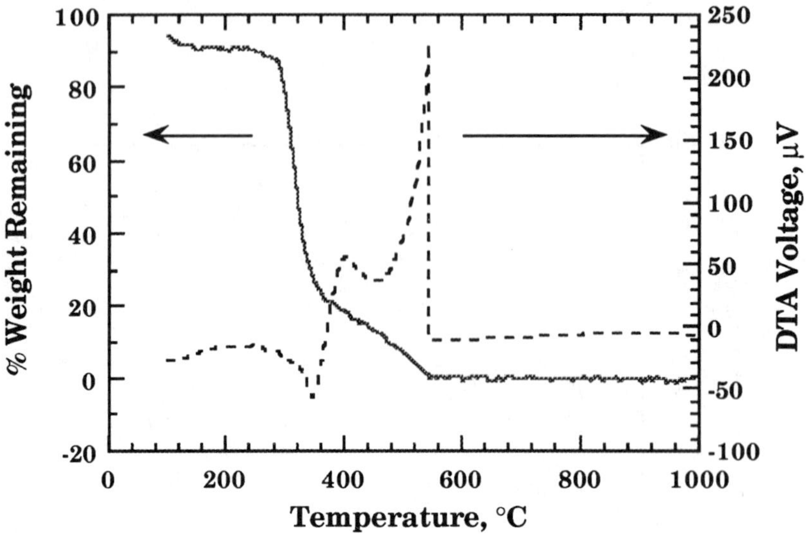

Figure 8: Thermogravimetric and Differential Thermal Analysis of Maltodextrin in Air. Decomposition begins at approximately 290°C and was completed at 550°C.

CONCLUSIONS

Our results indicate that oxalic acid produced the most clay-like plasticity in highly concentrated alumina suspensions. Maltodextrin exhibited similar behavior although to a lesser degree, and the glucose system did not perform as well. Plastic behavior at high densities is believed to be a result of reduced inter-particle attraction associated with steric hindrance of surface-adsorbed additives. These additives appear promising for practical shape-forming applications because they are aqueous and only small concentrations are needed to produce the desired effects.

FUTURE WORK

In order to adapt these formulations to practical applications a better understanding of how these systems behave under conditions of higher shear strain and higher strain rates must be developed. These issues are being addressed in ongoing research using extrusion and ram-pressing experiments. In addition, future work entails the development of practical processing maps that are derived from continuum plasticity models.

ACKNOWLEDGMENTS

The authors wish to thank the Office of Basic Energy Sciences, U.S. Department of Energy and the Iowa Corn Board for supporting this research. We also wish to thank J. Jane and P. Tomasik for valuable discussions.

REFERENCES

[1]V. K. Pujari, "Effect of Powder Characteristics on Compounding and Green Microstructure in the Injection-Molding Process," *J. Am. Ceram. Soc.*, **72** [10] 1981-84 (1989).

[2]V. K. Pujari, K. E. Admin, and P. H. Tewari, "Development of Improved Processing and Evaluation of Silicon Nitride," In *Proc. Int. Gas Turbine and Aeroengine Congr. and Exposition, ASME 91-GT-317*; Orlando, FL, 1991.

[3]V. K. Pujari and D. M. Tracey, "Processing Methods for High Reliability Silicon Nitride Heat Engine Components," In *Proc. Int. Gas Turbine and Aeroengine Congr. and Exposition, ASME 93-GT-319*; Cincinnati, OH 1993.

[4]G. C. Stangle and I. A. Aksay, "Simultaneous Momentum, Heat and Mass Transfer with Chemical Reaction in a Disordered Porous Medium: Application to Binder Removal from a Ceramic Green Body," *Chem. Eng. Sci.*, **45** [7] 1719-1731 (1990).

[5]L. Bergström, C. H. Schilling, and I. A. Aksay, "Consolidation Behavior of Flocculated Alumina Slurries," *J. Am. Ceram. Soc.*, **75** [12] 3305-14 (1992).

[6]T. Kramer and F. F. Lange, "Rheology and Particle Packing of Chem- and Phys-Adsorbed Alkylated Silicon Nitride Powders," *J. Am. Ceram. Soc.*, **77**, 922-928 (1994).

[7]C. H. Schilling, "Plastic Shaping of Colloidal Ceramics," Ph. D. Thesis, University of Washington, Seattle, 1992.

[8]J. N. Israelachvili and R. M. Pashley, "Molecular Layering of Water at Surfaces and Origin of Repulsive Hydration Forces," *Nature*, **306**, 249-250 (1983).

[9]J. N. Israelachvili, "Forces Between Surfaces in Liquids," *Adv. Colloid Interface Sci.*, **16**, 31-47 (1982).

[10]H. van Olphen, *An Introduction to Clay Colloid Chemistry*, John Wiley, New York, 1977.

[11]A. M. Homola, J. N. Israelachvili, M. L. Gee, and P. M. McGuiggan, "Measurements of and Relation Between the Adhesion and Friction of Two Surfaces Separated by Molecularly Thin Liquid Films," *J. Tribol.*, **111**, 675 (1989).

[12]R. M. Pashley and J. N. Israelachvili, "DLVO and Hydration Forces Between Mica Surfaces in Mg^{2+}, Ca^{2+}, Sr^{2+}, and Ba^{2+} Chloride Solutions," *J. Colloid and Interface Sci.*, **97**, 446-455 (1984).

[13]J. N. Israelachvili and R. M. Pashley, "Double-Layer, van der Waals and Hydration Forces Between Surfaces in Electrolyte Solutions," in Biophysics of Water (F. Franks, Ed.), John-Wiley, New York, 1982.

[14]R. M. Pashley and J. N. Israelachvili, "Molecular Layering of Water in Thin Films Between Mica Surfaces and Its Relation to Hydration Forces," *J. Colloid and Interface Sci.*, **101**, 511-523 (1984).

[15]J. N. Israelachvili, *Intermolecular and Surface Forces: With Applications to Colloidal and Biological Systems*, Academic Press, San Diego, 1985.

[16]C. H. Schilling, S. B. Biner, H. Goel, and J. Jane, "Plastic Shaping of Aqueous Alumina Suspensions with Sucrose and Maltodextrin Additives," *J. Environ. Poly. Degrad.*, **3** [3] (1995).

[17]R. M. Smith, "Ultrasonic Characterization of Green and Partially-Sintered Ceramics," M.S. Thesis, Iowa State University, Ames, IA, 1995.

[18]H. Goel, "Superplastic Shaping of Colloidal Ceramics with Saccharides and Dicarboxylic Acids," M.S. Thesis, Iowa State University, Ames, IA, 1995.

[19]F. M. Tiller and C.-D. Tsai, "Theory of Filtration of Ceramics I: Slip Casting," *J. Am. Ceram. Soc.*, **69** [12] 882-887 (1986).

[20]F. M. Tiller and C. S. Yeh, "Relative Liquid Removal in Filtration and Expression," *Filtration and Separation*, **March/April** 129-135 (1990).

[21]S. Chatraei and C. W. Macosko, "Lubricated Squeezing Flow: A New Biaxial Extensional Rheometer," *Journal of Rheology*, **25** [4] 433-443 (1981).

[22]P. R. Soskey and H. H. Winter, "Equibiaxial Extension of Two Polymer Melts: Polystyrene and Low Density Polyethylene," *Journal of Rheology*, **29** [5] 493-517 (1985).

[23]Annual Book of ASTM Standards, "Standard Test Methods for Apparent Porosity, Water Absorption, Apparent Specific Gravity, and Bulk Density of Burned Refractory Brick and Shapes by Boiling Water," 15.01, ASTM C20-87, 1987.

GELCASTING OF SILICON COMPOSITIONS FOR SRBSN

S. D. Nunn, J. O. Kiggans, Jr., R. E. Simpson II, and J. P. Maria
Oak Ridge National Laboratory, Oak Ridge, TN 37831-6087

ABSTRACT

Compositions of high purity silicon metal containing various oxides as sintering aids were gelcast to prepare green bodies for subsequent nitridation and sintering to form sintered reaction bonded silicon nitride (SRBSN). An acidic aqueous- and an alcohol-based gelcasting system were used. The solids content in the slurry affected the green density of the cast samples, however, there was no apparent correlation between the green density and the percent nitridation. Samples were nitrided under flowing, high-purity nitrogen in either a tungsten element furnace, a graphite element furnace, or in a microwave furnace. The samples nitrided in the graphite and microwave furnaces showed 9-16% higher levels of nitridation than in the tungsten furnace. A N_2-4% H_2 gas mixture was compared to pure nitrogen in the tungsten furnace. The addition of the hydrogen increased nitridation levels by 14-19%. The nitrided and sintered SRBSN ceramics had densities of 95-98% of theoretical, with no direct relationship being observed between the level of nitridation and the fired density. The average 4-point flexure strength of the samples ranged from 428 to 741 MPa.

INTRODUCTION

Sintered reaction bonded silicon nitride (SRBSN) which is formed by reacting silicon powder with nitrogen gas to form silicon nitride has been studied for many years.[1,2] The properties of SRBSN are usually inferior to those of silicon nitride materials formed from high quality silicon nitride powders, especially at elevated temperatures. This is because metal impurities (typically Fe) are usually added to the SRBSN composition to enhance the nitridation reaction. Reduced high temperature capability has limited the potential applications of SRBSN; however, on a cost basis, SRBSN is an attractive alternative to silicon nitride materials formed from silicon nitride powder.[3] The cost of high purity silicon powder is only about 20% of the cost of silicon nitride powder. In addition to lower raw material costs, the reaction of silicon with nitrogen to form silicon nitride results in a volume increase of about 21.6% over the original silicon powder volume. This normally occurs by the growth of the silicon nitride crystals into the pore structure of the powder compact and results in an increase in density without an accompanying reduction in the size of the part.[4] This can reduce the amount of shrinkage observed during the sintering of the component to form a densified material, and make net-shape forming and dimensional control more manageable. Recently, it was demonstrated that SRBSN could be fabricated from high purity silicon powders without metal additions by using microwave heating and that the resulting material had improved properties compared to conventional SRBSN.[3] Further studies of the fabrication of SRBSN compositions using high purity silicon powder are described in the present work.

Both conventional and microwave heating methods for firing SRBSN were examined in this study. The conventional process typically involves two distinct steps: the nitriding reaction and the sintering run.[4] This is primarily because typical sintering processes for silicon nitride containing low amounts of refractory sintering aids involve packing the parts in a bed of powder (usually silicon nitride) to minimize weight loss at high temperatures. If a silicon part is packed in a powder bed prior to nitridation, the powder can act as an insulating blanket holding in the heat generated by the exothermic reaction of the silicon metal converting to silicon nitride. Under these conditions, it is difficult to control the exothermic heating of the part which can result in melting of the silicon metal before complete nitridation can occur. In microwave heating, it has been demonstrated that due to the volumetric heating effect, silicon compacts can be nitrided and sintered in a single processing step.[3,5,6]

To take full advantage of the desirable characteristics which are offered by SRBSN materials, a viable and economical forming method for producing complex shapes needs to be developed. One such forming process which was utilized in this study is gelcasting. Gelcasting is a powder forming process which offers distinct advantages as an alternative to the more conventional ceramic forming methods such as dry pressing, slip casting, and injection molding.[7-10] Principal advantages include near-net-shape forming, high green density, and low organic levels in the dried green ceramics. A wide variety of ceramic materials have been prepared using the gelcasting process. These include silicon nitride, silicon carbide, alumina, and zirconia, among others.

Initial efforts to gelcast silicon metal compositions using the currently available process were unsuccessful. Two principal difficulties were observed, both of which could be traced to the fact that the typical gelcasting slurry is aqueous and basic (having a pH of about 8.4). The first was the poor dispersion characteristics of the silicon powder in the slurries. The second was the generation of gas bubbles in the slurry caused by reaction of the silicon metal with water. Attempts at surface modification of the powder to improve dispersion and reduce reactivity were unsuccessful, primarily because the slurries needed to be milled to obtain uniform mixing of the constituent powders and to achieve high solids loading. The milling action generated clean particle surfaces which negated any positive effects of the prior surface treatment.

Experimental investigations showed that these processing difficulties could be overcome in two ways: by reducing the pH of the aqueous slurry or by using isopropyl alcohol as the solvent system in place of water. However, the available gelcasting systems could not be used under these conditions. An effort was undertaken to identify chemicals which might be utilized to form gelcast bodies under the required processing conditions. Both an acidic aqueous system and an alcohol-based system were developed and successfully adapted for SRBSN gelcasting. The solids loading was limited to 35-40 vol % in the acidic aqueous system while slurries containing 50 vol % solids were prepared in the alcohol-based system. These new gelcasting systems are at present proprietary and will not be discussed in detail here. This paper describes the characteristics of the gelcast silicon green bodies, the nitriding and sintering behavior of a number of gelcast SRBSN compositions, and the preliminary evaluation of the mechanical properties of the sintered materials.

EXPERIMENTAL PROCEDURES

Batches were prepared by milling the constituents in an attritor mill (Union Process, Model 01) using a Teflon-lined jar, a high density polyethylene rotor, and high-density zirconia milling media. There was no measurable change in the weight of the media after the milling operation, indicating negligible wear and contamination of the slurries. The compositions which were prepared are listed in Table I, where both the as-prepared composition and the theoretical composition after nitriding are shown. High-purity silicon powder (Elkem, grade HQ, 99.95% purity) was used. The oxide sintering aids are typical of those being used in the preparation of high quality silicon nitride materials. The batches labeled Nd and Y were prepared in the acidic aqueous-based gelcasting system, and the slurries contained 35-40 vol % powder. Batches PM1, PM2, and PM3 were prepared in the alcohol-based gelcasting system and contained 48 vol % solids.

Table I. Compositions for Gelcasting of Sintered Reaction Bonded Silicon Nitride

Batch	Composition, wt %					
	Si	Si_3N_4	Al_2O_3	Y_2O_3	Nd_2O_3	SiO_2
As-prepared						
Nd	64.17	13.58	4.08		18.17	
Y	74.65	15.80	5.24	6.94		
PM1	84.39		6.71	8.90		
PM2	70.61	14.70	6.32	8.38		
PM3	68.68	14.57	6.26	8.30		2.19
After nitriding (theoretical)						
Nd		84.41	2.86		12.74	
Y		92.00	3.44	4.56		
PM1		90.00	4.30	5.70		
PM2		90.00	4.30	5.70		
PM3		88.50	4.30	5.70		1.50

Si: Elkem Metals Co., Buffalo, NY: Grade Si-HQ, 99.95%; 4.2 µm
Si_3N_4: Ube Industries, Japan; Grade E-10
Al_2O_3: Reynolds, Malakoff, TX; Grade RC-HP DBM
Y_2O_3: Molycorp, White Plains, NY; Grade 5600, 99.99%
Nd_2O_3: Molycorp, White Plains, NY; Grade 5410, 99.9%
SiO_2: U. S. Silica, Berkeley Springs, WV; Micronized silica, 5 µm

After milling, the slurries were vacuum deaired and cast in 101.6 x 254 x 9.53 mm (4 x10 x 0.375 in.) plate molds made of aluminum. The castings were then gelled by placing them in an oven at 65°C for 1h. The molds were cooled to room temperature and the gelled plates were removed. The plates were placed in a closed container to prevent rapid drying of the surface during the initial drying stage when the part shrinks until particle-particle contacts are formed. The initial drying stage typically lasted for about 12 h, after which no additional shrinkage was observed. The drying was then completed by allowing the remaining liquid to evaporate under ambient conditions.

The organic binders were removed from the dried parts by slowly heating the samples in air. Most of the organic was burned off at less than 400°C, but the parts were heated to 600°C to insure complete combustion of any residue. The green density of the samples was estimated by weighing the parts and measuring their dimensions. Most of the samples prepared using the aqueous gelcasting system were isopressed at 69-379 MPa (10-55 ksi) to increase the green density.

Nitriding of the gelcast samples was done in three different furnaces: a graphite element furnace, a tungsten element furnace, and a microwave furnace. The nitriding was done using a flowing atmosphere of high-purity nitrogen. One run in the tungsten furnace used a N_2-4% H_2 atmosphere. These nitriding conditions are summarized in Table II. The temperature profiles used for the nitriding runs are shown in Fig. 1. The profiles for the aqueous and the alcohol gelcastings varied slightly, with the aqueous system samples having an initial 1-h hold at 1250°C, while the initial hold for the alcohol system samples was at 1200°C. The extent to which the silicon was converted to silicon nitride during the nitridation cycle was calculated based on a theoretical weight gain for the silicon of 66.5% for 100% conversion. The presence of oxygen on the surface of the silicon particles was not taken into consideration. Knowing the amount of silicon in each sample before the nitriding run and the weight gain of the sample during the run, the percent nitridation was calculated. The samples which were nitrided in the microwave furnace were taken directly from the 1450°C hold temperature to the sintering temperature of 1800°C while maintaining the flowing nitrogen atmosphere. The estimate of the percent nitridation for these samples was affected by any weight loss which may have occurred during the sintering part of the run.

Table II. Nitriding of Gelcast Sintered Reaction Bonded Silicon Nitride

Sample	Green Density (kg/m^3)	Nitriding Furnace	Nitriding Atmosphere	Percent Nitridation*
		Acidic aqueous gelcasting		
Nd 1	1220 (43%)	Graphite	N_2	96.8
Nd 2	1540 (54%)†	Graphite	N_2	96.3
Nd 3	1670 (59%)†	Graphite	N_2	96.5
Nd 4		Graphite	N_2	87.8
Y 1	1610 (64%)†	Graphite	N_2	92.0
Y 2	1610 (64%)†	Graphite	N_2	91.5
Y 3	1610 (64%)†	Graphite	N_2	83.8
Y 4	1590 (63%)†	Graphite	N_2	83.8
Y 5	1600 (64%)†	Graphite	N_2	89.9
		Alcohol-based gelcasting		
PM1 1-2	1380 (55%)	Tungsten	N_2	78.9
PM2 1-2	1340 (51%)	Tungsten	N_2	82.2
PM3 1-2	1400 (53%)	Tungsten	N_2	79.6
PM1 3-2	1380 (55%)	Tungsten	N_2-4% H_2	89.7
PM2 3-2	1340 (51%)	Tungsten	N_2-4% H_2	96.0
PM3 3-2	1400 (53%)	Tungsten	N_2-4% H_2	95.0
PM1 4-2	1380 (55%)	Graphite	N_2	91.8
PM2 4-2	1340 (51%)	Graphite	N_2	91.2
PM3 4-2	1400 (53%)	Graphite	N_2	88.4
PM1 9-1	1380 (55%)	Microwave	N_2	90.3**
PM2 9-1	1340 (51%)	Microwave	N_2	92.0**
PM3 9-1	1400 (53%)	Microwave	N_2	87.2**

*Based on starting silicon content in the sample and the measured weight gain after nitriding.
†After isopressing.
**Microwave processing was done in one step which included both nitriding and sintering. The calculation to estimate the percent nitridation was affected by weight loss during the sintering portion of the firing cycle.

The samples which were nitrided in the tungsten and graphite furnaces were packed in coarse silicon nitride powder inside a graphite crucible and placed in a graphite resistance furnace for sintering. The samples were heated at 10°C/min to 1800°C and held at that temperature for 2h. The sintering took place under a static nitrogen pressure of 172 kPa (25 psia).

The sintered materials were machined to form B-size (MIL-STD-1942A) 4-point bend test bars. Densities of the machined bars were measured by Archimedes method using high purity ethyl alcohol. The test bar dimensions were 3 x 4 x 50 mm, with the edges of the bars chamfered at a 45° angle. The fixture used in testing the bars had an outer span of 40 mm and an inner span of 20 mm. The bars were loaded to failure in a screw-driven universal testing machine (Instron Model 4465). The fracture strength, σ_f, was calculated using the following equation:[11]

$$\sigma_f = \frac{3P(l_1 - l_2)}{2bd^2} \tag{1}$$

where P is the applied load at fracture, l_1 is the outer span of the test fixture, l_2 is the inner span, b is the width (or breadth) of the test bar, and d is the thickness (or depth) of the bar. The Weibull modulus, m, was estimated using the following equation:[12]

$$m = 1.21\left(\frac{\overline{\sigma}_f}{s}\right) - 0.47 \qquad (2)$$

where σ_f is the average fracture strength of a group of test specimens and s is the standard deviation of the fracture strength.

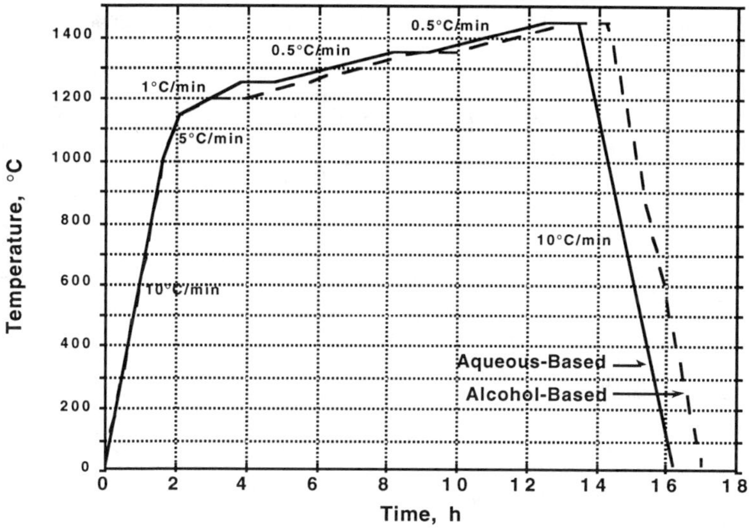

Fig. 1. Firing schedules used for nitriding of the SRBSN compositions.

RESULTS AND DISCUSSION

The green density of the as-cast samples showed a distinct difference between the aqueous- and the alcohol-based gelcastings. The lower solids content in the aqueous slurries resulted in green densities of about 43%; the alcohol-based samples had green densities of 51-55%. Most of the aqueous-based samples were isopressed prior to nitriding. Sample Nd 1 remained as-cast, Nd 2 was isopressed at 69 MPa (10 ksi), sample Nd 3 at 345 MPa (50 ksi), sample Nd 4 and all of the Y samples were isopressed at 379 MPa (55 ksi). The resulting green densities are shown in Table II.

A high green density can be detrimental in conventional nitriding processes due to the tendency of the nitriding reaction to start at the surface of the sample and progress inward. The volume expansion can close off the pore structure and prevent the nitrogen gas from reaching the unreacted silicon in the interior of the sample, especially in thick samples. However, as shown by the results in Table II, the percent nitridation of the gelcast samples showed no consistent dependency on the green density. There were, however, significant effects due to the gas composition and the type of furnace. The addition of 4% H_2 to the nitriding gas increased the nitridation by 14-19% in the tungsten furnace. This result was not surprising, since the hydrogen acts to decompose the oxide layer on the surface of the silicon particles making them more reactive. Both the graphite and the microwave furnaces showed improved nitridation over the tungsten furnace when using a pure nitrogen atmosphere. Samples in the graphite furnace showed 11-16% higher nitridation levels, while those in the microwave were 10-14% higher, even though the percent nitridation of the microwave samples was probably underestimated due to weight losses which typically occur during the sintering part of the cycle.

The density of the fired samples is summarized in Table III. Examination shows a slightly higher density for the samples fired in the one-step process in the microwave furnace. Selected data are plotted in Figs. 2a and 2b to aid in determining relationships between the process variables and the measured properties. There is no apparent correlation between the percent nitridation and the final density. Only the Nd samples show a direct relation between green density and fired density. For the PM samples, the density appears to depend more upon the starting composition than on any other factor.

Table III. Density and Strength Measurements of Gelcast SRBSN

Sample	Fired Density (kg/m^3)	Fired Density (%)	Flexure Strength (MPa)	Number Tested	Weibull Modulus*
		Acidic aqueous gelcasting			
Nd 1	2830	81.8			
Nd 2	3150	91.0			
Nd 3	3220	93.1			
Nd 4	3380	97.7	369 ± 43	13	9.9
Y 1	3150	96.5	497 ± 71	8	8.0
Y 2	3170	97.1	560 ± 88	8	7.3
Y 3	3170	97.2	489 ± 58	15	9.8
Y 4	3190	97.8	595 ± 74	17	9.3
Y 5	3210	98.3	430 ± 54	17	9.2
		Alcohol-based gelcasting			
PM1 1-2	3200	97.7	609 ± 39	10	18.6
PM2 1-2	3160	96.4	587 ± 46	8	15.2
PM3 1-2	3110	95.1	428 ± 111	10	4.2
PM1 3-2	3190	97.3	668 ± 59	6	13.2
PM2 3-2	3190	97.2	574 ± 184	5	3.3
PM3 3-2	3150	96.0	574 ± 133	6	4.8
PM1 4-2	3190	97.5	509 ± 54	6	10.9
PM2 4-2	3160	96.4	741 ± 75	3	11.5
PM3 4-2	3140	95.8	529 ± 34	5	18.2
PM1 9-1	3220	98.3	601 ± 94	10	7.3
PM2 9-1	3210	98.1	599 ± 128	10	5.2
PM3 9-1	3200	97.6	531 ±107	10	5.5

*Estimated using the equation: m = (1.21 x (Average Strength/Standard Deviation)) - 0.47.

The strength and Weibull modulus are shown in Table III and the strength is plotted in Figs. 2a and 2b. Again, the values show no correlation with the percent nitridation. For each of the alcohol-based gelcast batches, the strength shows a weak relationship with the fired density, except for the PM*n* 4-2 samples which show no correlation. The strength and Weibull modulus values obtained in the present study compare favorably with the results reported in earlier work.[3]

CONCLUSIONS

Compositions of high purity silicon metal containing various oxides as sintering aids were gelcast to prepare green bodies for subsequent nitridation and sintering to form sintered reaction bonded silicon nitride (SRBSN). An acidic aqueous- and an alcohol-based gelcasting system were used. In the aqueous system the solids loading was 35-40 vol %, while the alcohol-based slurries contained 48 vol % solids. The solids content in the slurry affected the green density of the cast samples, however, there was no apparent correlation between the green density and the percent nitridation. Samples were nitrided under flowing, high-purity nitrogen in either a tungsten

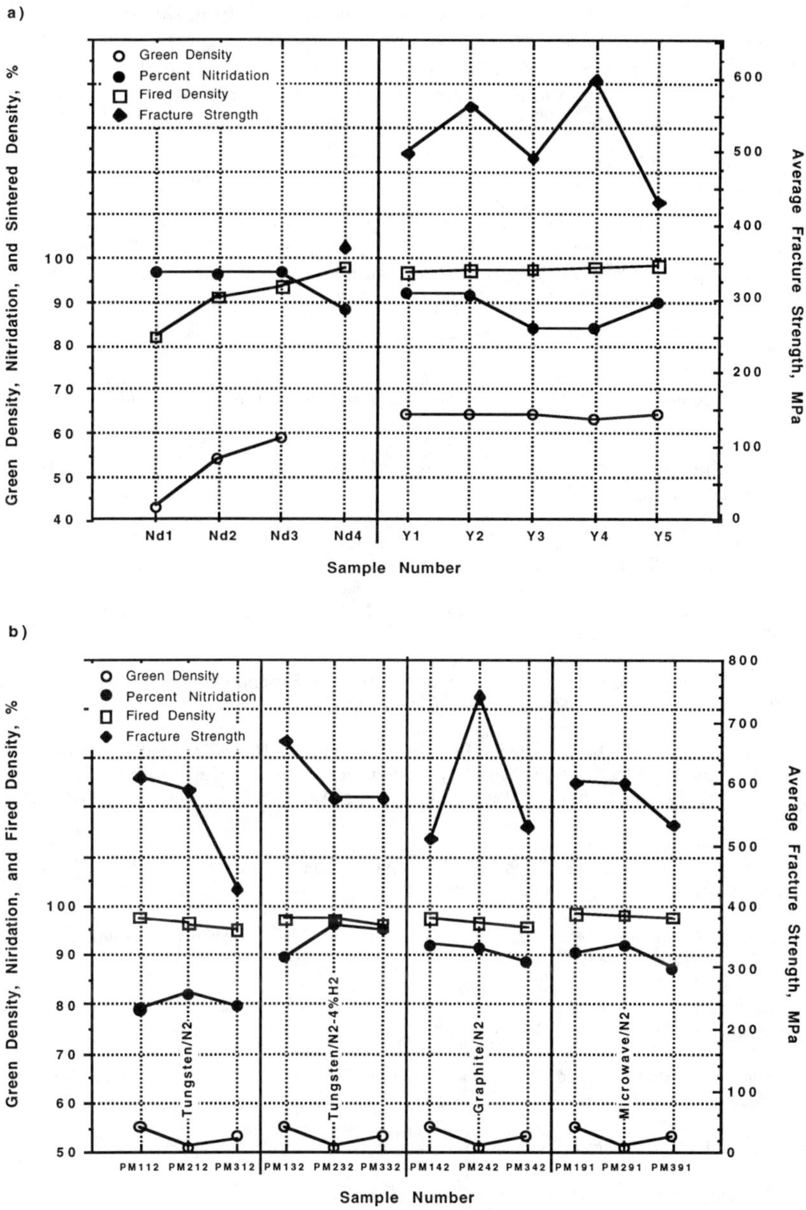

Fig. 2. Comparison of selected sample properties: a) acidic aqueous-based gelcastings, and b) alcohol-based gelcastings.

element furnace, a graphite element furnace, or in a microwave furnace. The samples nitrided in the graphite and microwave furnaces showed 9-16% higher levels of nitridation. In the tungsten furnace, a N_2-4% H_2 gas mixture was also examined in the nitriding runs. The addition of the hydrogen increased nitridation levels by 14-19%. The sintered ceramics had densities of 95-98% of theoretical, with no direct relationship being observed between the level of nitridation and the fired density. The average 4-point flexure strength of the samples ranged from 428 to 741 Mpa.

ACKNOWLEDGMENTS

Research sponsored by the U.S. Department of Energy, Assistant Secretary for Energy Efficiency and Renewable Energy, Office of Transportation Technologies, as part of the Ceramic Technology Project of the Propulsion System Materials Program, under contract DE-AC05-84OR21400 with Martin Marietta Energy Systems, Inc.

REFERENCES

1. A. J. Moulson, "Reaction-Bonded Silicon Nitride: Its Formation and Properties," *J. Mater. Sci.,* **14** 1017-1051 (1979).

2. F. L. Riley, "Nitridation and Reaction Bonding," pp. 265-288 in Nitrogen Ceramics, ed. F. L. Riley, Noordhoff, The Netherlands (1977).

3. T. N. Tiegs, J. O. Kiggans, Jr., and K. L. Ploetz, "Sintered Reaction-Bonded Silicon Nitride by Microwave Heating," *Mat. Res. Soc. Symp. Proc.,* **287** 283-288 (1993).

4. J. A. Mangels and G. J. Tennenhouse, "Densification of Reaction-Bonded Silicon Nitride," *Am. Ceram. Soc. Bull.,* **59** [12] 1216-1222 (1980).

5. J. O. Kiggans, Jr. and T. N. Tiegs, "Characterization of Sintered Reaction Bonded Silicon Nitride Processed by Microwave Heating," *Mat. Res. Soc. Symp. Proc.,* **269** 285-290 (1992).

6. J. O. Kiggans, Jr., C. R. Hubbard, R. R. Steele, H. D. Kimrey, C. E. Holcombe, and T. N. Tiegs, "Characterization of Silicon Nitride Synthesized by Microwave Heating," in Ceramic Transactions, Microwaves: Theory and Applications in Materials Processing, Am. Cer. Soc., Westerville, OH 267-272 (1991).

7. M. A. Janney and O. O. Omatete, "Method for Molding Ceramic Powders Using a Water-Based Gelcasting," U. S. Pat. No. 5 028 362, July 2, 1991.

8. A. C. Young, O. O. Omatete, M. A. Janney, and P. A. Menchhofer, "Gelcasting of Alumina," *J. Am. Ceram. Soc.,* **74** [3] 612-618 (1991).

9. O. O. Omatete, M. A. Janney, and R. A. Strehlow, "Gelcasting—A New Ceramic Forming Process," *Am. Ceram. Soc. Bull.,* **70** [10] 1641-1649 (1991).

10. M. A. Janney and O. O. Omatete, "Method for Molding Ceramic Powders Using a Water-Based Gelcasting Process," U. S. Pat. No. 5 145 908, Sep. 8, 1992.

11. Y. Murakami, Editor-in-Chief, Stress Intensity Factors Handbook, Pergamon Press, 1987.

12. Jeffrey Neil, GTE Laboratories Incorporated, unpublished work.

FILTER-PRESSING OF ALUMINA DISPERSIONS

R. Fries and B. Rand
School of Materials
The University of Leeds
Leeds, LS2 9JT, UK.

ABSTRACT

The filter-pressing characteristics of aqueous alumina dispersions containing either submicron or nano-sized particles have been compared with respect to ionic strength. The highest green densities for both systems were achieved at electrolyte concentrations < 0.01 mol dm^{-3} where long-range repulsive interparticle forces stabilize the slips. A slight increase in density with ionic strength in this range was attributed to an increase in the ratio of particle radius-to-double layer thickness, κa. At higher electrolyte concentrations, above the critical coagulation concentration, the green densities dropped to significantly lower values due to the onset of flocculation and the formation of open particle networks characterized by strong attraction which resisted rearrangement into a dense green microstructure. The green densities of the compacts consolidated from the submicron dispersions at ionic strength > 1 mol dm^{-3} were significantly higher than those prepared close to the critical coagulation concentration although the slips exhibited properties typical for a flocculated structure. The results may indicate the presence of short-range repulsive forces at high salt concentration for the submicron slips not accounted for by the classical DLVO-theory. In order to characterize the early stages of filter-cake consolidation the initial cake permeability was determined from the compaction curves.

INTRODUCTION

The general move towards the use of ultra-fine powders for engineering ceramics applications in order to reduce sintering times and temperatures, minimize grain growth or develop novel properties in the final products requires special processing techniques. Due to the presence of van der Waals forces all powders are attractive under atmosphere conditions and this gives rise to a number of problems when the surface-to-volume ratio of the powder is high (1). Submicron and especially nano-sized powders tend to form strong agglomerates which on consolidation lead to an irregular green structure characterized by large voids. Surface hydration effects can further increase the strength of the agglomerates and make their breakdown a difficult task. Therefore colloidal consolidation techniques such as filter-pressing, slip-casting or centrifugal sedimentation are the appropriate processing routes since the interparticle forces in a ceramic suspension can be controlled by the addition of suitable surface active agents (2).

In filter-pressing, a ceramic slip is forced against a filter to separate the particles from the suspending fluid and form a filter-cake which takes the shape of the filter. It has in recent years become a forming method in its own right using contoured porous molds for the automated production of various types of ceramics. Slip-casting is a related forming technique where the capillary suction of the porous mold provides the driving force for the liquid-solid separation. Filter-pressing, however, offers the opportunity for achieving a higher green density in the product and a more rapid consolidation than can be obtained by slip-casting.

The kinetics of filter-pressing are well established and usually follow an approach based on Darcy's law

$$\frac{1}{A}\frac{dV}{dt}=\frac{kP}{\eta x} \tag{1}$$

where the term on the left hand side represents the volume of fluid withdrawn from the slip per unit area and time, k the cake permeability, P the pressure, η the viscosity of the suspension fluid and x the filter-cake thickness. The properties of the slips and the way such slips are consolidated during formation of the filter-cake are influenced by a number of factors. The most important are interparticle forces, particle size distribution, particle shape and density and the imposed perturbations during or prior to consolidation (3).

The present study was aimed at comparing the filter-pressing behavior of electrostatically stabilized submicron and nano-sized alumina dispersions as a function of ionic strength, i.e. electrical double layer thickness. The main points of interest were how the ratio of particle radius, a, to double layer thickness, $1/\kappa$, i.e. the parameter κa, affects the filtration behavior and how the increased surface-to-volume ratio of the nanosized powders is reflected in different filter-pressing characteristics compared to submicron powder slips. Another aspect of interest was the filtration behavior at high electrolyte concentrations, above the critical coagulation concentration. A significant reduction in the attractive interparticle potential due to short range hydration repulsion which resulted in increased filter-cake densities at high ionic strength has recently been reported by Velamakanni et al. (4). A similar reduction in the thickness of the stabilizing layer could be particularly useful for nano-dispersions where the electrical double layer thickness is commensurate with the particle radius.

EXPERIMENTAL

The powders required had to be of uniform particle size and of relatively well understood surface chemistry. It was decided that alumina would be the most appropriate material. Ultra-fine aluminas characteristically comprise metastable phases, thus the nano-sized material chosen was a δ-alumina from Degussa, Al_2O_3 C, produced by flame hydrolysis of $AlCl_3$ (5). The fact that the phase transition to the α-modification inhibits sintering of the metastable forms was not a problem since the study was not designed to extend beyond the green consolidation stage. The BET surface area of the powder was 89 $m^2\ g^{-1}$ which corresponds to an average equivalent spherical particle radius of 10.7 nm. The powder tended to agglomerate strongly when exposed to the atmosphere due presumably to the effect of surface hydration. Therefore the powder was lightly calcined under vacuum to 300 °C and stored in a dry box. Powders from the box were immediately dispersed in the suspension fluid for characterization. The submicron powder was a chemically synthesized α–alumina, AKP-53, produced by Sumitomo. The BET surface area of the material was 13.9 m^2 g^{-1} which corresponds to an average equivalent spherical particle radius of 54.4 nm. The sizes of both powders were confirmed by TEM examination. All suspensions were ball-milled for at least 12 hours to break up agglomerates. The ionic strength was adjusted with KCl which was considered to be an ideal indifferent electrolyte since no sintering was to be attempted in this model investigation. Salt concentrations in the range from 0.001 to 2 mol dm^{-3} were investigated. The solids volume fraction of the slips for filtration were 0.073 and 0.2 for Al_2O_3 C and AKP-53, respectively. However, the slips of ionic strength > 0.1 containing nano-sized powder had to be somewhat diluted to enable effective milling. The amount of slip used for filter-pressing was therefore increased accordingly to achieve the same volume of solids in the compact. Two different pH-values, namely 3.6 and 11.5, achieved by adding, respectively, HCl or NaOH were investigated. These values correspond to a negative and a positive surface potential, respectively.

The filter-pressing device consisted of a die with a moving plunger of 2.54 cm diameter which was sealed from the outer cylinder by an o-ring. The KCl-solution was pressed through a cellulose nitrate filter of 0.1 μm pore size, backed by a steel disc containing small capillaries. The experiments were carried out using the 100 kN load cell of a floor-mounted Instron. A known volume of slip (25 ml) was filter-pressed at constant crosshead speed of 0.2 mm min^{-1} to a maximum pressure of approximately 35 MPa. The variation in displacement and pressure was recorded on a chart recorder.

RESULTS

The investigated range of ionic strength corresponds to double layer thicknesses from < 1 to > 10 nm which results in κa-ratios ranging from approximately 0.5 to > 15 for the nano-sized powder and 5 to > 50 for the submicron material. The small κa-ratios of the nano-sized suspensions are reflected in a strong dependence of their rheological behavior on double layer thickness. Below the critical coagulation concentration (> 0.01 mol dm^{-3}) the nano-sized suspensions showed a transition from viscous to elastic behavior with increasing solids fraction which was associated with the onset of double layer overlap and shifted to higher solids fractions as the ionic strength increased. The rheological properties of the slips are reported in detail elsewhere (6).

Figures 1 and 2 show the volume fractions of solids in the green compacts immediately after ejection from the filter-press and after drying as a function of ionic strength for the nano-sized Al_2O_3 C and the submicron AKP-53 compacts, respectively. Although the general trends of solids fraction against ionic strength are similar for both powders, the green densities of the compacts consolidated from the nano-sized powder reached only about two thirds of the values of the submicron compacts: the volume fractions of solids, ϕ, of the Al_2O_3 C slips ranged from 0.3 to 0.4 (Fig. 2) whereas the AKP-53 values ranged from 0.55 to 0.6 (Fig. 3). The highest densities were generally achieved from the deflocculated slips (ionic strength < 0.01 mol dm^{-3}) where relatively long-range repulsive forces stabilize the particles in suspension. Both figures show a slight increase in filter-cake density with increasing ionic strength, i.e. decreasing double layer thickness, within the low ionic strength regime. However, for the nano-sized powders (Fig. 2) this trend was reversed by the large shrinkage on drying of the sample of lowest ionic strength. The slip stabilized without any KCl addition at pH 3.6 shrank by more than 10 % compared to values of approximately 5 % for the other Al_2O_3 C filter-cakes. In contrast, the more dense submicron compacts did not shrink on drying but expanded by approximately 1 % indicating the slow release of strain elastically stored when compacting the filter-cake during the final stages of the filtration.

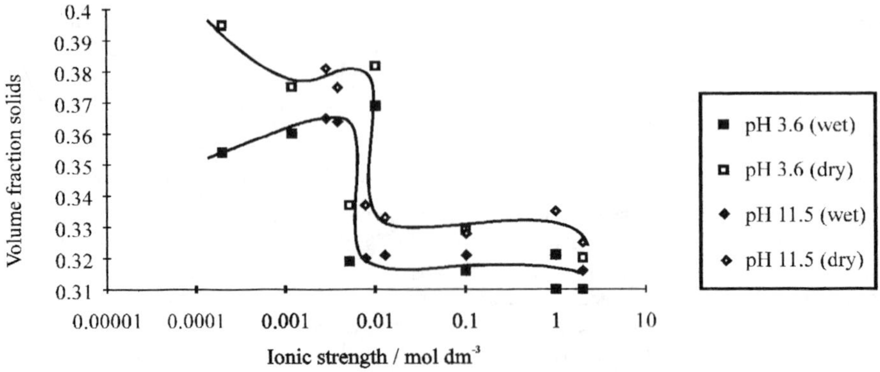

Fig. 1: Green densities of filter-pressed Al_2O_3 C compacts against ionic strength.

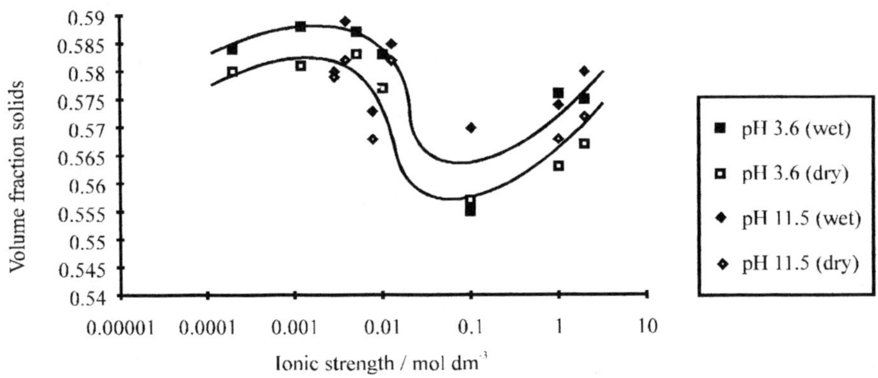

Fig. 2: Green densities of filter-pressed AKP-53 compacts against ionic strength.

When the ionic strength approached values close to the critical coagulation concentration (> 0.01 mol dm^{-3}) a sharp drop in filter-cake solids fraction was detected for both powders. The transition was more abrupt for the nano-sized samples were both high and low filter-cake solids fractions coincided near 0.01 mol dm^{-3} ionic strength. At this electrolyte concentration the slips exhibited minimum viscosities and storage moduli in rheological measurements but became prone to switching to an unstable condition on compaction as their double layers became too thin to provide a sufficient energy barrier against flocculation under the forces encountered during the filter-pressing operation.

Increasing the electrolyte concentration further beyond the critical coagulation concentration had little effect on the densities of the nano-sized powder compacts (Fig. 1). In fact, the filter-cake solids fraction of the Al_2O_3 C compacts slightly decreased with increasing ionic strength up to electrolyte concentrations of 2 mol dm^{-3} indicating strong flocculation in the primary energy minimum. No evidence for any additional short-range repulsive forces was found for the nano-sized Al_2O_3 C in this study. However, the high electrolyte concentration range of the submicron powder was characterized by a marked rise in filter-cake solids fraction when increasing the ionic strength from 0.1 to 2 mol dm^{-3} (Fig. 2). This is inconsistent with the strong flocculation in an infinitely deep potential energy minimum as predicted by the DLVO theory and possibly indicates the presence of additional stabilizing forces.

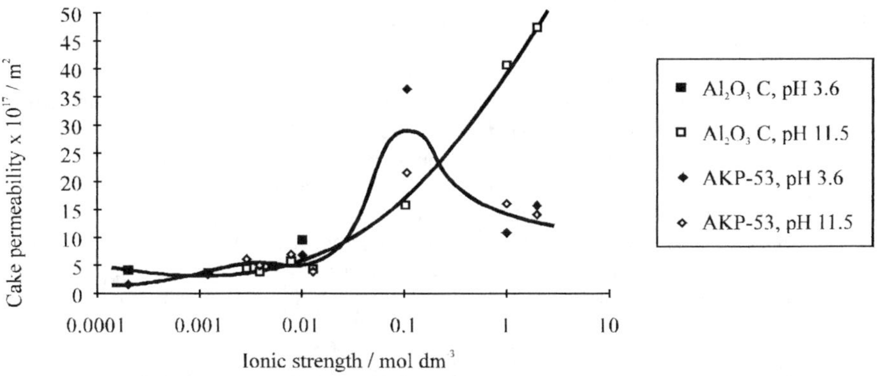

Fig. 3: Initial cake-permeability against ionic strength.

All compaction curves were characterized by a linear increase in pressure with displacement up to the point of particle contact after which the pressure increased exponentially upon further compaction of the filter-cake. A linear increase is expected for a constant rate of water removal, as arranged in the experiments, and implies relatively constant values of the permeability constant and volume fraction of solids in the cake. It was therefore possible to calculate the initial permeability, k, of the filter-cakes with Eq. 1 using the volume fraction of solids at the point of particle contact as ϕ_c and

$$x = L\left(\frac{\phi}{\phi_c - \phi}\right) \tag{2}$$

where L is the measured displacement of the plunger in the filter-press. Figure 3 shows the initial filter-cake permeability as a function of ionic strength. It can be seen that the same trends observed in the green densities of the compacts were already visible during the early compaction stages. Low cake permeabilities, indicating a dense cake structure, are found for the deflocculated samples with little indication of an influence of the actual double layer thickness. Beyond the critical coagulation concentration the permeability increased continuously with ionic strength for the Al_2O_3 C compacts giving no evidence for any additional repulsive forces in this range but showed a distinctive maximum around 0.1 mol dm^{-3} for the AKP-53 samples. The latter results are consistent with the possible presence of non-DLVO forces in the submicron system at high electrolyte concentrations.

DISCUSSION

The green densities of the two studied alumina powders are highest in the low ionic strength regime where long-range repulsion stabilizes the particles in the slips. The slight increase in solids fraction in the compacts with increasing ionic strength is believed to be associated with the decrease in double layer thickness which possibly allows a somewhat closer approach of some of the particles in the filter-cake. Up to the point of particle contact in the consolidation of a deflocculated slurry neighboring particles repel each other due to their electrical double layers. When pushed closer together the double layers act as a lubricant easing particle rearrangement. This leads to a high solids concentration in the filter-cake. Once a stable arrangement has been achieved throughout the filter-press, the force experienced by some of the particles will exceed the height of the potential energy barrier and these particles will flocculate in the primary minimum, i.e. form attractive links with neighboring particles and strongly resist further rearrangement. However, investigations into the consolidation behavior of powder compacts have shown that the particles in such a compact experience a wide range of forces (7). Thus, in filter-pressing operations some particles can be expected to be pushed into the primary minimum while others remain on the repulsive side of the DLVO energy barrier. This implies that when the double layer thickness is significant with respect to particle size it may even in the consolidated state have an influence on the solids fraction in the filter-cake.

When the double layers are large compared to the particle size, as is the case for the deflocculated nano-sized alumina suspensions, the point of particle contact is shifted to low volume fractions around 0.2. On further compression some of the particles will be pushed into the primary minimum in the energy curve while others maintain their large repulsive double layer. With more and more particles becoming attractive rearrangement becomes more difficult and the final green density is relatively low. The microstructures of these compacts are characterized on the one hand by particles maintaining their extended double layers and on the other hand by attractive particles forming an open network throughout the compact that can withstand high compressive stresses and lock in the open microstructure.

As the ionic strength increases, the force required to push particles into the primary minimum decreases with decreasing thickness of the double layer. This leads to an earlier onset of flocculation during the filter-pressing operation and more particles flocculating into the primary minimum. The resulting microstructure is increasingly characterized by an open network of aggregated particles that is difficult to rearrange due to the strong interparticle attraction. The decrease in green density of samples of ionic strength close to the critical coagulation concentration illustrates how limited particle rearrangement overrode the effect of the decreased double layer thickness in this range. It is interesting to note that in spite of their different κa-ratios the slips of both the nano-sized and submicron powders switched to an unstable condition at approximately the same ionic strength (~ 0.01 mol dm^{-3}).

Above the critical coagulation concentration all slips exhibited the rheological behavior of flocculated systems irrespective of their ionic strengths. The low solids concentration at the point of particle contact confirms those results and indicates the presence of an open network of attractive interparticle links at low solids concentrations. However, the relatively high filter-cake densities and the low initial cake permeabilities of the AKP-53 samples at high ionic strength (> 1 mol dm^{-3}) cast some doubt on the validity of the DLVO-theory for this system. Theory predicts the flocculation in an infinitely deep potential energy minimum which is inconsistent with the increasing cake density at high electrolyte concentration. The results seem to suggest that the slips of high ionic strength are only weakly flocculated and that particle rearrangement into a relatively dense green structure is possible under an externally applied force. The relatively low initial cake permeabilities in this range indicate that some particle rearrangement took place even at the moderate pressures of the early filtration stages.

Velamakanni et al. (4) found similar effects when filter-pressing aqueous alumina suspensions at high electrolyte concentration. They suggested hydration repulsion, that is repulsion due to a solvation layer in water, as the dominant mechanism. Assuming that hydrated cations will penetrate into the Stern layer at high ionic strength they suggested that water molecules outside the Stern layer may in turn adsorb onto the cationic surface giving rise to a structured solvation layer of several water molecules thickness extending into the liquid. The resulting short-range repulsion is believed to be associated with the energy required to desorb the hydrated cations from the particle surface. For such a system the deep minimum in the potential energy curve due to van der Waals attraction would be replaced by a moderate solvation minimum at small interparticle distances followed by strong interparticle repulsion as the hydration layers start to overlap. To date, this mechanism has only been reported for negatively charged alumina surfaces. Our findings, however, indicate similar mechanisms at both low and high pH. The fact that the nano-sized Al_2O_3 C slips did not show any hydration effects at high salt concentration is possibly due to the different surface chemistry of the δ-alumina.

CONCLUSIONS

The filter-pressing behavior of electrostatically stabilized suspensions strongly depends on both the surface-to-volume ratio of the powder and the ratio of particle radius-to-double layer thickness, κa. In general, the highest green densities were achieved at low ionic strength, where relatively large double layers stabilized the particles in suspension. For both the nano-sized and submicron slips only a slight increase in density was found as the κa-ratio increased. Two opposing mechanisms seem to affect the consolidation. The decreased double layer thickness allows a closer approach of neighboring particles which leads to a dense green compact. However, the higher ionic strength also leads to a lowering of the energy barrier against flocculation which limits particle rearrangement at high pressures as particles are 'pushed' more easily into the primary minimum. At high salt concentrations and both low and high pH the submicron slips exhibited the rheology of flocculated systems but could be consolidated to densities close to those achieved from deflocculated slips. The results are consistent with the concept of short-range lubricating forces not accounted for by the classical DLVO-theory. In contrast, the densities of the compacts consolidated from the nano-sized Al_2O_3 C powder decreased continuously with increasing ionic strength beyond the critical coagulation concentration giving no evidence for any hydration effects.

ACKNOWLEDGEMENTS

The authors gratefully acknowledge financial support by the CEC.

REFERENCES

(1) R.G. Horn, Surface Forces and Their Action in Ceramic Materials, J. Am. Ceram. Soc., 73, 1117-1135, 1990.
(2) F.F. Lange, Powder Processing Science and Technology for Increased Reliability, J. Am. Ceram. Soc., 72, 3-15, 1989.
(3) R. Fries and B. Rand, "Slip-Casting and Filter-Pressing", Chapter 7 in Materials Science and Technology Vol.17: Processing of Ceramics. Edited by R. Cahn. VCH, Weinheim, to be published in 1995.
(4) B.V. Velamakanni, J.C. Chang, F.F. Lange and D.S. Pearson, New Method for Efficient Colloidal Particle Packing via Modulation of Repulsive Lubricating Hydration Forces, Langmuir, 6, 1323-1325, 1990.
(5) G. W. Kriechbaum and P. Kleinschmidt, Superfine Oxide Powders - Flame Hydrolysis and Hydrothermal Synthesis, Adv. Mater., 10, 330-337, 1989.
(6) B. Rand and R. Fries, Viscoelasticity of Nano-Alumina Dispersions, in this Volume.
(7) L.T. Kuhn, R.M. McMeeking and F.F. Lange, A Model for Powder Consolidation, J. Am. Ceram. Soc., 74, 682-85, 1991.

Mechanism and Kinetics of Electrophoretic Deposition of Al_2O_3

Partho Sarkar and Patrick S. Nicholson

Ceramic Research Group
Department of Materials Science and Engineering
McMaster University
Hamilton, Ontario, Canada

ABSTRACT

The four main electrophoretic deposition (EPD) mechanisms are discussed and their shortcomings pointed out. The Hamaker constant for Al_2O_3 in ethanol suspension is determined by modelling the relationship between particle interaction energy and suspension stability. The Derjagun-Landau-Verwey-Overbeek (DLVO) interaction energy curve for Al_2O_3 particles in ethanol suspension is calculated and the minimum deposition voltage determined. Three probe dc measurements were conducted to explain discrepancies between the calculated and experimentally-observed voltage. A mechanism proposed is based on the DLVO theory and particle-lyosphere destortion/thinning. Kinetic equations for EPD are developed for constant current and constant voltage deposition using mass balance conditions and verified by experimental data.

I INTRODUCTION

In recent years it has become clear that colloid ceramic forming is the best way to reduce the flaw population in ceramic green bodies. Electrophoretic deposition(EPD) is a colloidal process wherein ceramic bodies are shaped directly from a stable colloid suspension by application of an electric dc field.

EPD is a combination of two processes, electrophoresis and deposition. Electrophoresis is the motion of charged particles in a suspension under the influence of an electric field gradient. It was discovered by the Indian scientist, G.M. Bose in the 1740's [1] in a liquid-siphon experiment. The Russian, Reuss (1807[2]), was the first to observe electric-field-induced motion of solid particles (clay) in water.

The second process is deposition, ie.; the coagulation of particles into a dense mass. The present paper discusses both processes, applies modified Derjaguin, Landau, Verwey and Overbeek (DLVO) theory thereto, to explain observed phenomena, and examines the mechanisms and kinetics of EPD.

II EPD OF SUSPENDED PARTICLES

EPD needs a stable suspension with particles charged so as to respond to an applied electric field. Therefore only electrostatic and electrosteric stabilization are appropriate for EPD. Since most of the literature on EPD uses electrostatically stabilized suspensions, the present discussion will be limited to this mechanisms.

Oxide particle surfaces are usually covered by amphoretic hydroxyl groups. These suffer reaction with the H^+/or OH^- of the suspension liquid, which ionize depending on the suspension pH. The reactions are as follows:

$$MOH_{(surface)} + OH^- = MO^-_{(surface)} + H_2O \quad \text{(high pH)} \qquad .....(1)$$

$$MOH_{(surface)} + H^+ = MOH^+_{2(surface)} \quad \text{(low pH)} \qquad ......(2)$$

These equations suggest, there is an intermediate pH where the number of positively and negatively-charged surface groups are equal thus resulting in zero surface charge.

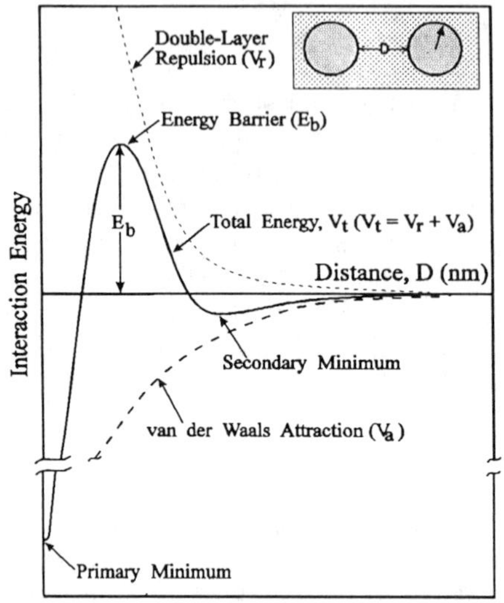

Figure 1. Schematic of the interaction energy as a function of separation between two particles.

Counter ions form a charge cloud around the particles masking their surface charge. This ionic atmosphere is called the "diffuse-double-layer" or "lyosphere". When the particle/lyosphere system moves, fluid mechanics dictates some ions in the lyosphere "shear off" and counter-ions are lost. A moving particle and its remnant double-layer will thus be charged with sign the same as the particle surface charge. The residual potential at the shear plane is called the shear or zeta(ζ)-potential and is experimentally accessible.

Derjaguin and Landau [3] and Verwey and Overbeek [4] developed a theory (DLVO) to describe the stability of a colloid suspension using attractive- and repulsive-energy concepts.

When two particles approach, double layers overlap and a repulsive force is experienced. The origin of the repulsive force between two similarly charged particles is osmotic, not electrostatic. The force at any point between the particles consists of two components ie; an electrostatic component which is always attractive and an osmotic pressure component, always repulsive. In a double-layer, the repulsive osmotic pressure between counter-ions forces them away from the particle surfaces and from each other to increase their configurational entropy. The bringing of two particles together therefore forces the counter-ions from their preferred equilibrium state ie., against the osmotic repulsion but favoured by electrostatic interaction. Since the former dominates the latter, the net force is repulsive. To determine this force, it is necessary to formulate the ionic distribution in the double-layer. There is no analytical solution for the Poisson-Boltzmann equation describing the ionic potential of a spherical double layer, so the repulsive energy is calculated in the present study via an exact numerical solution of the second order Poisson-Boltzmann equation following the algorithm of Chan et al [5].

A van der Waals attraction always exists between similar particles and is given by:

$$V_a = -(A/6D)\{(r_1 r_2/(r_1+r_2)\} \quad .....(3)$$

where A:: Hamaker constant, D:: separation between the particle and r_1, r_2:: radius of particle.

Since this attraction relationship follows a power law, it dominates at small separation. A primary minimum is therefore observed at the position of particle/particle contact (D=0;Figure 1). As separation increases, double-layer repulsion starts to dominate and, as a consequence, the total-interaction-energy curve exhibits a peak the height of which, E_b, constitutes the energy-barrier against coagulation. A secondary minimum develops at large distances thus the attractive energy dominates.

If particles overcome the primary energy barrier(E_b), they will adhere strongly and irreversibly due to the primary minimum. The secondary minimum can also give rise to coagulation but particle adhesion is weak and reversible. The attractive energy is insensitive to the suspension pH and ionic concentration for a given suspension but the double-layer-repulsion is sensitive to these parameters. Therefore, by controlling the pH/ionic concentration of the suspension, it is possible to control the double-layer repulsion, ie., design the total interaction energy curve to induce stability.

III THE DEPOSITION OF SUSPENDED PARTICLES

III-1 Mechanism of Deposition

The literature suggests four mechanisms of deposition during EPD. According to the first, particles undergo an electrode reaction which neutralizes them and a deposit forms. Benson et al [6] proposed that secondary processes at the electrode produce hydroxides which polymerize and absorb on the particles, holding them together in the deposit. Koelman [7] proposed particles accumulate near the electrode due to the applied field then adhere by flocculation associated with the increased ionic concentration and consequent reduced ζ-potential near the electrode. In 1940 Hamaker and Verwey [8] proposed particles are brought to the electrode by the applied field which exerts sufficient force to overcome their mutual repulsion to give coagulation.

III-2 The Kinetics of Electrophoretic Deposition

The rate at which a deposit forms during EPD is very important to control the thickness of the deposited layer and the spacing between the layers of lamellar composites.

If 'A' is the area of the depositing electrode, 'w(t)' the weight of the deposit, 'u' the velocity of the particle, C(t) the concentration of the suspension and 't' the deposition time, then, the deposited weight is given by:

$$dw/dt = AuC(t) \quad .....(4)$$

If the mass balance condition is imposed then at time t=0; $C(0)= w_o/V$ and at t=t; $C(t)= (w_o-w(t))/V$, where w_o is the initial weight of the powder in the suspension and V is the volume of the suspension. The solution using this boundary condition is:

$$w(t) = w_o(1 - e^{-kt}) \quad .....(5)$$

where k is called "kinetic parameter": k = (Au)/V .....(6)

IV EXPERIMENTAL METHOD

Electrophoretic mobility was measured using a doppler-electrophoretic-light scattering analyzer. The ζ-potential was calculated from the mobility using Henry's expression. Optical absorbance was recorded in a UV spectrometer at 450nm wave length as a function of time.

V RESULTS AND DISCUSSION

V-1 Examination of the Mechanism of EPD

The depositing electrode ie.; the cathode, was separated from the suspension by a dialysis membrane. A liquid, the same as the suspension liquid, was filled into the cavity between the cathode and the membrane to provide a path for ions to reach the cathode. The particles stopped at the membrane but the current continued to pass. Thus ions permeate the membrane and complete the electrical path during the EPD process. A deposit formed on the membrane. This clearly demonstrates the particle/electrode reaction is not necessary for electrophoretic deposition as proposed by Mechanism One.

Since deposition occurs at the anode or cathode depending on the particle sign, and, noble metals, carbon etc, can be deposited, a mechanism involving hydroxide formation as proposed by Benson et al is unlikely.

Stable suspensions provide a dense, homogeneous deposit which cannot involve flocculation with its low density structure. Dialysis membrane deposition was observed at any location between the anode and cathode thus eliminating the mechanism proposed by Koelman deposition due to local increases of ionic concentration.

The Hamaker mechanism starts by assuming a particle has reached its fixed position (the origin electrode, in Figure 1) on application of an electric field. A second particle now approaches. It must overcome the energy barrier between it and the stationary particle. If the applied field exerts a force greater than the mutual repulsion force, the second particle will coagulate with the first. The maximum slope on the right hand side of the energy barrier peak represents the minimum force required to coagulate two particles and the minimum electric field required for coagulation can be calculated therefrom by equating the Stoke velocity and the electrophoretic mobility. Several researchers have calculated this electric field and compared it with observed values. They found the experimental field required is much lower than the calculated values. The unexpected discrepencies cast doubt on the Hamaker mechanism and other mechanisms were proposed. Present results explain the discrepency.

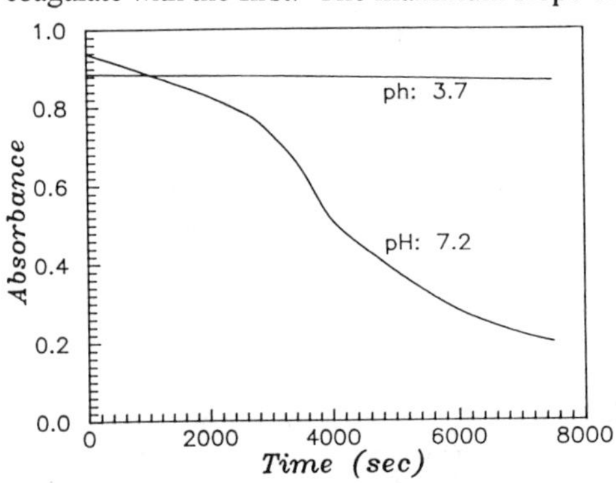

Figure 2. Optical absorbance at 450nm wave length as a function of time, for an Al_2O_3 suspension.

The van der Waals attractive energy is given by Equation (3). The only unknown

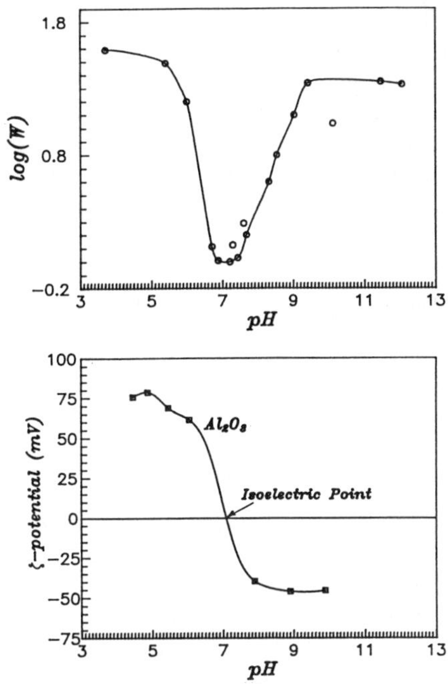

Figure 3. Al_2O_3 suspension (a) stability ratio (W) and (b) ζ-potential, as a function of pH.

parameter is the Hamaker constant, A which can be determined from the suspension stability-ratio and the ζ-potential. The suspension stability-ratio, W, is a measure of the effectiveness of the potential barrier in preventing particle coagulation. It is defined as the ratio of the rate of slow to fast coagulation. Optical measurements were used to determine these rates of coagulation and Figure 2 is a plot of the optical absorbance of an Al_2O_3 suspension at different pH values as a function of time.

The optical absorbance curves exhibit an initial linear region of slope directly proportional to the coagulation rate of the suspension. There is rapid change of absorbance with time at pH 7.2 whereas little change shows at pH 3.7. This indicates the suspension is more stable at pH 3.7 than pH 7.2. This result is not surprising since, pH 7.2 is the isoelectric point. All measurements were thus performed assuming pH 7.2 had the fastest coagulation rate. The slope at pH 7.2 is defined as is R_f and all slopes for pH $\neq$ 7.2 are divided by this value. Figure 3 is a plot of logarithm W, and ζ-potential as functions of pH. The suspension stability is high at low and high pH. At intermediate pH values the suspension stability reaches a minimum between pH 6.89 and 7.4which corresponds to ζ-potential values of 20mV and -18mV respectively.

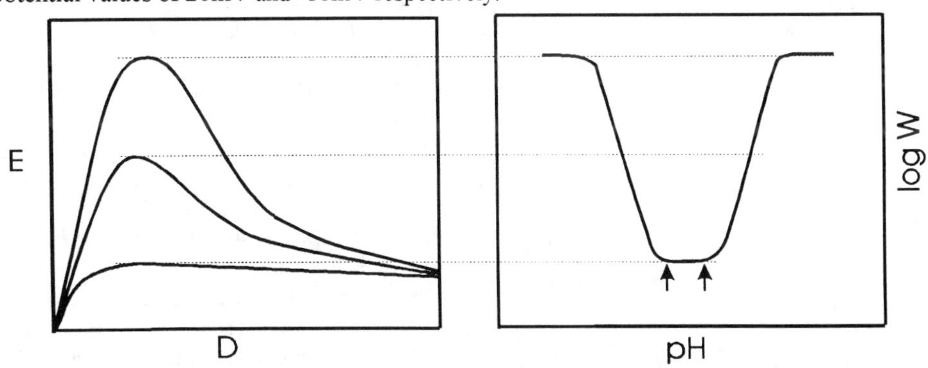

Figure 4. Schematic relationship between the interaction energy and the stability ratio of a suspension.

Since the particle-surface-potential is not experimentally accessible, the surface potential is usually replaced by the ζ-potential in repulsive energy calculations. This substitution is reasonable for organic suspending liquids since the ionic concentration is low. As a result the difference between the surface and ζ-potentials is low.

Figure 4 is a schematic plot of the interaction energy .vs. the suspension stability. When the suspension stability is maximum, the energy barrier, E_b is maximum. As stability decreases, the height of the energy barrier decreases. The energy barrier, E_b is ≤ 0 between the two points shown in the Figure 4. To determine the Hamaker constant, the energy barrier, E_b, is assumed to be zero at point 'a'. The total interaction energy curve was calculated using a ζ-potential value of 20mV and a value of Hamaker constant (fitting parameter) such that, $E_b = 0$ at point 'a'. The identified Hamaker Constant value was 1.1×10^{-19}J which agrees with values reported in the literature (1.12×10^{-19}J).

Figure 5 is a plot of the calculated interaction energy, V_t as a function of the separation between two spherical Al_2O_3 particles at ζ-potentials 80 and 60mV. The total energy, V_t, is expressed in units of kT. The height of the energy barriers are ~110 kT and ~45 kT at ζ-potentials 80 and 60 mV respectively. To determine the maximum force between the two particles, the interparticle force (ie., $-dV_t/dD$) was calculated as a function of distance. The maximum interparticle force represents the minimum force between particles necessary for deposition. The resulting minimum required electrical potentials are 55V/cm and 35V/cm respectively. Experimentally observed values are as low as 1V/cm. This discrepancy is significant and will now be explained.

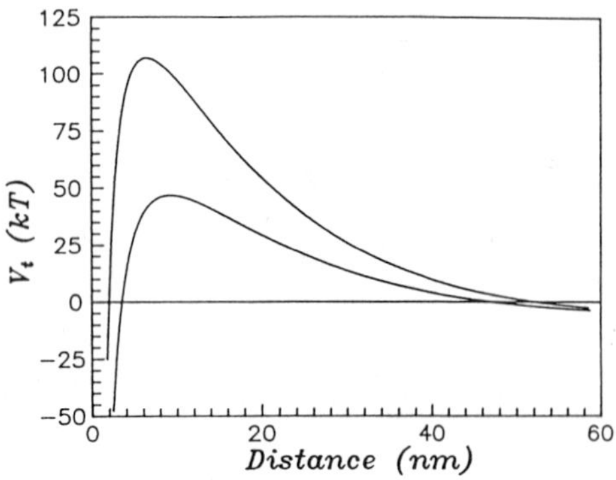

Figure 5. Calculated interaction energies between two Al_2O_3 particles in ethanol suspension at 80mV and 60mV ζ-potential.

The experimental voltage is the average voltage drop per cm between the electrodes ie., the total voltage drop between the anode and cathode divided by the distance between them. Other researchers calculated the experimental voltage drop using this assumption. In the present work three-probe dc measurements were made to precisely profile the voltage around the depositing electrode (the cathode). The third probe was attached to a digital micrometer and its distance from the cathode determined. A voltmeter was connected between the cathode and the third probe and the voltage drop recorded. Figure 6 is a plot of the voltage profile near the cathode under constant current deposition conditions of $0.1 mA/cm^2$ and an electrode separation of 2.3cm. This shows the

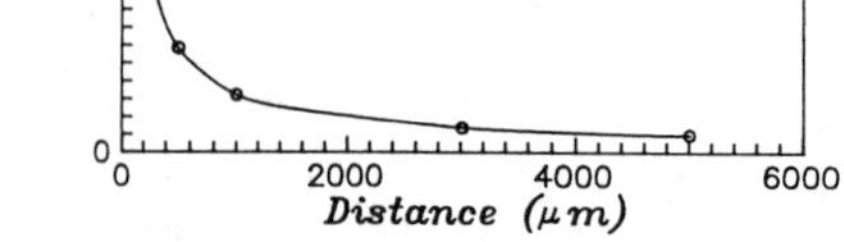

Figure 6. Deposition voltage (volts/cm) as a function of distance from the depositing electrode.

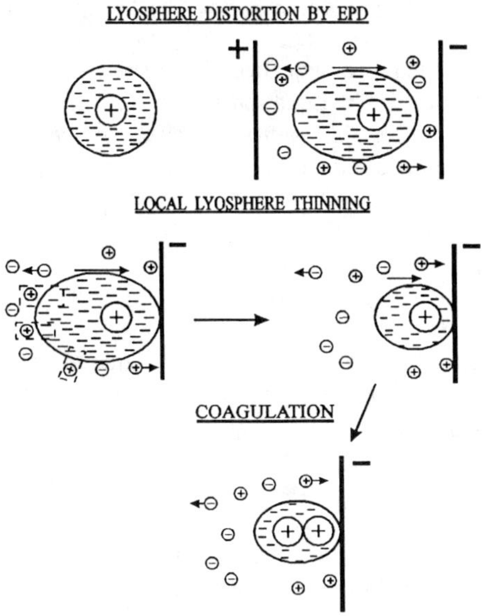

Figure 7. Schematic of the deposition mechanism.

voltage drop is ~140V/cm close to the cathode and it falls off very rapidly with distance approaching a value of the average voltage between the electrodes (~3V/cm). The actual voltage drop per cm near the cathode is ~3X higher than the calculated value. This difference is the source of the discrepancies reported in the literature.

Figure 7 schematically illustrates the modification of the DLVO picture by the EPD process. Application of an electrical field deforms the spherical lyosphere as the particle moves towards the cathode and the anions (the diffuse layer) are attracted to the anode. Not only do particles reach the cathode but cations also do. The latter species increase the local positive-ion concentration at the cathode and these cations will react with the lyosphere anions thus to re-establish chemical equilibrium, initiating its destruction, decreasing the repulsive barrier between particles and inducing coagulation.

VI-2 The Kinetics of Electrophoretic Deposition

An Al_2O_3 suspension (0.76 v/o in absolute alcohol, pH =3.4) was electrophoretically deposited under constant current conditions ($0.56A/cm^2$). Figure 8 is plot of the deposited weight fraction as function of time. The experimental points fit the curve calculated from Equation (5) using k=0.0005. The velocity of the particles during deposition was determined as 64.3μm/sec using this value of k. The particle velocity is the product of electrophoretic mobility (μ_e) times the voltage drop/cm (E) in the suspension.

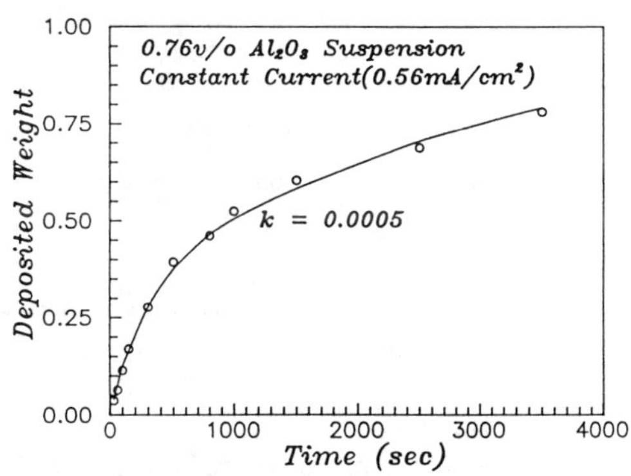

Figure 8. Deposited weight as a function of time for an Al_2O_3 suspension, under constant current conditions.

In the present case, E ≈ 54V/cm thus $\mu_e = 1.2(\mu m\ sec^{-1})/(Vcm^{-1})$, a value that agrees with that measured by laser zeta-meter. Figure 9 is a similar plot for constant voltage deposition conditions. There is considerable disagreement between the calculated and experimental results. The experimental data initially fits with the calculated results, using k = 0.00135 but the calculated weight is higher after longer times. Thus constant voltage data cannot be fitted by a constant k-parameter value. The resistivity of the deposit increases with depositing time thus, under constant voltage the current decreases and the voltage drop/cm decreases with time.

This result demonstrates the need to modify the kinetic equation if constant voltage deposition conditions are utilized. The concentration C(t) in Equation (4) was considered exclusively a function of time. This is valid for constant current conditions. The velocity is a function of time under constantvoltage conditions. The modified equation for constant voltage conditions is;

$$\begin{aligned} dw/dt &= A\,u(t)\,C(t) = A\,\mu_e\,E(t)\,C(t) \\ &= A\mu_e\,E(t)\,((w_o - w(t))/V) = k'\,E(t)(w_o - w(t)) \qquad .....(7) \end{aligned}$$

where $k'=(A\mu_e)/V$, is a modified kinetic parameter with universal application. Equation (7) contains two time-dependent parameters, E(t) and w(t). These are combined and the solution of the equation is:

$$R'w(t) + (R'w_o + 1)\log[(w_o - w(t))/w_o] + k'V_{app.}\,t = 0 \qquad .....(8)$$

where R' = (resistivity ratio of deposit to suspension - 1)/(deposit density x deposit area). The constant-voltage, deposited-weight was recalculated via Equation (8) and is plotted in

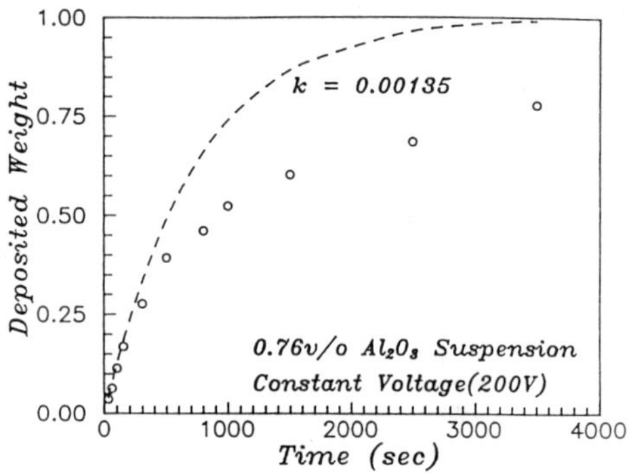

Figure 9. Deposited weight as a function of time for an Al_2O_3 suspension, under constant voltage conditions.

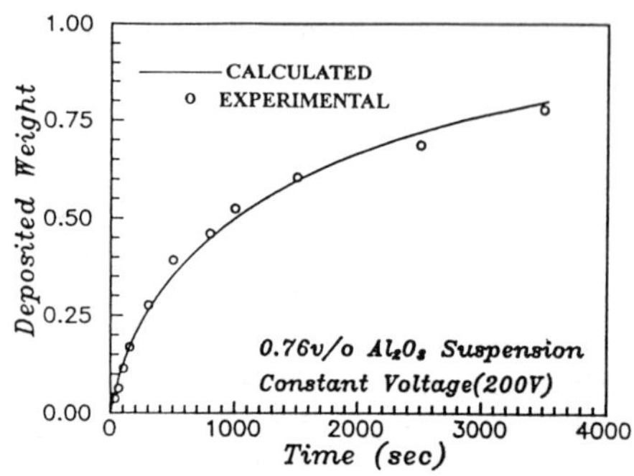

Figure 10. Fit of the experimental data via Equation (8) (for constant voltage condition).

Figure 10 with the experimental data. Good agreement is observed between the experimental and calculated data.

VI SUMMARY

EPD mechanisms have been discussed. A new mechanism is proposed based on DLVO theory and lyosphere distortion/thinning. A method was developed to determine the Hamaker constant and mass balance conditions were used to develop the kinetic equations for constant-current and constant-voltage deposition conditions. The equations were experimentally verified.

References

1. G.M. Bose, Phil. Trans. Roy. Soc. **43** 419 (1745).
2. J.J. Shyne and H.G. Scheible, Modern Electroplating, Ed. by P.A. Lowenheim, John Wiley and Sons, New York, p714 (1963).
3. B.V. Derjaguin and L. Landau, Acta Physicochim, **14** 633 (1941).
4. E.J.W. Verwey and J.Th.G. Overbeek, Theory of Stability of Lyophobic Colloids, Elsevier, Amsterdam (1948).
5. D.Y.C. Chan, R.M. Pashley and L.R. White, J. Colloid Inter. Sci. **77** 283 (1980).
6. P. Benson, Electrochim. Acta, **9** 275 (1964).
7. H. Koelmans, Philips Res. Rep. **10** 161 (1955).
8. H.C.Hamaker and E.J.W Verwey Trans. Fara. Soc. **36** 180 (1940).

ACRYLIC BINDERS FOR ALUMINA AND OTHER NON-CLAY CERAMIC EXTRUSIONS

X.L. Kevin Wu and Tony T. Donato
Rohm and Haas Company, Spring House, PA 19477

The role of binder in non-clay ceramics extrusion processes is to provide plasticity for the forming process and green strength for better handling. Recently, an aqueous acrylic emulsion binder system for extruding alumina and other non-clay ceramics has been developed. The influence of binders on the mechanical properties of the extrudates and the burnout properties of the binders will be discussed.

INTRODUCTION

Extrusion processes have been used extensively in the manufacture of ceramic parts with constant cross-sections, such as hollow tubes, rods, and honeycomb catalytic substrates. A fundamental requirement for extrusion is the plasticity of the feed material. When additives are combined with water or other liquid, they greatly contribute to the plasticity of the feed material for extrusion. Other than clay, which is the most common additive[1], cellulose is widely used as a binder for ceramic extrusion.[2,3] Recent studies have shown that certain synthetic polymers, such as polyethylene glycol, polyethylene oxide,[4] and polyvinyl alcohol,[5] can also be used as binders for ceramic extrusion.

Cellulosic binders are manufactured by chemical reactions between natural cellulose and a variety of chemicals. Methyl cellulose binders provide high plasticity for the feed materials and high green strength for the extrudates.[3,6] When these binders are used at temperatures above their gellation points, their cohesive properties are satisfactory for high-quality surface finishes. In almost all cases, however, it is not practical to extrude ceramics at elevated temperatures. Heat generated during high-shear mixing of ceramic powder, binder, and liquid often induces local gellation of the cellulose binders, which results in an inhomogeneous system with domains of high binder concentration. Cracks occurred during drying significantly increase the scrap rate. Drying the extrudates slowly in a controlled humidity environment might overcome this effect; however, long drying, especially for parts of large cross-section, could significantly decrease the production efficiency. In addition, cellulosic binders produce a high char yield when burned out in an inert atmosphere, prohibiting their use with Si_3N_4, SiC, and other non-oxide ceramics, which must be inert-fired.

An ideal binder for ceramic extrusion should provide high plasticity to the paste or particulate feed materials. The binder should also impart the appropriate green strength for post machining. If the elastic modulus is too high, machining green extrudates using conventional grinding wheels or cutting tools may be difficult or inefficient. On the other hand, if the green strength of the extrudates is too low, a high scrap rate will be produced.

Our studies of using aqueous-based polyacrylic binders for ceramic extrusion will be discussed in this article. Emphasis is placed on the performance of aqueous-based polyacrylic binders, how they influence the mechanical properties of the extruded ceramic materials, and the burnout characteristics of these acrylic binders.

EXPERIMENTAL PROCEDURE

(1) Raw Materials and Extrusion

The alumina powder was Alcoa A-16SG (Alcoa Chemicals, Bauxite, AK) with a median particle size of 0.48 μm and specific surface area of 8.2 m^2/g. The SiC (800 grit) was from Washington Mills Electro Minerals Corporation, and the partially (Y_2O_3) stabilized zirconia powder was HSY-3.0 [Zirconia Sales (America) Inc.] with an average particle size of 0.7 μm and a specific surface area of 6.5 m^2/g. Polyacrylic binders, Duramax™ B-1039 binder ("B-1039"), Duramax™ B-1051 binder ("B-1051"), and Duramax™ B-1052 binder ("B-1052") were from the Rohm and Haas Company. All acrylic binders were aqueous emulsions and were used as supplied. Organic base AMP-95™ ("AMP-95", 95% 2-amino-2-methyl-1-propanol solution) was from the Angus Chemical Company (Buffalo Grove, IL). Carbowax™ polyethylene glycol 400 was used as a lubricant ("PEG-400", Union Carbide Co., Danbury, CT). The cellulose binder was Methocel™ A4M (Dow Chemical Company, Midland, MI). Deionized water was used in the formulation.

All the raw materials were admixed in a blender (Cuisinart Classic Model). When acrylic binders were used, the solution of AMP-95™ was added into the alumina powder during the mixing. After 5 min. of mixing, a measured amount of the premixed acrylic emulsion binders was added and subsequently mixed for another five minutes. The intensive mixing was to ensure homogeneity of the binder, liquid, and powder. Finally the remaining water was added and mixed thoroughly. The particulate paste was further mixed with a Ross Double Planetary mixer (Charles Ross & Son Company, Hauppauge, NY).

The final materials were rolled through a three-roll mill (Anthony Mfg. Co., Lindenhurt, NY) before extrusion. A flake-like material was obtained after milling and was broken down by the blender to small particulates before extrusion. The alumina rods were extruded using a ram extruder (2" diameter barrel with 12" stroke, Loomis 232-20L) at room temperature through a round die with a diameter of 0.792 centimeters.

(2) Characterization of the Extrudates

The extruded alumina rods were dried in air for more than 16 hours at room temperature and then were placed in an oven at 75°C for over 12 hours before mechanical testing. The results of the mechanical testing of the extruded green rods were generated by a universal testing machine (Instron Model 1122) at a crosshead speed of 2.5mm/min. A three point bending method was used to determine the green strength of the rods.

(3) Binder Characterization

Samples of binder were dried at 70°C for at least 72 hours in a forced-air oven and then cooled to room temperature. Burnout residue was determined gravimetrically to an accuracy of ± 0.0001g. An ashing furnace and a tube furnace were used to conduct the binder's pyrolysis in air and in nitrogen, respectively. The burnout profiles of the binders were evaluated using thermal gravimetrical analyses (TGA) at a heating rate of 20°C/min. from room temperature to 800°C. Glass transition temperatures were determined by differential scanning calorimetry (DSC). To minimize the loss of absorbed moisture, DSC samples were placed in sealed sample boats and tested without thermal pre-annealing.

RESULTS AND DISCUSSION

(1) Influence of the Binders on the Extrusion

Two typical formulations of the extrusion pastes are shown in Table 1. The extruded alumina rods using the acrylic binders are shown in Figure 1. The plasticity of the paste was attributed to three acrylic binders, B-1039, B-1051, and B-1052. Binders B-1051 and B-1039 mainly contribute the high-yield stress for the paste, which consequently imparts the wet green strength and dimensional stability for the extrudate. Binder B-1052, on the other hand, mainly provides plasticity for the paste and ductility in the dried extrudate. The dried extrudates produced from the formulation labeled "high modulus" have higher green strength and elastic modulus than the extrudates produced using the formulation labeled "low modulus" in Table 1. In contrast, using the formulation labeled "low modulus", the dried extrudates have higher ductility and plasticity than the extrudates produced using the formulation labeled "high modulus" in Table 1. We believe that for most ceramic extrusion, the formulation labeled "low modulus" will perform better than the formulation labeled "high modulus" , except for those ceramic parts with very thin cross-sectional areas, such as honeycomb and thin-wall tubes.

Table 1. Typical Paste Formulation

Component	Weight (high modulus)	Weight (low modulus)	Function
A-16SG Al_2O_3	100.0	100.0	ceramic powder
PEG-400	0.5	0.5	lubricant
B-1039	2.7*	---	binder
B-1051	1.3*	1.3*	binder
B-1052	---	2.7	binder
AMP-95	1.0	1.0	pH control
water	20.5‡	19.1‡	media

*net polymer weight of the aqueous emulsions

‡includes water from the aqueous emulsions

The glass transition temperatures determined by differential scanning calorimetry (DSC) of binders B-1039, B-1051, and B-1052 are 169°C, 81°C, and 6°C, respectively. The glass transition temperature (Tg) is the temperature at which a noncrystalline polymer takes on the characteristic properties of the glassy state--brittleness, stiffness, and rigidity. Like many other polymeric materials, when a binder is used at a temperature well above its Tg, the binder will be soft, flexible, very ductile, and elongated. On the other hand, when the application temperature of a binder is well below its Tg, the binder will be rigid and brittle. Two differentiated Tgs will be found when two polymer binders with distinguishable Tgs are blended together. However, for ceramic application, a blended binder could be considered as one binder with one "apparent" Tg, which is determined by the ratio of the binders used in the blend.

AMP-95 raises the pH of the paste to enhance the performance of both binders, especially B-1051. The advantage of using AMP-95™ is that it volatilizes completely at low temperatures; the boiling point is 164°C. Other inorganic bases, such as ammonium hydroxide (NH_4OH), ammonium carbonate ($(NH_4)_2CO_3$), and sodium hydroxide (NaOH), should have the same function as AMP-95. The negative effects of these bases are either a very strong irritating odor released during the mixing or alkaline metal residues left in the final ceramics.

The paste formulations shown in Table 1 are very cohesive, producing ceramic sheets with smooth surfaces. The other benefit of the cohesiveness is that extruded sheets will not stick to each other. No phase separation was observed during the extrusion, which is probably due to the strong interaction between water and the acrylic binders. The amount of water in the formulation can be adjusted according to the working pressure of the extruder and other production requirements. A low water-content paste requires higher extrusion pressure; therefore, the extrudates will have greater dimensional retention.

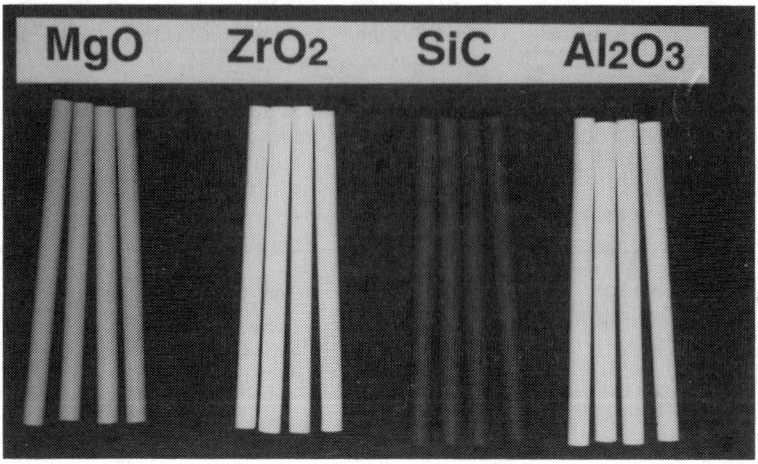

Figure 1. Dried ceramic extrudates using acrylic binders.

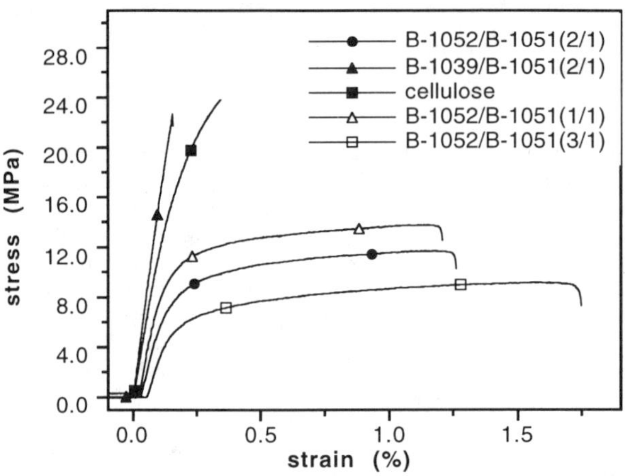

Figure 2. Strain and stress curves obtained from the flexural tests of dried ceramic extrudates produced using acrylic and cellulose binders.

Generally the pastes formulated with acrylic binders are very cohesive, which is highly desirable for a ceramic extrusion process using a ram extruder. To obtain a paste with the maximum cohesive property, the amount of water added to the formulation should be tightly controlled. Too much water often decreases the cohesive properties. On the other hand, it is desirable to attain a high-packing density of ceramic particles in the binder system while maintaining acceptable extrusion pressure. High-packing density requires high-solid loading; however, low-extrusion pressure demands higher levels of binder and of higher water content or additional effective lubricant. A rule of thumb is that if the extruder can tolerate high-extrusion pressure and if the resulting green strength is adequate, small amounts of binder and water should be used in the formulation.

2. Mechanical Properties of the Green Extrudates

For non-clay ceramics, the mechanical properties of the dried green extrudate are primarily controlled by the binders. One of the significant advantages of using the acrylic binder system is that the mechanical properties (elastic modulus, green strength, and ductility) of the extrudate can be readily varied by different binder combinations and by binder concentrations used in the feed material formulations.

The results in Figure 2 represent the acrylic binders that offer tunable mechanical properties for dried green extrudates with sufficient extrudability. At the 4% binder level, the mechanical behavior of the extrudates varied from very brittle with a high-elastic modulus to high ductile with a low-elastic modulus. The elastic moduli of acrylic binders can be either higher or lower than the widely used cellulose binder. The variation provides broad process latitude, which enables manufacturing of a variety of ceramic parts by post-machining of extrudates. A blend of B-1039 and B-1051 binders will produce high elastic and high green strength extrudates that are suitable for grinding and simple cutting. A combination of B-1052 and B-1051 can be used to produce a low to medium elastic modulus extrudate.

Figure 3a and 3b plot the binder concentration dependence of the green strength and the elastic modulus of extrudates. Both the green strength and the elastic modulus of the extrudate increase with an increase in the binder level, indicating that binders contribute to the mechanical properties of extrudates. Figure 3 also shows that the elastic modulus and the green strength of the extrudate produced using combinations of B-1039 and B-1051 are significantly higher than those extrudates using combinations of B-1052 and B-1051.

The elastic modulus and green strength of extrudates can also be varied by different binder combinations as shown in Figure 3c. The elastic moduli and the green strength of the extrudates using B-1052 and B-1051 are significantly lower than those using B-1039 and B-1051. It is much easier to form a homogeneous feed material when binder B-1052 and B-1051 are employed. When B-1039/B-1051 is used with high-intensity mixing, a homogeneous feed material may also be obtained.

Many complex-shaped ceramic parts are manufactured through post-machining of green extrudates. The ductility and elasticity required for each machining process are different. The ductility of the extrudates using acrylic binders can be tuned by different concentrations, by different binder combinations, or by changing the ratio of the binder in each combination. The plots in Figure 4a,b demonstrate that the ductility of the extrudates may be changed either by an increase in binder concentration, by different binder combinations, or by a change in binder ratio. It is obvious that the most practical way of increasing ductility is to use the binder combination of B-1052/B-1051. Therefore, these results indicate that the extrudates with B-1051/B-1052 are more ductile than the extrudates with B-1039/B-1051.

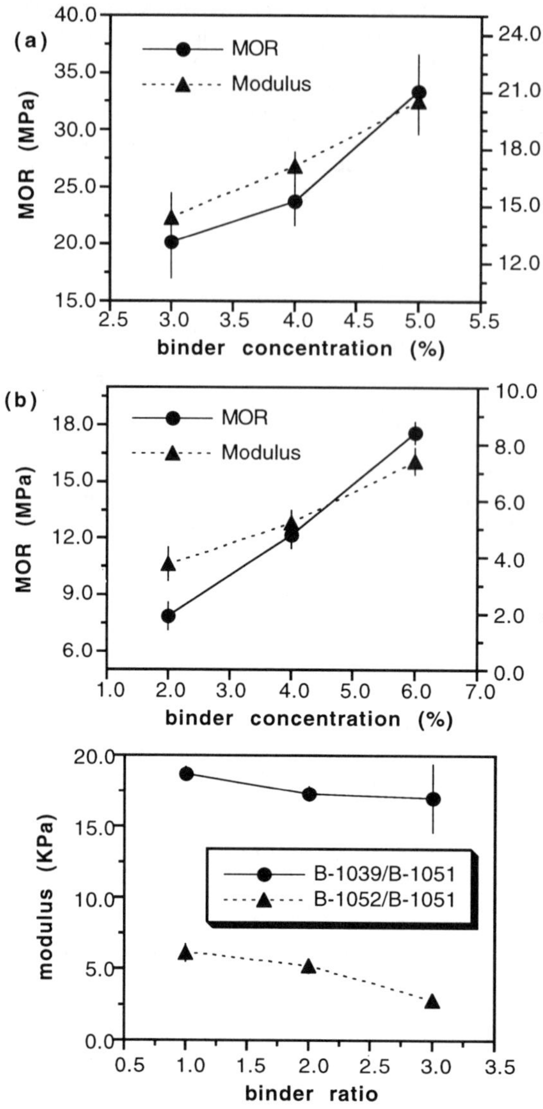

Figure 3a,b,c. Green strengths and elastic moduli of dried ceramic extrudates produced using acrylic binders. (a) B-1039/B-1051 at the ratio of 2/1, (b) B-1052/B-1051 at the ratio of 2/1, (c) 4% binder.

*(3) **Burnout Characteristics***

Burnout characteristics of the binders were studied by thermogravimetric analyses (TGA) and pyrolysis of bulk samples. During the firing, the binders were removed by thermolysis either in oxidizing, reducing, or inert atmospheres. Both binder B-1039 and B-1052 started to lose weight at approximately 220°C and 300°C, respectively, and continuously lost weight with a relative sharp transition between 400°C and 500°C. The burnout temperature ranges for B-1039 and B-1052 were 280°C and 200°C, respectively. In contrast, binder B-1051 showed a relatively broader temperature range of burnout (350°C) and a sharper weight loss at similar temperatures. When both binders are pyrolized in nitrogen, the burnout profiles are similar to the ones obtained in air. However, there is a shift to higher temperatures. The similarity in pyrolysis profiles in air and nitrogen suggests that the binders decompose by analogous mechanisms in both atmospheres.

The small sample size and limited dynamic range of TGA limit the accuracy in determining burnout residues. The burnout residues of two acrylic binders and a cellulosic binder are listed in Table 2. The small amount of residue left after air pyrolysis are ash by-products from salts used in the manufacturing of the binders. Residues after nitrogen firing include this ash and carbonaceous material left from incomplete binder burnout. The high residue after firing cellulose under nitrogen indicates that oxygen is critical for complete removal of this binder. It is well known that acrylic polymers undergo depolymerization or "unzipping" with no need for added oxygen, explaining that the depolymerization does not strongly depend on the atmospheres.[7]

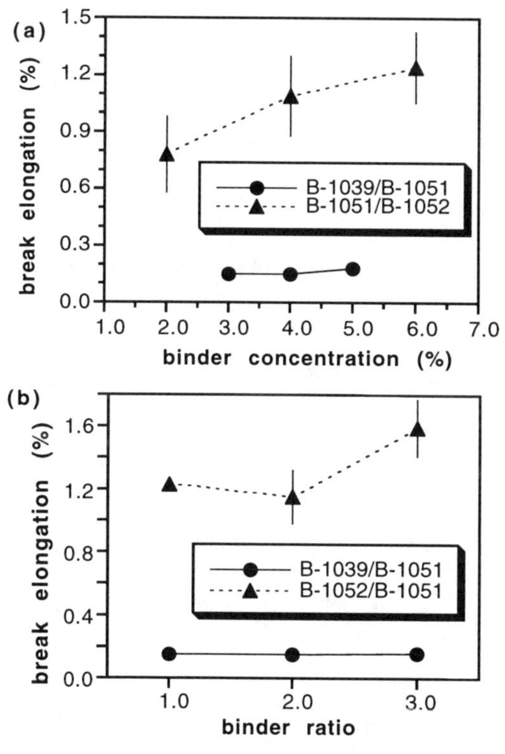

Figure 4a,b. The ductility of dried ceramic extrudates produced with acrylic binders. (a) Binder ratios of B-1039/B-1051 and B-1052/B-1051 are 2/1 and (b) 4% binder.

Table 2. Burnout Residues of Polymeric Binders

Binder	Residue (%, air)	Residue (%, N_2)
B-1039	0.23±0.02	2.1±0.1
B-1051	0.26±0.02	4.4±0.1
B-1052	1.3±0.3	5.5±0.2
cellulose	0.22±0.02	13.3±0.5

CONCLUSIONS

High quality alumina and other non-clay ceramics were extruded using acrylic binders. These acrylic binders show high cohesive and plastic properties in the feed paste. The mechanical properties (strength, elastic modulus, brittleness, and plasticity of the extrudate) can be varied by using the appropriate combination of acrylic binders. The burnout profiles of acrylic and cellulose binders were reported and discussed. The thermal pyrolysis results obtained in air show that the residue of acrylic binder B-1052 is comparable or slightly higher than that of the cellulose binder. However, the residue of cellulose in nitrogen is significantly higher than that of the acrylic binders.

REFERENCES

[1]W. Gleißle, J. Graczyk, "Formbarkeit keramischer Massen durch Extrusion," *Chem.-Ing.-Tech.*, **65**(10) 1206-13 (1993).

[2]J.S. Reed, Introduction to the Principles of Ceramic Processing; pp.359-60. Wiley, New York, 1988.

[3]L.E. Schuetz, "Methylcellulose Polymers as Binders for Extrusion of Ceramics," *Am. Ceram. Soc. Bull.*, **65**[12] 1556-59 (1986).

[4]J. Zheng, J.S. Reed, and S.K. Verma, "Ceramic Extrusion with PEG," Am. *Ceram. Soc. Bull.*, **73**[12] 61-66 (1994).

[5]A.Y. Chen and J.C. Cawley, "Extrusion of Al_2O_3 Ceramics with Low Organic Content," *Ceram. Eng. Sci. Proc.*, **10**[7-8] 773-783 (1989).

[6]H. Böhm and S. Blackburn, "Agglomerate Breakdown in Fine Alumina Powder by Multiple Extrusion," *J. Mat. Sci.*, **29**, 5779-86 (1994).

[7]Y.N. Sun, M.D. Sacks, and J.W. Williams, "Pyrolysis Behavior of Acrylic Polymers and Acrylic /Ceramic Mixtures," *Ceras. Trans.*, **1**, 538-48 (1988).

BINDERS FOR FLEXIBLE CERAMIC SHEET EXTRUSION

X.L. Kevin Wu and Tony T. Donato
Rohm and Haas Company, Spring House, PA 19477

Flexible ceramic tapes are typically manufactured by tape casting or roll compaction. The green extrudate currently produced via conventional binders are brittle after drying. Recently, aqueous acrylic emulsion binder systems for extruding flexible ceramic and metal tapes have been developed. Green tapes made with these acrylic emulsion binder systems have high strength and flexibility both at wet and dry stages. The influence of these binder systems on extrusion parameters, mechanical properties of extrudates, and burnout characteristics are discussed.

INTRODUCTION

Extrusion processes have been used extensively in the manufacture of ceramic parts with constant cross-sections, such as hollow tubes, rods, and honeycomb catalytic substrates. A fundamental requirement for extrusion is the plasticity of the feed material. When additives are combined with water or other liquid, they greatly contribute to the plasticity of the feed material for extrusion. Other than clay, which is the most common additive, cellulose is widely used as a binder for ceramic extrusion.[1,2] Recent studies have shown that certain synthetic polymers, such as polyethylene glycol, polyethylene oxide,[3] and polyvinyl alcohol,[4] can also be used as binders for ceramic extrusion.

Thick ceramic sheets are important materials for electronic substrates and sensors. Roll compaction is a widely employed process for making thick ceramic sheets. This process has the advantages of high production rate and low processing cost. However, high differential shrinkage and inadequate surface finish are undesirable features associated with roll compaction. There are other cases[5,6] of using an extrusion process to fabricate ceramic sheets; however, this process has not been widely used in a commercial scale partially due to the unavailability of a high-performance binder. An ideal binder for ceramic sheet extrusion should provide high plasticity to the feed material. The binder should also impart high ductility to the green sheet to which post processing (such as thermal lamination, cutting, punching, and rolling) will be applied. A low ductile or brittle sheet will either fail during post processing or generate a high scrap rate. High-performance binders are critical for commercially manufacturing thin or thick ceramic sheets for extrusion processes.

In this article we will report our studies of using aqueous-based polyacrylic binders for ceramic sheet extrusion. Emphasis is given to the performance of these aqueous-based polyacrylic binders including the parameters of controlling the extrusion behavior of the binders, the mechanical properties of extruded ceramic green sheets using these acrylic binders, and the burnout characteristics of these binders.

EXPERIMENTAL PROCEDURE

(1) Raw Materials and Extrusion

The alumina powder was Alcoa A-14 (Alcoa Chemicals, Bauxite, AK) with median particle size of 5.0 μm and specific surface area of 0.5 m^2/g. The polyacrylic binders, Duramax™ B-1051 binder ("B-1051") and Duramax™ B-1052 binder ("B-1052") were from the Rohm and Haas Company. Both acrylic binders were aqueous emulsions and were used as supplied. The organic base was AMP-95™ ("AMP-95", 95% 2-amino-2-methyl-1-propanol solution) and was from the Angus Chemical Company. Deionized water was used in all formulations. The cellulose binder was Methocel™ A4M (Dow Chemical Company, Midland, MI).

All the raw materials were admixed in a blender (Cuisinart Classic Model). When acrylic binders were used, a solution of AMP-95™ was added into the alumina powder while mixing. After five minutes of mixing, a measured amount of the premixed acrylic emulsion binders was added and subsequently mixed for another five minutes. The intensive mixing was to ensure the homogeneity of the distribution of the binder, liquid, and powder. Finally, the remaining water was added and mixed thoroughly. The particulate paste was rolled through a three-roll mill (Anthony Mfg. Co., Lindenhurst, NY) before extrusion. Air pockets might be trapped inside the sheet-like paste. Small blisters would appear on the surface of the sheets if air pockets were trapped in the paste. It is very important to remove the air pockets in the paste by complete de-airing. After the three-milling, the dough-like material obtained was broken down to small particulates by the blender before extrusion. The alumina sheets were extruded at room temperature through a slit die using a ram extruder (2" diameter barrel with 12" stroke, Loomis 232-20L). De-aring was done for one minute at the pressure of 10 torrs.

(2) Characterization of the Extrudates

The extruded alumina sheets were dried in air for more than 16 hours and then placed in an oven at a temperature of 75°C for two hours. The results of the mechanical studies of the extruded green sheets were generated by a universal testing machine (Instron Model 1122) at a crosshead speed of 5.1mm/min.

(3) Binder Characterization

Samples of binder were dried at 70°C for at least 72 hours in a forced-air oven and then cooled to room temperature. Burnout residue was determined gravimetrically to an accuracy of ± 0.0001g. An ashing furnace and a tube furnace were used to conduct the binder's pyrolysis in air and in nitrogen, respectively. The burnout profiles of the binders were evaluated using thermal gravimetrical analyses (TGA) at a heating rate of 20°C/min. from room temperature to 800°C. Glass transition temperatures were determined by differential scanning calorimetry (DSC). To minimize the loss of absorbed moisture, DSC samples were placed in sealed sample boats and tested without thermal pre-annealing.

RESULTS AND DISCUSSION

(1) Influence of the Binders on the Extrusion

A typical formulation of the extrusion paste is shown in Table 1. The extruded alumina green sheets are shown in Figure 1. The plasticity of the paste was primarily from two acrylic binders, Duramax™ B-1051 and B-1052. B-1051 binder mainly contributes to the high-yield stress for the paste, which consequently imparts wet green strength and dimensional stability to the extrudate. The B-1052 binder, on the other hand, mostly provides the plasticity for the paste and ductility of the dried extrudate.

The glass transition temperatures determined by differential scanning calorimetry (DSC) of B-1051 and B-1052 are 81°C and 6°C, respectively. The glass transition temperature (Tg) is the temperature at which a noncrystalline polymer takes on the characteristic properties of the glassy state--brittleness, stiffness, and rigidity. Like many other polymeric materials, when a binder is used at temperatures well above its Tg, the binder is soft, flexible, and ductile. In contrast, when the application temperature of a binder is well below its Tg, the binder will be rigid and brittle. Often, differentiated Tgs are observed when two polymer binders with distinguishable Tgs are blended together.

Table 1. Typical Paste Formulation

Component	Quantity (g)	Function	Volume Percent
alumina	100.0	ceramic powder	40.4
B-1052	10.8*	binder	17.3
B-1051	1.2*	binder	2.0
AMP-95	3.3	pH control	5.3
water	21.8‡	media	34.9

*net polymer weight of the aqueous emulsions
‡includes water from the aqueous emulsions

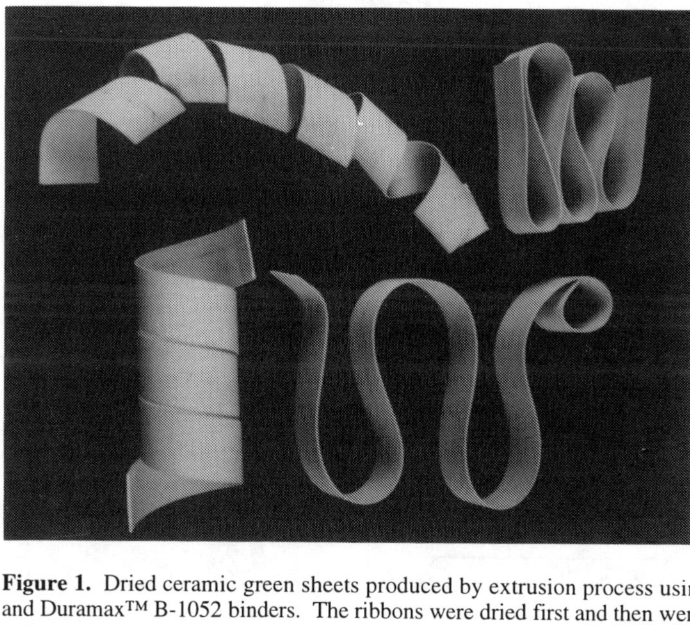

Figure 1. Dried ceramic green sheets produced by extrusion process using Duramax™ B-1051 and Duramax™ B-1052 binders. The ribbons were dried first and then were curled. (The numbered unit on the ruler is centimeters.)

AMP-95 raises the pH of the paste to enhance the performance of both binders, especially B-1051. The advantage of using AMP-95™ is that it volatilizes completely at low temperatures; the boiling point is 164°C. Other inorganic bases, such as ammonium hydroxide (NH_4OH), ammonium carbonate ($(NH_4)_2CO_3$), and sodium hydroxide (NaOH), should have the same function as AMP-95. The negative effects of these bases are either a very strong irritating odor released during the mixing or alkaline metal residues left in the final ceramics.

The formulated paste shown in Table 1 is very cohesive and consequently produces ceramic sheets with smooth surfaces. Also, the extruded sheets tend not to stick to each other. Phase separation was not observed during extrusion. This is believed to be a result of the strong interaction between water and these acrylic binders. The amount of water in the formulation may be adjusted according to the working pressure of the extruder and other production requirements. Low water-content pastes require high-extrusion pressure; therefore, the extrudates have higher dimensional retention.

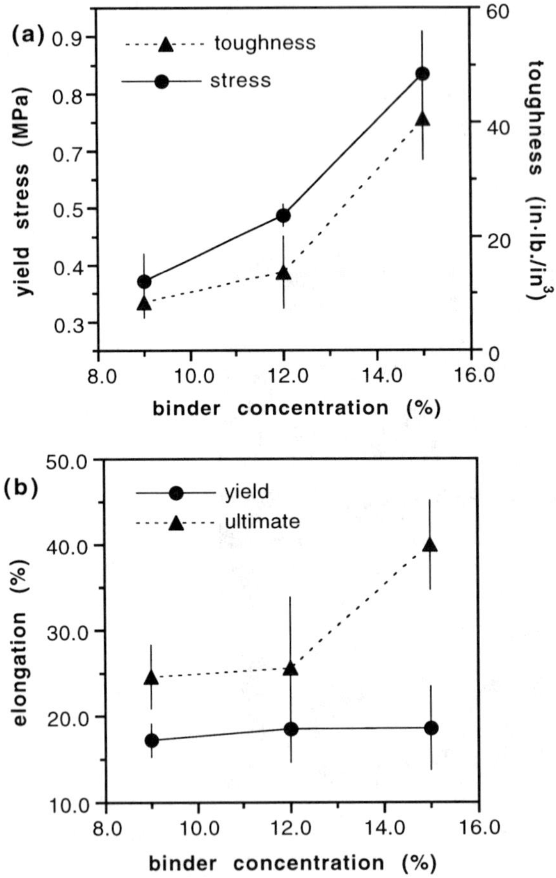

Figure 2a,b. Binder concentration dependence of mechanical properties of extruded ceramic green sheets using B-1051 and B-1052 binders.

It is desirable to attain high-packing density of ceramic particles in a binder system while maintaining acceptable extrusion pressure. High-packing density requires high-solid loading; however, low-extrusion pressure demands higher levels of binder and of water content or an additional effective lubricant. A rule of thumb is that if the extruder can tolerate high-extrusion pressure and if the green strength is adequate, a small amount of binder and water should be used in the formulation.

(2) Mechanical Properties of Extrudates

The mechanical properties of extrudates are strongly influenced by the binders used in the extrusion process. Alumina pastes formulated with three different binder levels with the same ratio of B-1052/B-1051 were studied. The extrusion pressure, the preparation procedure, and the composition of these pastes were similar, except the binder levels were varied. The results of the strain-stress measurements are shown in Figure 2. At a fixed binder ratio, the yield stress and toughness of the ceramic green sheets increase with increased binder levels. This behavior is similar to those ceramic green parts produced by other processes. The higher green strength is usually associated with higher binder levels, indicating that the green strength is attributed to the binder. The plots in Figure 2a imply that the tensile strength is primarily contributed by polymeric binders and suggests that higher green strength could readily be achieved by adding more binder to the formulation. Higher binder levels are needed for fabricating ceramic parts with thin sheets or multi-step post-extrusion processes.

The performance of these binders was further examined to determine the binder level dependence on other mechanical properties of the extruded ceramic green sheets. The results in Figure 2b show that the yield elongation is independent of the binder level, but the break elongation increases with the increase of the binder levels. These results illustrate that higher ductility may be obtained by addition of binders to the paste formulations. However, it is not very efficient to use higher binder levels to enhance ductility. For instance, strain at break increases from 25% to 30% at binder levels of 12% and 15%, respectively.

The ductility of the green sheets may be changed more effectively by varying the ratio of binder B-1051 and B-1052. The plot in Figure 3a shows the relationship between the peak and ultimate elongation for the green sheets extruded using the same total binder level but different binder ratios. The ultimate elongation is 43.9% for green sheets extruded with the binder ratio of 6:1, B-1052/B-1051. However, the yield elongation reaches 148.6% with a binder ratio of 12. These results indicate that the major factor affecting the plasticity of the green extrudate is binder B-1052.

Figure 3b shows that the yield stress of the green sheets decreases and the toughness increases rapidly with higher ratios of B-1052/B-1051. At the lower ratio of B-1052/B-1051, a high-strength but low-extensible green sheet is obtained. The plot also shows that higher ratios of B-1052/B-1051 impart lower strength but higher extensibility. As discussed before, the polymer chains of binder B-1052 are much more flexible than the polymer B-1051 chains at room temperature. The ultimate elongation values of the extruded green sheets are within the range of typical engineering plastics.

(3) Burnout Characteristics of the Acrylic Binders

Burnout characteristics of the binders were studied by thermogravimetric analyses (TGA) and pyrolysis of bulk samples. During ceramic firing, binders are removed by thermolysis in oxidizing, reducing, or inert atmospheres. Binder B-1051 starts to lose weight at approximately 180°C and continues to lose weight with a relatively sharp transition around 400°C until 550°C. The burnout temperature range is approximately 350°C. In contrast, binder B-1052 shows a narrower temperature range of the thermolysis, i.e., 250°C, and a sharper weight loss at similar temperatures.

When both binders are pyrolized in nitrogen, the burnout profiles are similar to the ones obtained in air. However, there is a shift to higher temperatures. The similarity in pyrolysis profiles in air and nitrogen suggests that the binders decompose by analogous mechanisms in both atmospheres.

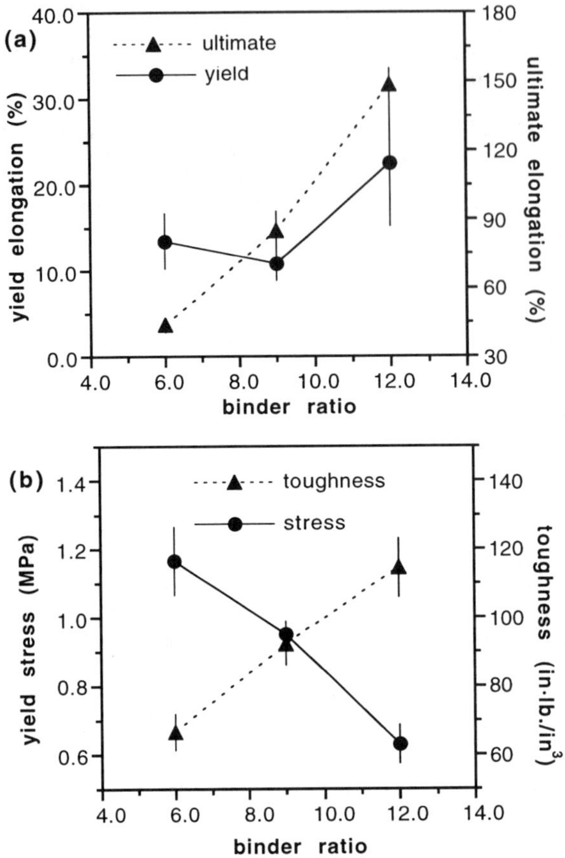

Figure 3a,b. Binder ratio (B-1052/B-1051) dependence of mechanical properties of extruded ceramic green sheets.

Table 2. Burnout Residues of Polymeric Binders

Binder	Residue (%, air)	Residue (%, N_2)
B-1051	0.26±0.02	4.4±0.1
B-1052	1.3±0.3	5.5±0.2
cellulose	0.22±0.02	13.3±0.5

The small sample size and limited dynamic range of TGA limit the accuracy in determining burnout residues. The burnout residues of two acrylic binders and a cellulosic binder are listed in Table 2. The small amount of residue left after air pyrolysis are ash by-products from salts used in the manufacturing of the binders. Residues after nitrogen firing include this ash and carbonaceous material left from incomplete binder burnout. The high residue after firing cellulose under nitrogen indicates that oxygen is critical for complete removal of this binder. It is well known that acrylic polymers undergo depolymerization or "unzipping" with no need for added oxygen, explaining that the depolymerization does not strongly depend on the atmospheres.[7]

CONCLUSIONS

There is an inherent difference between B-1051 and B-1052 binders. This difference is indicated by glass transition temperatures. The harder binder (higher glass transition temperature binder) imparts tensile strength whereas the soft binder (lower glass transition temperature binder) provides ductility. Therefore, this versatile binder combination offers broad process latitude for manufacturing ceramic sheets by extrusion processes.

Highly ductile ceramic green sheets were extruded using aqueous acrylic emulsion binders. The mechanical properties, such as tensile strength, ductility, and yield strength, of the green sheets are strongly dependent on the binders used in the extrusion process. The tensile strength of the sheets could readily be increased by adding more binder to the formulation. However, ductility is not strongly dependent on binder level. In addition, higher ductility could be achieved effectively by an increased ratio of B-1052/B-1051. The major factor affecting the plasticity of the green extrudate is the ratio of soft to hard binders.

The burnout profiles of acrylic and cellulose binders were reported in this study. The thermal pyrolysis results obtained in air show that the residue of acrylic binder B-1052 is comparable or slightly higher than that of the cellulose binder. The residue of cellulose in nitrogen is significantly higher than acrylic binders, which makes the cellulose binder undesirable for non-oxide ceramics. Compared with B-1052, B-1051 binder has a broader burnout profile.

REFERENCES

[1]J.S. Reed, Introduction to the Principles of Ceramic Processing; pp.359-60. Wiley, New York, 1988.

[2]L.E. Schuetz, "Methylcellulose Polymers as Binders for Extrusion of Ceramics," *Am. Ceram. Soc. Bull.*, **65**[12] 1556-59 (1986).

[3]J. Zheng, J.S. Reed, and S.K. Verma, "Ceramic Extrusion with PEG," Am. *Ceram. Soc. Bull.*, **73**[12] 61-66 (1994).
[4]A.Y. Chen and J.C. Cawley, "Extrusion of Al_2O_3 Ceramics with Low Organic Content," *Ceram. Eng. Sci. Proc.*, **10**[7-8] 773-783 (1989).
[5]H. Takebe, M. Yoshida, K. Hayashi, and K. Morinaga, "Fabrication of Alumina Sheets by Extrusion," J. *Ceram. Soc. Jpn.*, **100**[5] 750-754 (1992).
[6]J.J. Thompson, "Forming Thin Ceramics," Am. *Ceram. Soc. Bull.*, **42** [9] 480-81 (1963).
[7]Y.N. Sun, M.D. Sacks, and J.W. Williams, "Pyrolysis Behavior of Acrylic Polymers and Acrylic /Ceramic Mixtures," *Ceras. Trans.*, **1** 538-48 (1988).

EFFECTS OF PARTICLE SIZE DISTRIBUTION AND EXTRUSION RATE ON THE DIMENSIONAL STABILITY OF EXTRUDED ALUMINA RODS

K.J. Woodard and D.R. Dinger
Clemson University
Clemson, SC 29634

ABSTRACT

A major objective of the extrusion process is to precisely control green and fired densities and associated shrinkages. In this study, the properties of extruded alumina rods were examined as functions of particle size distribution and extrusion rate.

INTRODUCTION

Much of the research involving extrusion of high alumina bodies has focused on the prediction and modeling of paste extrusion characteristics (1-7) or on the influences of binder phases on extrusion rheologies (8,9). Very few studies have been conducted to examine the effects of particle size distribution on extrusion behaviors and product properties. Benbow and Bridgwater (10) examined the effects of particle size distribution by blending materials according to Furnas' packing theory and observed an increase in green and fired densities. However, no mention was made of the extrusion behaviors of these materials. Travnickova, et.al., (11) studied the relationships between extrusion behaviors and changes in the overall sizes of very narrowly distributed alumina powders without any mention of the properties of the finished products.

In a production situation, changes in particle size distribution of the feedstock will manifest themselves as variations in pressures required for extrusion as well as fluctuations in the densities and shrinkages of the final product. Therefore, the objective of this research was to study the effects of particle size distribution and extrusion rate on the pressures required for extrusion and on the properties of the finished products.

EXPERIMENTAL PROCEDURE

Three Bayer processed calcined aluminas, A-14, -325 mesh, A-152SG, and Primalox-10, were provided by Alcoa.* Their particle size distributions and specific surface areas were measured using a Sedigraph 5100 particle size analyzer and a Gemini 2160 surface area analyzer **. The results from the particle size analyses were input into the MIX10N computer program. This program calculates percentages of the raw materials necessary to achieve particle size distributions with particular distribution moduli (n) in the Dinger-Funk packing equation. Different distribution moduli correspond to different packing efficiencies. A distribution modulus of 0.25 was chosen for this research because size distributions having a modulus in the range of 0.1 to 0.25 have been demonstrated to have satisfactory plastic forming properties (12). The two remaining distributions were chosen from the individual raw materials. Alcoa A-14 was chosen as a standard reference material because of its widespread use in industry, and Primalox-10 was chosen as an example of a very fine grained, monodisperse material. Raw materials of this type have traditionally been regarded as necessary for manufacturing highly dense, alumina ceramics. Particle size distributions are shown in Figure 1.

Each distribution was prepared for extrusion first by mixing all of the dry alumina raw materials with hydroxyethylcellulose binder in a Hobart mixer for five minutes. Oleic Acid was added to the batch as a lubricant and Darvan C as a dispersant. Distilled water was then added while mixing until the batch changed from a loose granular mix into a cohesive plastic mass. This is known as the plastic transition point of the material, and it is different for each particle size distribution. After the transition point was observed each distribution was pre-extruded three times using a piston extruder to improve homogeneity. Samples from each distribution were then extruded using the same piston extruder with a die cone angle of 60° at three different extrusion rates, 1.69 mm/s, 3.39 mm/s and 6.77 mm/s. The loads on the piston during extrusion were recorded . Samples from each batch were dried after which their dimensions and bulk densities were measured. All of the samples were then fired to 1600°C with a two hour soak. Fired dimensions, shrinkages, and bulk densities were measured.

*Aluminum Company of America, Bauxite, AR
**Micromeritics Instrument Corporation, Norcross, GA

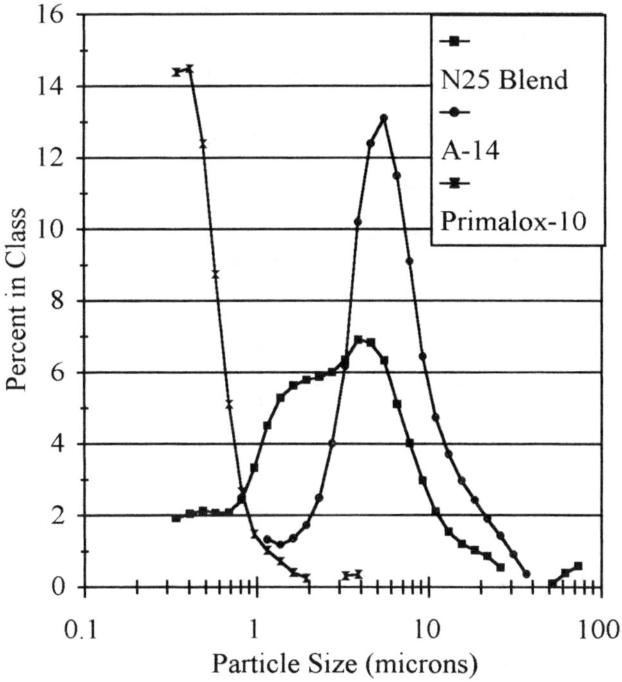

Figure 1. Particle size distributions of alumina extrudate bodies.

RESULTS AND DISCUSSION

The average pressure drop across the extruder as a function of shear rate, particle size distribution, and solids content for a 60° die is shown in Figure 2. These results show that the blended particle size distribution exhibited the lowest extrusion pressures at each of the extrusion rates, even considering that this body was extruded at the lowest water content of the three. This indicates that blending of the raw materials effectively increased the packing efficiency of the body. Less pore volume was present in the packed powder and more water was available for separation of the particles resulting in lowered extrusion pressures. This figure also shows that extrusion pressures generally increased as extrusion rates increased, as expected.

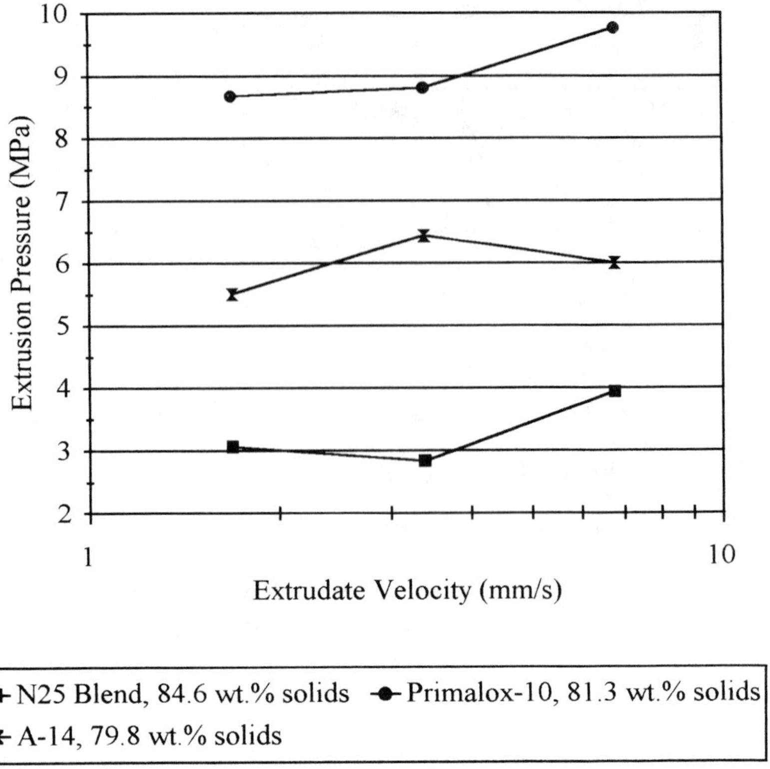

Figure 2. Extrusion Pressure as a Function of Extrudate Velocity and Particle Size Distribution.

Excessive increases in extrusion pressures with increasing velocities can be indicative of the onset of dilatancy, where the structure of the powder compact begins to open due to interparticle interactions. Such expansions produce increased pore volumes and reduce the fraction of carrier fluid available to separate particles.

Lower product densities, because of increased pore volumes, are typical of unrelaxed dilatant structures in powder compacts (12). Table 1, however, which shows dry densities of the extruded bars, gives no indication of dilatancy. Dry densities for each of the particle size distributions remained approximately constant as functions of extrudate velocity. This suggests that the observed extrusion behaviors were not associated with dilatancy.

Table I. Dry Density Results

Extrudate Velocity	A-14	N25 Blend	Primalox-10
1.69 mm/s	1.96 g/cm^3	2.39 g/cm^3	2.13 g/cm^3
3.39 mm/s	1.99 g/cm^3	2.41 g/cm^3	2.16 g/cm^3
6.77 mm/s	1.97 g/cm^3	2.39 g/cm^3	2.16 g/cm^3

The greater dry densities for the blended particle size distribution reinforces the fact that this distribution packed more efficiently than either A-14 or Primalox-10. Similar behavior was also observed by Benbow (10), although his methods for fabrication of the experimental size distributions differed from those in this study.

Fired densities and shrinkages for each of the test bodies are shown in Table 2. The results indicate that Primalox-10 yielded the highest fired densities and shrinkages, followed by the N25 Blend and A-14, respectively. Although one would normally expect bodies with greater dry densities to sinter well, the single firing schedule used in this study did not indicate the maximum densities achievable by any of the test bodies. It is evident, however, that the highest surface area body (Primalox-10) sintered faster and to a higher density, due to increased driving forces, than the other two bodies.

Table II. Fired Densities and Shrinkages

Property	A-14	N25 Blend	Primalox-10
Fired Density	2.29 g/cm^3	2.94 g/cm^3	3.87 g/cm^3
Fired Shrinkage	6.31%	7.73%	19.10%
Particle Crowding Index	6.70 X 10^9	3.05 X 10^{12}	1.89 X 10^{13}
Specific Surface Area	.768 m^2/g	3.00 m^2/g	8.16 m^2/g

The particle crowding index, PCI, which is the number of particles per true cm^3 of powder calculated from particle size distribution and surface area data (12), increased as fired densities and shrinkages also increased. Increased PCI values indicate that more particles were present per unit volume of powder. The PCI values for these three bodies correspond with higher surface areas and increased sintering reactivities.

CONCLUSIONS

This work has shown that blending of alumina raw materials can lead to lower extrusion pressures at higher solids loadings. By lowering extrusion pressures, typical processing problems such as bleeding, binder migration, and particle size segregation can be avoided. An increase in dried densities was also observed as a result of blending which could lead to cost savings for manufacturers through lower firing temperatures needed to achieve required product densities and dimensions.

Primalox-10 produced the greatest fired densities and shrinkages, despite having lower dry densities, than the blended size distribution. This was attributed to its significantly higher reactivity during sintering as indicated by the large values of specific surface area and particle crowding index, and by the extremely fine particle size distribution. Fired densities and shrinkages in general increased as PCI values increased.

During processing in the plant, changes in particle size distribution of the raw materials will show themselves as variations in the fired densities, shrinkages, and sintering rates, unless the manufacturer changes process procedures to accommodate such variations. One way to achieve this is to blend different raw materials to achieve constant particle size distributions with constant PCI values. This would insure that densities and shrinkages could be held to tight tolerances by manufacturing a product with consistent packing densities and sintering reactivities.

REFERENCES

1. J.J. Benbow and J. Bridgwater, "The Role of Frictional Forces in Extrusion," pp.80-90, Tribology in Particulate Technology, B.J. Briscoe and M.J. Adams, Adam Hilger, Bristol U.K., 1987.

2. J.J. Benbow, T.A. Lawson, E.W. Oxley, and J. Bridgewater, "Prediction of Paste Extrusion Pressure," Bull. Am. Cer. Soc. 68 [10] p.1821-1824, 1989.

3. J.J. Benbow, S.H. Jazayeri, and J. Bridgwater, "Ceramic Extrusion Mechanics: The Effect of Paste Formulation and Liquid Phase Rheology on Die Flow Resistance," Ceramic Powder Science II, G.Messing, American Ceramic Society, Westerville, Ohio, 1987.

4. J.J. Benbow, "The Dependence of Output Rate on Die Shape During Catalyst Extrusion," Chem. Eng. Sci. 26 p.1467-1473, 1971.

5. A. Ovenston and J.J. Benbow, "Effects of Die Geometry on the Extrusion of a Clay Like Material," Trans. Br. Cer. Soc. 67 p.543-567, 1968.

6. D.B. Price and J.S. Reed, " Boundary Conditions in Electrical Porcelain Extrusion," Bull. Am. Cer. Soc. 62 [12] p.1348-1350.

7. W.B. Carlson, J. Zheng, and J.S. Reed, "Numerical Simulation of the Extrusion of Plastic Bodies, Forming Science and Technology for Ceramics, Ceramic Transactions 26 p.132-140, 1992.

8. J. Zheng, J.S. Reed, and S.K. Verma, "Ceramic Extrusion with PEG," Bull. Am. Cer. Soc. 73 [12] p.61-66, 1994.

9. J.F. White and A.L. Clavel, " Extrusion Properties of Non-Clay Oxides," Bull. Am. Cer. Soc. 42 [11] p.698-702, 1963.

10. J.J. Benbow and J. Bridgwater, " The Influence of Formulation on Extrudate Structure and Strength," Chem. Eng. Sci. 42 [4] p.753-756, 1987.

11. J. Travnickova, J. Havrda, and F. Ouijiri, "The Effect of Particle Size on the Rheological Properties of Ceramic Paste," Silikaty 33 [1] p.11-16, 1989.

12. J.E. Funk and D.R. Dinger, Predictive Process Control of Crowded Particulate Suspensions, Kluwer Academic Publishers, Boston, 1994.

INJECTION MOLDING OF CERAMICS USING A POLYACETAL BASED BINDER SYSTEM

Dr. J.S. Ebenhöch
BASF Corporation, 1609 Biddle Ave. Wyandotte, MI 48192;

J.H.H. ter Maat, Dr. W. Hesse, Dr. P. Truebenbach
BASF AG, RC/P - I 522, 67056 Ludwigshafen, GER;

Abstract

Among the production routes to small complex ceramic parts, powder injection molding is the most attractive alternative. It combines near net shape capability with good surface finish and is easy to automate. With the development of the catalytic debinding process for poly-acetal binders, the main impediments for the acceptance of ceramic injection molding as a mass production method can be overcome. The use of this system ensures short molding cycle times, high green strength and fast debinding without deformation. Ready to mold compounds are commercially available for various oxide and non-oxide ceramic materials as well as the equipment needed for this process.

1. Introduction

The most common manufacturing techniques for advanced ceramics are uniaxial pressing, cold- and hot isostatic pressing, slip casting, machining and ceramic injection molding (CIM). Powder injection molding is most suited for small, complex precision parts, that are produced in large quantities. In spite of its obvious advantages, near net shape capabitity, good surface finish and high production rates, CIM is still limited to niche applications. The reason for the hesitating acceptance of CIM has been the unrealiability of the process, inconsistant quality and poor yield of good parts. Defects, which reduce the strength of the material and eventually lead to failure of the part, can originate from any processing step [1]:

Compounding:
- inhomogeneities in the feedstock
- impurities, inclusions of foreign matter
- powder agglomerates

Molding:
- weld lines
- voids
- defects due to ejection of the moldings
- cracking, blistering after mold release

Debinding:
- deformation due to slumping or stress relaxation
- cracking or bloating due to internal gas pressure

These defects can not be healed by heat treatment. In most cases they will fully develop during sintering due to differences in densification.

2. The Polyacetal Binder System

Research at BASF has been aimed at developing a thermoplastic binder with improved moldability, which can be removed quickly from the molded part without leaving a char residue. Polyoxymethylen (POM), a commonly used engineering resin, combines many favorable properties for this purpose [2]. The strongly polar POM-chain adheres to the powder surface, which prevents demixing during molding. The high molecular weight and the crystallinity of the polymer ensure short molding cycles, high melt elasticity and high green strength. Most important of all, POM decomposes easily and cleanly into its monomer units, gaseous formaldehyde. Normal, commercially available POM has to be modified to meet the needs of powder injection molding. Due to the high powder loading of the compounds, the resistance against thermal and shear induced degradation has to be improved by additives, that leave no residue after pyrolysis. Copolymers and amorphous polymers, that mix with POM, reduce the viscosity and the shrinkage

due to crystallization. Organic buffers neutralize acidic or alkaline powder surfaces, that could attack the polymer chain. Surfactants stabilize the dispersion of the fine ceramic powder and prevent reagglomeration of the particles. The composition of the compound, powder loading, binder, additives and dispersants is optimized for each ceramic powder.
BASF offers ready to mold feedstocks for a variety of materials. The compounding technique and quality control ensure consistant composition, good homogeneity and very little impurities in the feedstocks.

3. The Catalytic Debinding

Commercialization of powder injection molding has been seriously impeded by the slow and difficult to control debinding process. To remove the binder from the green part, the polymer has to be decomposed into volatile products. At the same time, a pore network has to be formed, that allows the gas to diffuse out of the part without building up an internal pressure. The formation of this porous structure was found to be rate limiting.
The depolymerization of POM is catalyzed by gaseous acids and proceeds quickly above 100°C as a direct solid-gas transition.

$$\underset{\text{mp: } 165^{\circ}\text{C}}{(CH_2\text{-}O\text{-})_n} \xrightarrow[>100^{\circ}C]{\text{acid}} \underset{\text{bp: } -21^{\circ}\text{C}}{n\ CH_2{=}O}$$

The catalytic debinding process [3] is performed well below the melting point of the polyacetal. This low debinding temperature prevents the formation of a liquid phase and thus deformation of the green part due to gravitational distortion or stress relaxation. Capillary forces have no effect. Since the polymer is not penetrated by the vaporized acid, the decomposition is catalyzed exclusively at the gas-binder interface. Debinding proceeds according to a shrinking core mechanism. The gas exchange is limited to the already porous shell and no pressure is built up within the green part. The interface travels inward at a speed of 1-2 mm/h, depending on powder particle size and stage of debinding.
For most materials 100% nitric (b.p. = 85°C) acid is the most suitable catalyst. The vaporized acid is not corro-

sive. It does not react with the ceramic powders, which might deteriorate the properties of the sintered part. Since the process involves highly concentrated nitric acid and formaldehyde gas, the equipment has to comply to the industrial standards for safe operation and the ecological regulations.

The debinding process is performed in a steel lined, gas tight oven, that has to be equipped with a fan to achieve thorough mixing and proper heat tranfer. The flue gas, mainly formaldehyde in the carrier gas nitrogen, is passed through a burner. Nitric acid and NO_x are decomposed in the primary (reducing) combustion chamber by excess fuel gas, formaldehyde is oxidized in the second (excess air) stage [4]. Its concentration after the burner is below 0.5 ppm. The emissions during a debinding run (6 h) for 25 kg of typical CIM-parts are well below 50 g of NOx and 1 g formaldehyde, which is much less than the emission of a car or a domestic gas furnace.

Hardwired safety interlocks ensure sufficient purge of the oven before and after the process, prevent debinding without the flare burning, keep the door locked during the run and reset the oven into the purge mode (acid pump off), in case of a temperature- or gas flow high/low-alarm.

Ovens for the catalytic debinding, that meet all these requirements, have been developed jointly by BASF and various furnace manufacturers and are commercially available.

4. Materials and Properties

Ready to mold feedstocks for ceramic injection molding are available for a variety of materials (tab. I).

The engineering ceramic materials, alumina, zirconia, silicon carbide and silicon nitride are used for a variety of structural applications ranging from nozzles, cutting tools and sliding seals to pump parts and engine components. The colored zirconias are employeded for decorative purposes only. Common applications are watch cases, designer pens and jewelery.

The optimum solids loading of a feedstock is determined by the powder properties, mainly the particle shape and size distribution. Compounds with too high volume loading are difficult to mold and highly abrasive. Too low powder content causes problems during sintering, high shrinkage, poor precision and possibly distortion.

Feedstock	Material	Powder loading
AO H 59	Al_2O_3, 96%	59 vol.-%
AO F 56	Al_2O_3, 99.6%	56 vol.-%
TZP A 49	ZrO_2, 5% Y_2O_3, coprecip.	49 vol.-%
TZP B 49	ZrO_2, 5% Y_2O_3, fused	49 vol.-%
ZTA FB 54	Al_2O_3/ZrO_2, 80/20	54 vol.-%
SSC W 49	SiC (C, B)	49 vol.-%
SSN M 49	Si_3N_4 (5%Y_2O_3, 5% Al_2O_3)	49 vol.-%
TZP A 47 c	TZP A, black, colors	47 vol.-%

Table I: Composition of commercial CIM-feedstocks

The decorative ceramics require uniformity in color and surface finish, the technical ceramics are judged by their mechanical, thermal and chemical properties, e.g. high and low temperature strength, fracture toughness and corrosion resistance. The properties of these materials have mostly been published for HIP'ed samples. Table II summarizes some mechanical properties for CIM-parts from various feedstocks.

The flexural strength (4-point bending test, 40 mm/20 mm) and the fracture toughness (parallel notch) were obtained from test bars 3 * 4 * 60 mm^3. The Weibull modulus is based on at least 25 samples.

Often, failure of a sample is caused by surface defects. Polishing will remove most of these defects, resulting in a higher strength. HIP increases the density, which improves the strength, especially of the transformation toughened materials.

The test samples were molded under optimized conditions. The molding process greatly influences the qualtity of ceramic parts. Poor design of component or tool as well as improper selection of the molding parameters will cause inferior mechanical properties [5].

Material Finish	Density [g/ml]	4-pt Bend strength [MPa]	Fracture toughness [$MPam^{1/2}$]	Weibull
AO H 59 sintered	3.75	410	4.2	20
AO F 56 sintered polished	3.91 same	400 460	3.5 same	11 14
TZP A 49 sintered polished HIP,pol.	6.07 same 6.08	700 970 1310	8.7 same	11 12 11
TZP B 49 sintered polished HIP,pol.	6.05 same same	610 940 1290	7.6 same	12 11 13
ZTA FB 54 sintered polished	4.22 same	480 630	7.0 same	12 14
SSC W 49 sintered	3.16	410	3.0	14
SSN M 49 sintered polished	3.23 3.25	700	7.3	

Table II: Properties of injection molded samples

5. Conclusion

Ceramic parts from various materials can be manufactured via powder injection molding in good quality at high yield. The ready to use polyacetal based feedstocks are easy to process and tolerate a sufficiently wide variation of molding and debinding parameters. The catalytic debinding, although a chemical process, can be handled safely for personal and environment. This system was introduced into the market in 1992 and is commercially applied by several European parts producers.

6. References

[1] M.J.Edirisinghe, Ceram. Bull. 70(5), 824 (1991).

[2] J.H.H. ter Maat, J.S. Ebenhoech, H.J. Sterzel, Proc. 4th Int. Symp. Ceramic Materials and Components for Engines, Goeteborg (S) June 10 - 12, 1991.

[3] J.H.H. ter Maat, W. Schuette, H.J. Sterzel, EP A 413231.
J.S. Ebenhoech, J.H.H. ter Maat, P. Truebenbach, Powder Injection Molding Symposium, June 21 - 26, 1992, ed. by P.H. Booker, J. Gaspervich, R.M. German, MPIF, Princeton, NJ, p. 385 (1992).

[4] C. Hechler, Annual Meeting of DKG, Oct. 6 - 8, 1993.
D.C. Krueger, Adv. Powder Metallurgy & Particulate Materials 1994(4), p. 4/47.

[5] J.S. Ebenhoech, J.H.H. ter Maat, Adv. Powder Metallurgy & Particulate Materials 1993(5), p. 5/45.

CHEMICALLY BONDED CERAMIC PROCESSING OF MONO-CALCIUM ALUMINATE

Bradley R. Johnson, Mehmet A. Gülgün, and Waltraud M. Kriven, Department of Materials Science and Engineering, University of Illinois, 105 S. Goodwin Ave, Urbana, Illinois 61801, (217) 333-5258

ABSTRACT

A ceramic matrix composite consisting of mono-calcium aluminate as the matrix phase and calcia stabilized zirconia as the second phase was fabricated by warm-pressing. The hydraulic properties of calcium aluminate (CA) were exploited in order to form strong green bodies. Pressureless sintering of composites with zirconia loadings of 60 wt% (44 vol%) were accomplished with no evidence of constrained sintering. Novel processing techniques enhanced homogenization of the starting powders, thereby facilitating the elimination of agglomeration and constrained sintering. Samples were characterized using SEM, XRD, dilatometry and helium pycnometry. Vickers hardness testing was done to evaluate mechanical properties.

INTRODUCTION

Calcium aluminate cements (CAC) in addition to being hydraulic (i.e. reacting with water to form chemically bonding hydrate phases) have the unique property of being refractory. That is, the chemical bonds formed from reacting with water can be subsequently replaced (through a complex series of reactions) by diffusional bonds during sintering. As a result, CAC have found widespread use in the refractories industry for the formulation of castable refractory products. The key deleterious feature of this material, however, is that the chemically bonding hydrate phases initially formed at room temperature are metastable, and with time and temperature, they will convert to a more stable, denser hydrate phase. This conversion reaction is accompanied by a decrease in the bound water in the system, an increase in porosity and a concomitant decrease in strength[1].

Mono-calcium aluminate (CA) is the primary phase in CAC and is responsible for its rapid setting and refractory characteristics[1]. Borglum et al.[2] demonstrated the ability to form a chemically bonded ceramic with promising electrical properties using CAC. It was based on the direct formation of thermodynamically stable hydrate phases of CA. Thus, it may be possible to use pure CA to fabricate a wide array of ceramic matrix composites for electrical and mechanical purposes

The objective of this research is to design a processing route based on CA so as to use it as a generic matrix material for the fabrication of ceramic matrix composites (CMC). The second phase material used in this study to make CMC was stabilized zirconia. Some specimens were made with calcia stabilized cubic zirconia; others with zirconia that was partially stabilized with yttria.

EXPERIMENTAL PROCEDURES

Chemically synthesized CA was used as the binding phase and different types of stabilized zirconia (70% calcia stabilized cubic zirconia: Zirox CS -325 mesh, TAM Ceramics, Niagara Falls, NY and partially stabilized zirconia with 3 mole % yttria: HSY-3.0, Zirconia Sales (America), Inc.) were used as the secondary phase.

Previous work[3] demonstrated the difficulties of homogeneously mixing sub-micron powders. Preferential agglomeration was frequently encountered due to the high surface forces associated with such small particle sizes. Thus a process was developed for the synthesis of coated powders. The concept of coating powders has been reported in other places in the literature as well[4-8]. A flow diagram of the procedure used is shown in Figure 1. The process depicted is similar to one used for the synthesis of CA reported by Gülgün[9]. The major variations are that different molar ratios of ethylene glycol and citric acid were used (1:1), and the molar ratio of the citric acid-ethylene glycol resin (1 mole of resin = 1 mole of citric acid + 1 mole of ethylene glycol) to cations was 4:3. Additionally, zirconia powder (60 wt% based on the final CA yield) was added to the polymeric precursor solution containing calcium and aluminum ions. This polymeric precursor solution containing zirconia powder was heated on a hot plate with constant stirring, driving off both water and NO_x gases. The solution became

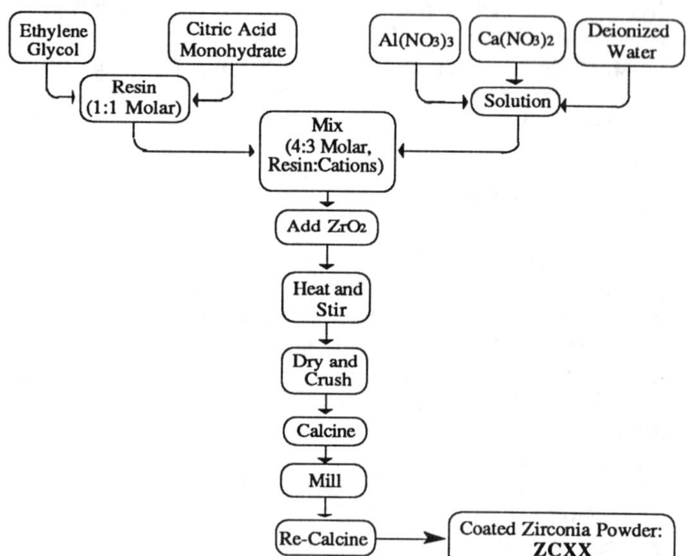

Figure 1. Flow diagram for coating zirconia powder with calcium aluminate.

increasingly viscous and sticky, eventually turning into a doughy mass which was placed in a box oven at 175°C for approximately one hour. The doughy mass expanded by roughly a factor of 25 into a large, hollow sphere with a thick porous crust. The dry porous material was broken up into chunks and ground into a powder with a mortar and pestle. Calcination of the dried polymeric precursor containing material was accomplished by placing the crushed powder on a platinum tray and heating at 10°C/min to 900°C and holding there for 1 hour.

A milling step was included to break up the coated powder agglomerates. The coated powder was placed in a polyethylene bottle with isopropyl alcohol (to form a suspension) and spherical zirconia milling media and milled for 60 minutes using a random orbital paint mixer (SPEX Industries, Inc., Scotch Plaines, NJ). After milling, the suspension was poured into a crystallization dish and the alcohol was evaporated using a hot plate. An additional calcination step (with the same heating schedule as before) was included, in order to eliminate potential contamination from the polyethylene bottle used in milling. The fabricated coated zirconia powder was designated "ZC60" where "ZC" stands for coated zirconia and "60" indicates that it contained 60 wt% of zirconia.

Composite specimens were fabricated from ZC60 powder through warm pressing. Water was added to the powder to form a paste which was vigorously mixed with a mortar and pestle (Figure 2.). The paste was then placed in a cylindrical die which was lubricated with a solution of graphite suspended in isopropyl alcohol. Carbon foil paper was used on the die faces and as a rim strip around the perimeter of the paste in the pressing cavity. This was done to facilitate removal of the specimen after pressing. The paste was pressed at 110 MPa while simultaneously being heated to 400°C using a cylindrical band heater around the die. A heat up rate of approximately 15°C/min was used, and the sample was held at 400°C for approximately 1 hour. Specimens were removed from the die prior to cooling down.

After removal from the die, the specimens were sectioned with a diamond wafer blade saw. Portions of each sample were sintered to 1350°C and held there for one hour.[10] Hardness testing was accomplished using a Vickers indenter (Nikon QM, Tokyo, Japan). A minimum of ten indents were made and average values are presented. Phase identification and hydrate decomposition behavior was accomplished using simultaneous thermal analysis (Netzsch Simultaneous Thermal Analyzer STA409, Selb, Germany) and X-ray diffraction (XRD, Rigaku, Danvers, MA, USA). Microstructural analysis was done using a scanning electron microscope (SEM, Hitachi S-530, Osaka, Japan) equipped with a back scattered electron

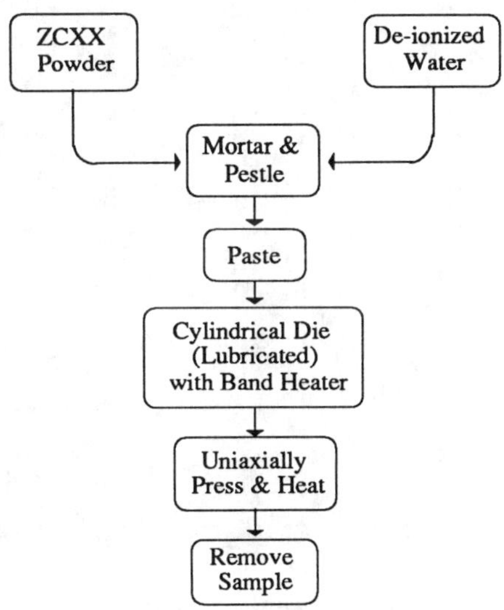

Figure 2. Flow diagram for warm pressing CA composite specimens.

(BSE) detector and energy dispersive spectroscopy (EDS).

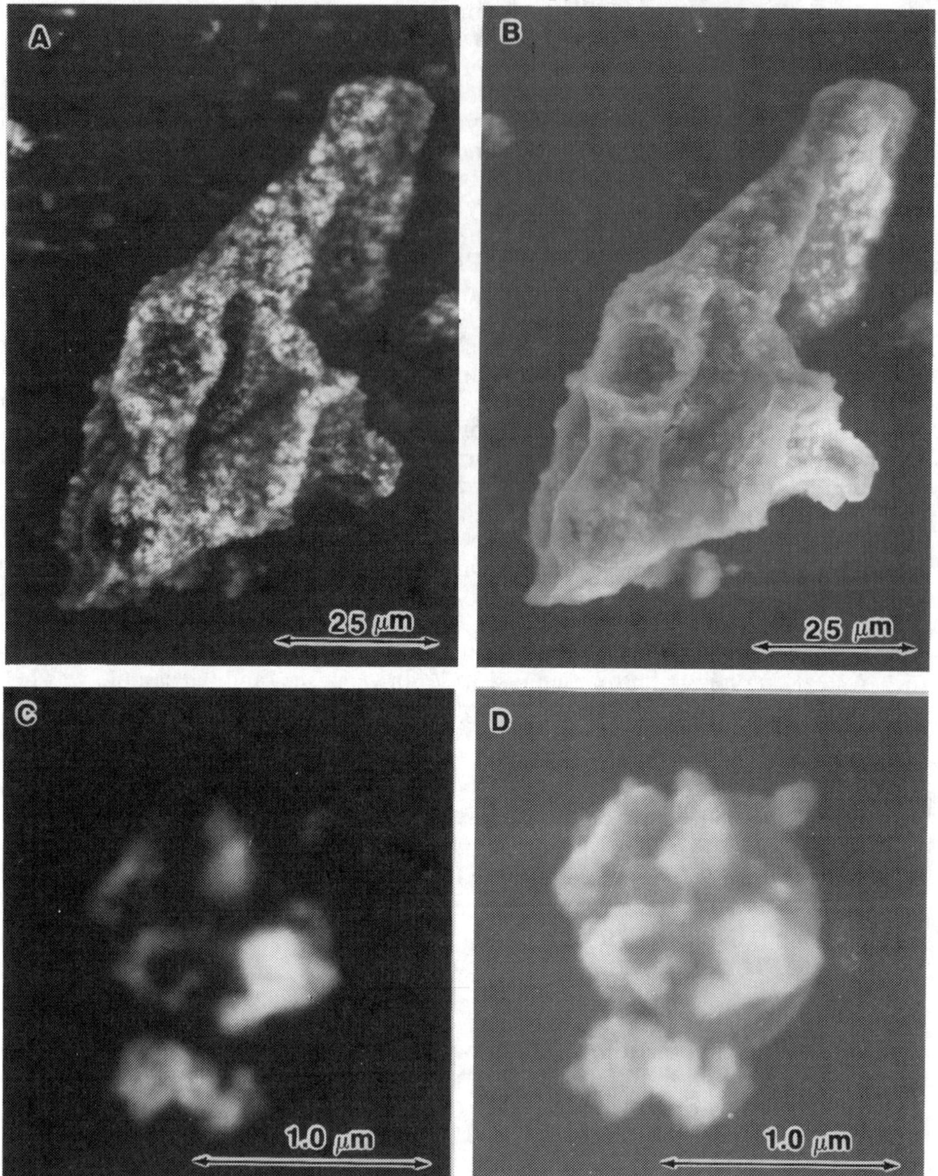

Figure 3. SEM micrographs of ZC60 powders made with yttria PSZ. (a) Pre-milled agglomerate in BSE imaging mode (b) Pre-milled agglomerate in SE imaging mode (c) Milled granule in BSE imaging mode (d) Milled granule in SE imaging mode.

RESULTS AND DISCUSSION

Evidence for the effectiveness of the powder coating can be seen from the series of micrographs shown in Figure 3. Since the BSE detector provides atomic number contrast, the zirconia (Z=40), having a larger atomic number than either calcium (Z=20) or aluminum (Z=13), shows up as the brightest phase. Thus the skeletal structures seen in white (Figures 3a, c) are composed of zirconia. The same areas were then imaged in secondary electron (SE) mode (detects both secondary and back scattered electrons) which provides better topographical imaging and is thus better able to image calcium and aluminum. The SE+BSE images (Figures 3b, d) therefore provide evidence that the areas filling out the zirconia skeleton are composed of calcium aluminate.

The agglomerated structure of the pre-milled ZC powders is clearly evident in Figures 3a, b. Since the CA coating is the result of pyrolysis of polymeric precursors, the formation of such agglomerates is to be anticipated. Hardness testing of pre-milled and post-milled ZC60 powders reflected the effects of poorer packing efficiency in the lower hardness values observed for the agglomerated, pre-milled powders.

SEM analysis of milled ZC60 powders revealed that the large agglomerates had been eliminated. Micrographs of a milled ZC60 powder granule (Figure 3b, d) were again taken in BSE and SE imaging modes. They reveal the presence of a calcium aluminate coating on sub-micron zirconia particles for the same reasons as stated before. Thus, it can be concluded that milling of coated zirconia powders not only effectively breaks up the agglomerates, but also leaves the coating intact.

Vickers hardness testing of warm pressed and sintered specimens was conducted to evaluate mechanical properties. Specimens made from both the calcia stabilized zirconia ("CS", 5 - 44 μm diam.) and the yttria partially stabilized zirconia ("PSZ", ~ 0.7 μm diam.) were evaluated. The average hardness values are shown in Figure 4. By way of comparison, the VHN for conventionally sintered PSZ is approximately 1200 kg/mm^2.[11] For warm pressed specimens, the average hardness of the PSZ composite was less than the CS specimen. This can be explained by the much greater zirconia surface area in the PSZ specimen (~ 50X due to smaller particle size, but same wt%) which would result in a thinner coating of

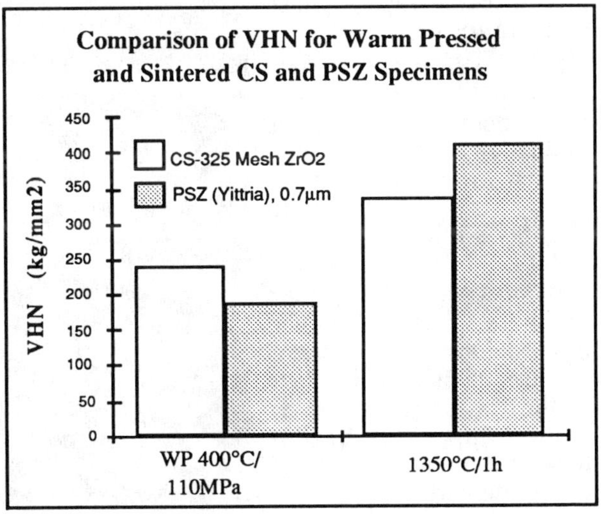

Figure 4. Effects of zirconia particle size and shape on VHN.

CA, and thus a lower strength matrix i.e. the zirconia loading is too high. A similar result was observed with warm pressed CS specimens when the zirconia loading was increased to 70 wt%.[3] Additionally, the much smaller particles for the PSZ specimen are easier to displace during indentation than the larger particles for the CS specimen, resulting in lower hardness. For sintered specimens, the trend is reversed - the PSZ composite is harder. Since it has a much smaller and a more uniform particle size, it has a much smaller and a more uniform pore size, and thus it sinters better, and so would be expected to be harder.

SEM micrographs of warm pressed CS and PSZ specimens are shown in Figure 5a and b. The key feature common to both is the generally uniform dispersion of zirconia in the CA matrix. This result is attributed to using the powder coating process previously described. For the PSZ composite, significantly less porosity and generally smaller pores were observed by SEM. For the CS specimen, the porosity was generally larger and more prevalent. This result is attributed to the better packing efficiency of the more uniformly shaped PSZ.

Simultaneous thermal analysis and XRD was conducted on CS specimens at several different temperatures. After warm pressing, the hydrate $Ca_4Al_6O_{13}\cdot 3H_2O$ ($C_4A_3H_3$) was formed. Examination of a pseudo-binary phase diagram for $CaO - Al_2O_3 - H_2O$ at 103 MPa (15 000 psi) indicates that this is to be expected.[12] There is indication that it remained present in the system from up to 700° (by TGA) to possibly 900°C (by XRD of a bulk specimen) which correlates with reports in the literature.[13] The advantage of producing this hydrate of CA is that it contains less water, is denser, and is stable up to higher temperatures than lower temperature hydrates.

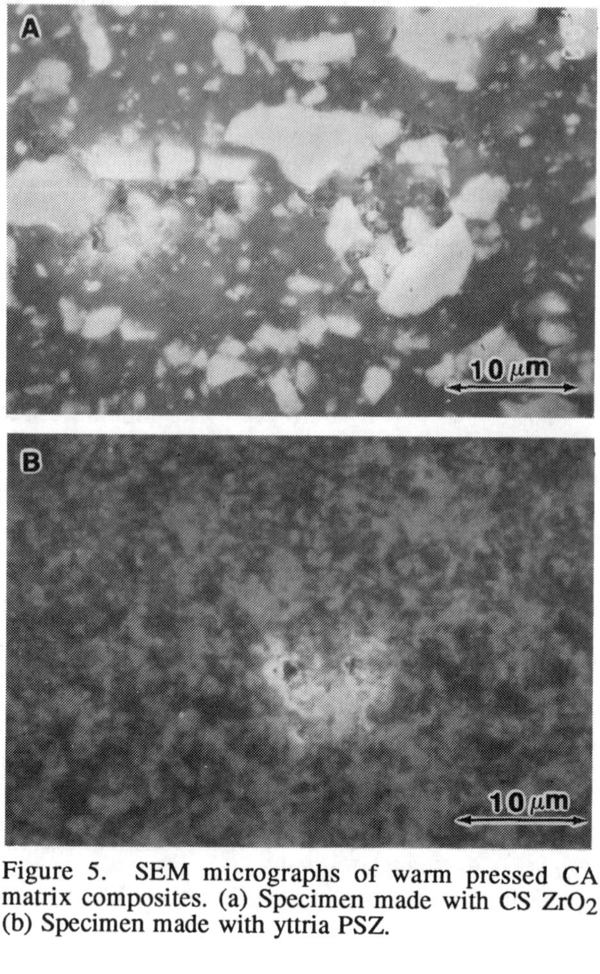

Figure 5. SEM micrographs of warm pressed CA matrix composites. (a) Specimen made with CS ZrO_2 (b) Specimen made with yttria PSZ.

By 1250°C there is evidence of a solid-state reaction having occurred between CA and ZrO_2 to form $CaAl_4O_7$ (CA_2) and $CaZrO_3$ (CZ). The XRD peaks for these compounds are pronounced and the endothermic tail on the DTA plot is in accordance with thermodynamic calculations.

One of the serendipitous features of this reaction is that the very hydraulically reactive mono-calcium aluminate is converted to the significantly less hydraulically reactive, and more refractory calcium di-aluminate.[13] The post-sintering water sensitivity of such a composite would be greatly reduced. This is suggestive of a cementitious analog of fugitive liquid phase sintering in ceramics.

CONCLUSIONS

1. In-situ coating of oxide powders with a second oxide powder can easily be accomplished by adding the "core" powder to a polymeric precursor solution of the second, outer coating powder. This coating is adherent and even persists after milling of the coated powder.

2. Coating of powders is an effective and efficient means of mixing sub-micron powders and circumventing the deleterious effects of preferential agglomeration caused by strong surface forces. It is a convenient method for obtaining a homogeneous mixture in mixed oxide ceramic matrix composites.

3. Warm pressing of CA ceramic matrix composites at 400°C and 110 MPa forms a hydrothermal hydrate ($C_4A_3H_3$) which is stable at temperatures above 700°C. Additionally, such composites can successfully be sintered, converting the chemical bonds to diffusional bonds and increasing hardness.

4. It appears possible (for some systems) that CA could be used as a transient species for forming strong green bodies at near ambient temperatures. It could then be eliminated by solid state reactions during sintering as it reacts with another compound to form the less hydraulic CA_2.

REFERENCES

[1]C.M. George, "Industrial Aluminous Cements", Structure and Performance of Cements, ed. P. Barnes, Applied Science Publications, London, (1983), pp. 415-470.

[2]B.P. Borglum, J.F. Young, and R.C. Buchanan, "Electrical Properties of Chemically Bonded Ceramics Based on Calcium Aluminate", *Advn. Cem. Bas. Mat.* , **1** 47-50 (1993).

[3]B.R. Johnson, M.A. Gülgün, W.M. Kriven, "Chemically Bonded Ceramic Matrix Composites: Densification and Conversion to Diffusion Bonding", in Ceramic Matrix Composites - Advanced High Temperature Structural Materials, ed. R.A. Lowden, J.R. Hellmann, M.K. Ferber, S.G. DiPietro and K.K. Chawla, (Mater. Res. Soc. Proc. **365**, Pittsburgh, PA 1995). In press.

[4]P.G. Desai, Z. Xu, and J.A. Lewis, "Synthesis and Characterization of $CaAl_2O_4$-Coated Al_2O_4 Microcomposite Particles," *J. Am. Ceram. Soc.*, submitted.

[5]R.W. Rice "Processing of Advanced Ceramic Composites" in Better Ceramics Through Chemistry, ed. C.J. Brinker, D.E. Clark and D.R. Ulrich (Mater Res. Soc. Proc. **32**, Elsevier Science Pub. Co., New York, 1984) pp. 337-343.

[6]C.L. Hu and M.N. Rahaman, "Dense Al_2O_3/ZrO_2 Particulate Composites by Free Sintering of Coated Powders", *J. Am. Ceram. Soc.*, **77** [3] 815-819 (1994).

[7]C.L. Hu and M.N. Rahaman, "Factors Controlling the Sintering of Ceramic Particulate Composites: II, Coated Inclusion Particles", *J. Am. Ceram. Soc.*, **75** [8] 2066-2077 (1992).

[8]F.A. Selmi an V.R.W. Amarakoon, "Sol-Gel Coating of Powders for Processing Electronic Ceramics", *J. Am. Ceram. Soc.*, **71** [11] 934-937 (1988).

[9]M.A. Gülgün, O.O. Popoola, I. Nettleship, W.M. Kriven and J.F. Young, "Preparation and Hydration Kinetics of Pure $CaAl_2O_4$", in Advanced Cementitious Systems: Mechanisms and Properties, ed. F.P. Glasser, P.L. Pratt, T.O. Mason, J.F. Young and G.J. McCarthy (Mat. Res. Soc. Proc. **245**, Pittsburgh, PA, 1992) pp. 199-204.

[10]M.A. Gülgün, B.R. Johnson and W.M. Kriven, "Chemically Bonded Ceramics as an Alternative to High Temperature Ceramic Processing" in Better Ceramics Through Chemistry VI, ed. A.K. Cheetham, C.J. Brinker, M.L. Mecartney and C.Sanchez (Mat. Res. Soc. Proc. **346**, Pittsburgh, PA, 1994) pp. 511-516.

[11]Zirconia Sales (America), Inc. literature on HSY 3.0 Partially Stabilized Zirconia, 814-C Livingston Court, Franklin Forest Business Park, Marietta, GA 30067, (404) 590-0239.

[12]M.S. Crowley, "Effect of Starting Materials on Phase Relations in the System $CaO-Al_2O_3-H_2O$", *J. Am. Ceram. Soc.*, **47** [3] 144-148 (1964).

[13]A.J. Majumdar and R. Roy, "The System $CaO-Al_2O_3-H_2O$", *J. Am. Ceram. Soc.*, **39** [12] 434-442 (1956).

[14]K.M. Parker and J.H. Sharp, "Refractory Calcium Aluminate Cements", *Trans. J. Brit. Ceram. Soc.*, **81** [2] 35-42 (1982).

THERMAL STABILITY OF Y- α '-SIALON AT HIGH TEMPERATURE

Akihiko Suda and Yoshio Ukyo
Toyota Central Research and Development Labs. Inc.
Nagakute-cho, Aichi-gun, Aichi-ken 480-11, Japan

ABSTRACT

Sialon composite composed of Y- α '-sialon and β '-sialon were annealed at high temperature to investigate the thermal stability of Y- α '-sialon. The amount of Y - α '-sialon was reduced and it was transformed into β '-sialon. Furthermore, the lattice constant (solubility) of Y- α '-sialon was increased, while that of β '-sialon was constant. It can be concluded that Y- α '-sialon is less stable than β '-sialon at high temperature.

INTRODUCTION

We have previously shown that the sialon composites containing Y- α '- sialon and β '- sialon can be obtained by hot-pressing mixtures of Si_3N_4-Y_2O_3-AlN powders(1-2). The mechanical properties of such sialon composites have been found to depend on the amount of Y- α '-sialon, the maximum strength achieved being as high as 1.3 GPa at room temperature and 1.0 GPa at 1400℃(3-4).

α '-sialon ceramics have attracted much attention because of their excellent properties, such as thermal shock resistance and high temperature strength. Extensive studies have been carried out on the processing, phase relations and mechanical properties of α '-sialon ceramics by many authors, and good reviews were reported by Ekstrom et al. (5) and Cao et al. (6).

Previous works have also shown that the sialon composites containing Y- α ' - sialon with various degree of solid solubility and β '-sialon with nearly constant solid solubility are formed by sintering mixtures of Si_3N_4, Y_2O_3 and AlN powders. When the amounts of Y_2O_3 and AlN were small, compounds containing SiO_2 such as $10Y_2O_3 \cdot 9SiO_2 \cdot Si_3N_4$ or $4Y_2O_3 \cdot SiO_2 \cdot Si_3N_4$ were formed in addition to Y-α'-sialon and β'-sialon(7). These results showed disagreement with those reported by other authors (8-9).

We have also investigated the thermal stability of Y-α'-sialon phase coexisting with β'-sialon at high temperatures, and found that the Y-α'-sialon with lower solubility is unstable and transforms into β'-sialon(10-11). This fact indicates that the sialon composite obtained by short time sintering for 1 or 2 hours are not in thermal equilibrium(12). In this study, sialon composite composed of Y-α'-sialon and β'-sialon were annealed at high temperatures for 130 hours and the changes in the crystalline phases and solubilities of Y-α'-sialon and β'-sialon were examined.

EXPERIMENTAL PROCEDURE

The powder mixture of Si_3N_4 (UBE, SN-E10)-1.5 mol% Y_2O_3-13.5 mol%AlN was sintered by hot-pressing at 1850℃ for 1h under 20 MPa pressure. The ratio of AlN to Y_2O_3 was 1 to 9 because all of the Y_2O_3 and AlN can be dissolved ideally into Si_3N_4 grains according to the following equation;

$$(4-3a)Si_3N_4 + a(Y_2O_3+AlN) = Y_{2a}(Si_{12-9a}, Al_{2a})(O_{3a}, N_{16-3a})$$

The sialon composite composed of about 24 vol% Y-α'-sialon and 76 vol% β'-sialon was obtained. The sialon composite was annealed between 1450℃ and 1850℃ for 130 hours. After annealing, the amount of Y-α'-sialon and the lattice constants (solubility) of Y-α'-sialon and β'-sialon were determined by the X-ray diffraction method. The microstructural change was observed by SEM.

RESULTS AND DISCUSSIONS

Change in Amount of Y-α'-sialon

The crystalline phases found in the specimen after annealing were also Y-α'-sialon and β'-sialon only. Figure 1 shows the dependence of the amount of Y-α'-sialon on annealing temperature. The amount of Y-α'-sialon already started to decrease from 1450℃, that is, some part of Y-α'-sialon was transformed into β'-sialon. This fact suggests that Y-α'-sialon is less stable than β'-sialon at high temperatures as reported in our previous work (12).

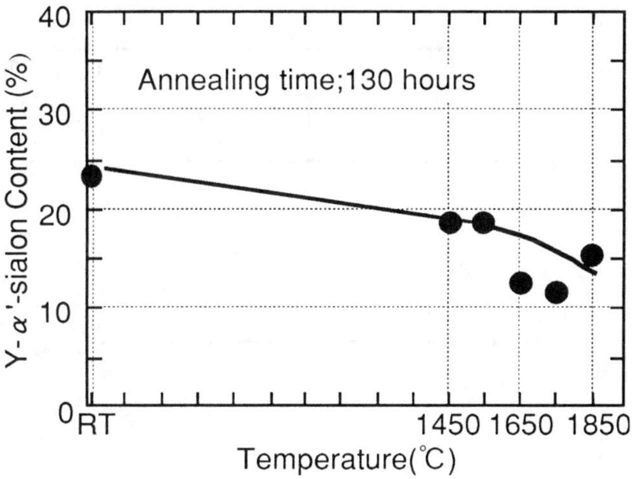

Fig.1 Temperature dependence of Y-α'-sialon content

Change in Lattice Constant (solubility) of Y-α'-sialon and β'-sialon

Figure 2 shows the dependence of the lattice constants of Y-α'-sialon and β'-sialon on the annealing temperature. The change in the lattice constants of Y-α'-sialon and β'-sialon corresponds to that of the solubility of Y-α'-sialon and β'-sialon. The lattice constant of Y-α'-sialon increased during annealing at all temperatures, that is, the solubility of Y-α'-sialon increased during annealing. As decribed in our previou report (12), the lattice constant of Y-α'-sialon still tends to

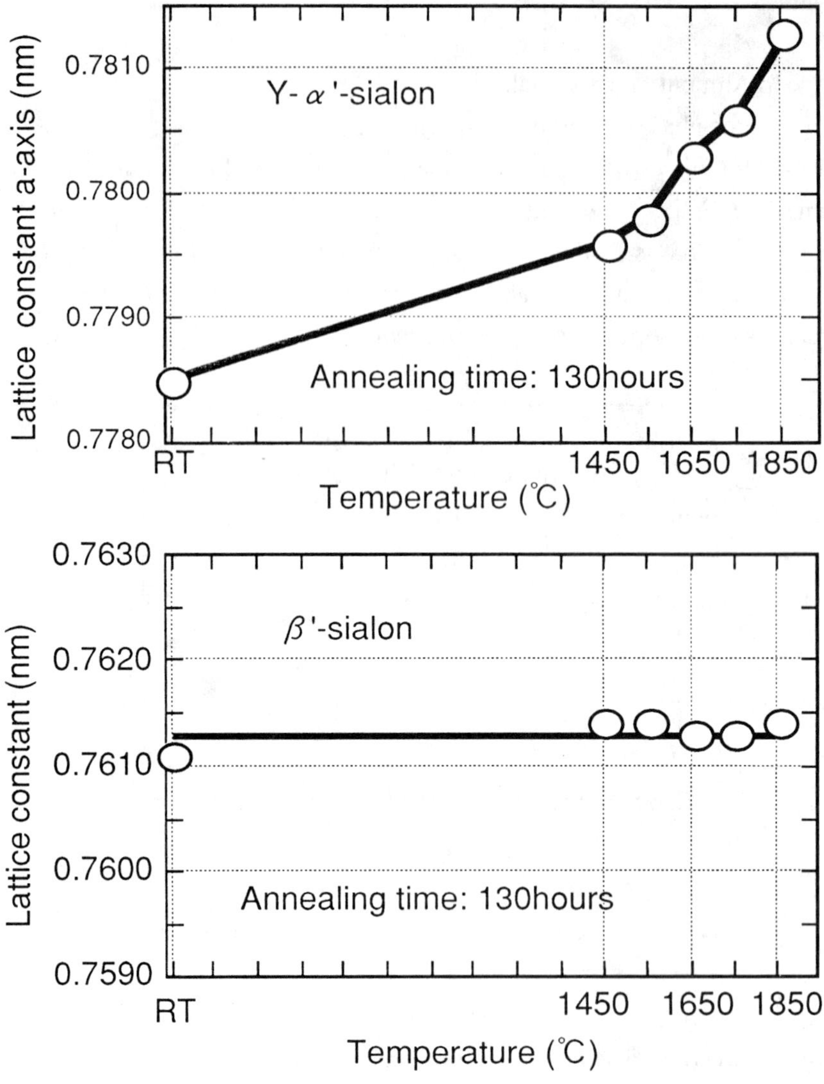

Fig.2 Temperature dependence of lattice constants of Y- α '-sialon and β '-sialon after annealing.

increase though amount of Y-α'-sialon became almost constant. On the other hand, the lattice constant of β'-sialon was almost constant during annealing, indicating that the solubility of β'-sialon was constant during annealing.

Observation of Microstructure

Figure 3 shows the microstructure of sialon composite before and after annealing at various temperatures. Before annealing, the sialon composite was composed of very fine grains smaller than 1 or 2 μm. These fine grains have grown into large grains during annealing. The grain growth was remarkable above 1750℃ and the grains with lengths of 30-50 μm were found after annealing at 1850℃. The grains grown were mainly prismatic, and it is, therefore, thought that the β'-sialon grains have grown during annealing.

Transformation Mechanism

The transformation mechanism from Y-α'-sialon into β'-sialon during annealing is estimated by taking our results obtained before into account as follows. At the initial stage of annealing, some part of the more unstable Y-α'-sialon transforms into β'-sialon. When the transformation from Y-α'-sialon to β'-sialon proceeds, the elements, such as Y and N, will be excluded to grain boundaries and the concentration of the elements at the grain boundaries will increase with time. These elements will diffuse to the ramaining Y-α'-sialon and dissolve again. Therefore, the lattice constants of the remaining Y-α'-sialon increased with time and become more stable. On the other hand, the lattice constants of β'-sialon were constant and its grains have grown into large grains with time.

CONCLUSIONS

The sialon composite composed of Y-α'-sialon and β'-sialon were fabricated by hot-pressing mixture of Si_3N_4, Y_2O_3 and AlN powders and annealed between 1450 ℃ and 1850℃ for 130 hours. Changes in crystalline phases and lattice constant (solubility) were examined. The follwing results were obtained.

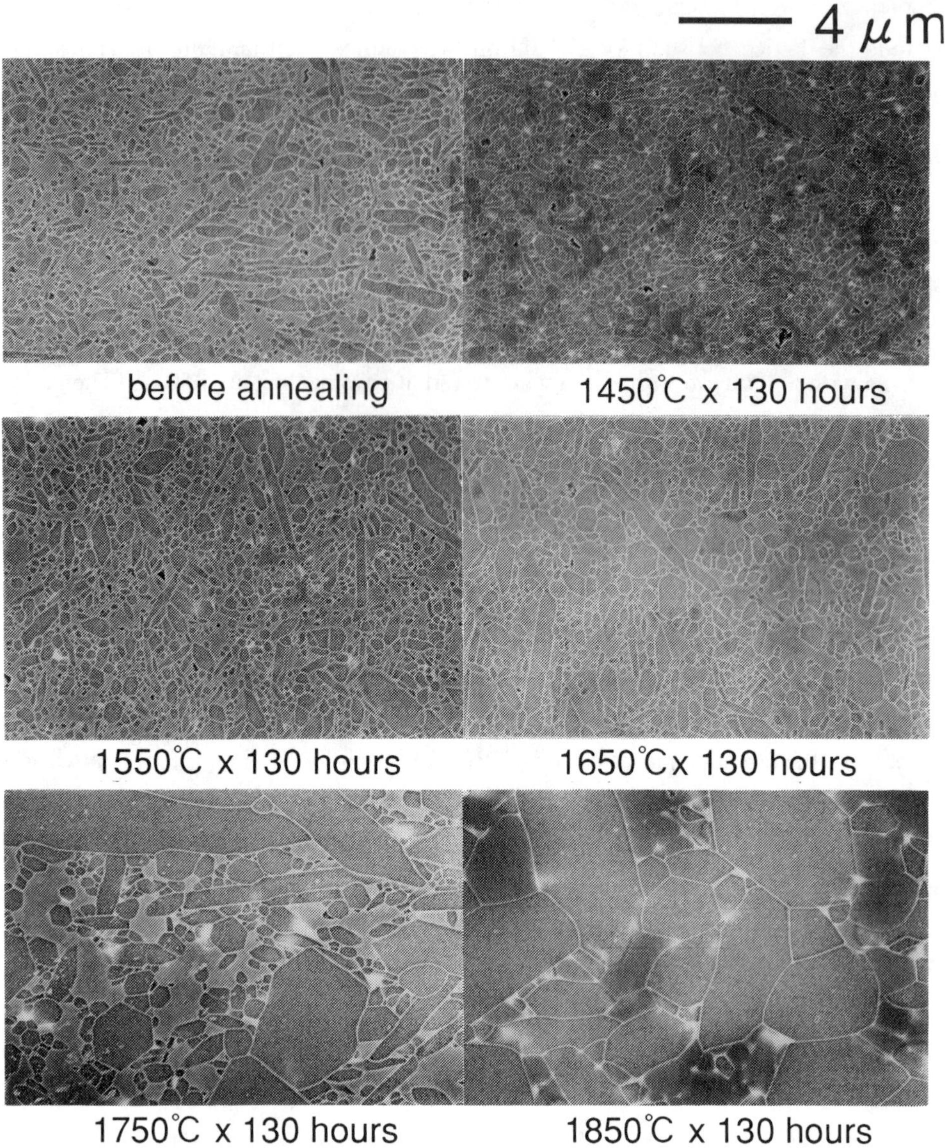

Fig.3 Microstructure of sialon composite before and after annealing between 1450 and 1850℃ for 130 hours.

(1) Y- α '-sialon transformed into β '-sialon during annealing at all temperatures investigated.
(2) The lattice constants (solubility) of Y- α '-sialon increased during annealing, while that of β '-sialon was almost constant during annealing.
(3) β '-sialon grains have grown, and Y or N excluded from Y- α '-sialon dissolved into the remaining Y- α '-sialon during annealing.

References

(1) Y. Ukyo and S. Wada, "Formation and Stability of Y- α '-sialon co-existing with β '-sialon" pp567-571 in Euro-Ceramics. Edited by G. de With, R. A. Terpstra and R. Metaselaar. Elsevier Applied Science, 1989.
(2) Y. Ukyo and S. Wada, "Grain Boundary Controlled Properties of Fine Ceramics" JFCC Workshop Series; Materials Processing and Design, Ed. by K. Ishizaki, K. Niihara, M. Isotani and R.G. Ford, Elsevier Applied Science (1992) pp.112-124
(3) Y. Ukyo and S. Wada, "High Strength Si_3N_4 Ceramics" Nippon Seramikkusu Kyokai Gakujutu Ronbunshi, 97[8] 872-874 (1989)
(4) S. Wada and Y. Ukyo,"Microstructure and Properties of Y- α '/ β 'Sialon Composite" pp29-33 in Proc. of the 34th Japan Congress on Materials Research(Kyoto, Japan,1991). Published by The Society of Materials Science, Kyoto, Japan, 1991.
(5) T. Ekstrom and M. Nygren,"SIALON Ceramics" J. Am. Ceram. Soc., 75[2] 259-276 (1992)
(6) G. Z. Cao and R. Metselaar, " α '-Sialon Ceramics: A Review"Chem. Mater., 3, 242-252 (1991)
(7) N. Sugiyama, Y. Ukyo and S. Wada,"Phase Relations between α ' and β ' -Sialons in $Si_3N_4(SiO_2)$-AlN-Y_2O_3 Syatem" J. Ceram. Soc. Jpn., 101[7] 729-73 (1993)
(8) M. Mitomo, F. Izumi, Y. Bando and Y. Sekikawa, "Characterization of α ' -SiAlON Ceramics" pp377-386 in Proc. Int. Symp. on Ceram. Component

for Engines, Edited by Somiya et al. KTK Science Publishers, Tokyo, Japan,1983.

(9) Z-K. Huang, P. Greil and G. Petzow, "Formation of α-Si_3N_4 Solid Solutions in the System Si_3N_4-AlN-Y_2O_3" J. Am. Ceram. Soc., 66[5] c96-97(1983)

(10) Y. Ukyo, N. Sugiyama and S. Wada, " Themal Stability of Y-α'-Sialon co-existing with β'-sialon" pp.141-146 in Proc. Int. Symp. on Science of Engineering Ceramics, Edited by S. Kimura and K. Niihara, The Ceramic Society of Japan, Tokyo, 1991.

(11) N. Sugiyama, Y. Ukyo and S. Wada,"Thermal Stability of Phases in $Si_3N_4(SiO_2)$-AlN-Y_2O_3 System" J. Ceram. Soc. Jpn., 102[2] 145-148 (1994)

(12) Y. Ukyo and S. Wada," Thermal Stability of Y-α'-sialon Prepared from Si_3N_4-Y_2O_3-AlN" J. Ceram. Soc. Jpn., 102[8] 713-717 (1994)

KEYWORD AND AUTHOR INDEX

3-aminopropyltriethoxysilane, 133

Abraham, Thomas, 3
Acrylics, 125, 281, 289
Adsorption, 125
Advanced ceramics, 15, 21
Aerosols, 83
Agostine, Sharon E., 125
Akashi, Kazuo, 91
Al_2O_3, 173, 241, 281, 297
  dispersions, 263
Aluminum, 91
Aluminum nitride, 21, 91
Aminosilanes, 133
APS, 133
Aqueous colloidal processing, 233
Attraction, 165

Berry, Amanda K., 125
Binders, 29, 125, 141, 193, 225, 281, 289, 305
  acrylic emulsion, 157
  crosslinkable, 157
  tape casting, 157
Biner, S.B., 241
Blakely, Keith A., 21
Blending, 193
Blum, Yigal D., 67
Bogan, Leonard E., 125
Breakpoint, 203
Bryd, Hayes, Jr., 157
Buchta, Michael A., 133

Cake permeability, 263
Carburization, 99
Ceramic injection molding, 305
Ceramic powders, 3
  nanosized, 3
Ceramic sheets, 289
Chemical vapor deposition, 107
Chemically bonded ceramics, 313
Coating of powders, 313
Colloidal suspension, 271
Combustion, 83
Commercialization, 15
Compaction, 203
Crimp, Melissa J., 173
Cumbers, Donald I., 193

Debinding, 305
Deevi, Sarojini, 99
Deevi, Seetharama C., 99
Deflocculants, 141
Diametral compression, 203
Dicarboxylic acids, 241
Dinger, D.R., 297
Dispersants, 117, 125
DLVO theory, 271
Donato, Antonio T., 157, 281, 289
Dry pressing, 141, 193, 217, 225

Ebenhöch, J.S., 305
Elasticity, 165
Electrolytic etching, 91
Electronic structural coatings, 3
Electrophoretic deposition, 271
Electrostatic repulsion, 165
Extrusion, 233, 281, 289, 297

Fabrication, 29
Farrell, D.J., 233
Filter pressing, 263
Finch, William C., 225
Fluidized-bed reactors, 99
Fries, R., 165, 263

Geiger, Greg, 15
Gel casting, 29, 255
Glass, S. Jill, 203
Glass transition, 225
Goel, H., 241
Granules, 141, 203
Green body, 217
Green densities, 263
Green microstructure, 263
Gülgün, Mehmet A., 57, 313

Hackley, Vincent A., 117
Hajaligol, Mohammad R., 99
Hematite, 37

Hesse, W., 305
High-temperature strength, 321
Hlavacek, V., 107
Horn, J.A., 183
Hydration, 313
Hydrothermal processing, 37
Hydrothermal synthesis of ceramics, 75
Hydroxyapatite, 37

Ibbitson, Scott A., 157, 193
Iron carbides, 99
Isoelectric point, 149
Ito, Shigeru, 91

Johnson, Bradley R., 313
Johnson, Sylvia M., 67

Kiggans, J.O., Jr., 255
Kim, Chong Hee, 47
Kim, Do Kyung, 47
Kinetics, 271
Komarneni, Sridhar, 37
Kriven, Waltraud M., 57, 313

Li, Qing Hua, 37
Lograsso, B.K., 241

Magnetic ferrites, 75
Magnetic media, 99
Magnetic properties, 99
Malghan, Subhas G., 117
Manganese-zinc ferrites, 75
Maria, J.P., 255
Markets, 3, 15, 21
McCauley, R.A., 149
McDermott, Gregory A., 67
Mechanical properties, 305, 321
Menon, V.C., 37
Microwave-hydrothermal processing, 37
Microwave processing, 37
Mixed oxide powders, 57
Moeggenborg, Kevin J., 217
Molybdenum powders, 107
Monocalcium aluminate, 313
Monodisperse spherical powders, 47
Moore, J.A., 241

Nano-alumina dispersions, 165
Nanosized particles, 83
Newton, Clay, 203
Nicholson, Patrick S., 271
Niobium powders, 107
Nitriding, 91
Nonaqueous suspensions, 173
Nonchelating polymer, 57
Nonoxides, 3
Nunn, S.D., 255

Ogasawara, Tsuneharu, 75
Oppermann, Dean A., 173
Organics, 149
Oxides, 3

Park, Hong Kyu, 47
Patterson, B.R., 183
PCI, 297
Plasma chemical reactor, 107
Plasticity, 241
Poly(acrylic acid), 117
Polyethylene glycol, 141
Polymers, 183
Polysilazanes, 67
Powder forming, 29
Powder suspensions, 183
Powder synthesis, 313
Powders, 15, 47
Pratsinis, Sotiris E., 83
Preceramic polymers, 67
Pressing, 233
Pressure filtration, 241
Prices, 3
Processing, 15, 313
Pyrolysis, 67

Rand, B., 165, 263
Rangel de Figueiredo Teixeira, Ana Maria, 75
Reed, J.S., 141
Rheology, 117, 165, 183, 241

Saccharides, 241
Saito, Shin-ichi, 91
Sarkar, Partho, 271
Schilling, C.H., 241
Shape forming, 233
Sheppard, Laurel M., 29

Shih, Wei-Heng, 133, 233
Shih, W.Y., 233
Shrinkage, 297
SiAlON, 321
SiC whiskers, 173
Silica, 83
Silicon, 255
Silicon nitride, 67, 117, 133, 141, 149, 233
Simpson, R.E., II, 255
"Soft" particle interactions, 165
Sol-gel coating, 233
Solution-polymerization route, 57
Spinel ferrites, 75
Spray drying, 141, 193, 225
Springback, 217
SRBSN, 255
Stability, 173
Suda, Akihiko, 321
Suppliers, 21
Surface chemistry, 117
Surface modification, 149
Synthesis, 3, 15, 37, 47

Tang, Xun, 157, 193
Tantalum powders, 107
Tape casting, 193
Tensile properties of green tapes, 157
ter Maat, J.H.H., 305
T_g, 225
Thermal plasma, 107
Thermal shock resistance, 321
Thermal stability, 321
Thermodynamic analysis of ceramic synthesis, 75
Thick films, 289
Titania, 83
Truebenbach, P., 305
Tungsten powders, 107

Ukyo, Yoshio, 321

Verma, S.K., 141
Viscoelasticity, 165
Viscosity, 125, 183

Walker, W.J., Jr., 141
Wang, H.M., 149
Warm pressing, 313
Whiskers, 133
Whitman, David W., 193, 225
Woodard, K.J., 297
Wu, Xianliang Kevin, 225, 281, 289

Yoshida, Kazuharu, 91
Yttrium oxide, 91

Zeta potential, 173, 271
Zhang, Mingli, 173
Zhao, G.Y., 107
Zhou, J., 183
Zhu, Wenhua, 83
$ZrTiO_4$, 47